Inland Waters - Ecology, Limnology, and Environmental Protection

Edited by Mohamed Nageeb Rashed

Published in London, United Kingdom

Inland Waters - Ecology, Limnology, and Environmental Protection
http://dx.doi.org/10.5772/10.5772/intechopen.1002161
Edited by Mohamed Nageeb Rashed

Contributors
Aarefa Jan, Abenezer Wendimu, Abnet Woldesenbet, Amy M. Villamagna, Andreea Voina, Ben J. Nugent, Brigid C. O'Donnell, Dean G. Fitzgerald, Diego García de Jalón, Donovan C. Kotze, Edineldo Lans-Ceballos, Erik Mosselman, Erwin J.J. Sieben, Fatma Dhaouadi, Gabriela Circiumaru, Jared B. Lamy, Joshua C. Hoekwater, Kandé Bangoura, Lotfi Sellaoui, Lucia-Andreea El-Leathey, Lynn S. McCarty, Marie Serena McConnell, Mario Mariglia, Massimo Rinaldi, Mohamed Lamine Komara, Rajendra Chouhan, Rares-Andrei Chihaia, Saima Syed, Subrata Pani, Suchitra Banerjee, Thomas Shahady, Tyson R. Morrill, Wondimagegnehu Tekalign

First published in London, United Kingdom, 2024 by IntechOpen
IntechOpen is the global imprint of INTECHOPEN LIMITED, registered in England and Wales, registration number: 11086078, 167-169 Great Portland Street, London, W1W 5PF, United Kingdom

British Library Cataloguing-in-Publication Data
A catalogue record for this book is available from the British Library

Additional hard and PDF copies can be obtained from orders@intechopen.com

Inland Waters - Ecology, Limnology, and Environmental Protection
Edited by Mohamed Nageeb Rashed
p. cm.

This title is part of the Environmental Sciences Book Series, Volume 21
Topic: Water Science
Series Editor: J. Kevin Summers
Topic Editor: Yizi Shang

Print ISBN 978-0-85466-641-6
Online ISBN 978-0-85466-640-9
eBook (PDF) ISBN 978-0-85466-642-3
ISSN 2754-6713

For EU product safety concerns:
IN TECH d.o.o., Prolaz Marije Krucifikse Kozulić 3, 51000 Rijeka, Croatia,
info@intechopen.com or visit our website at intechopen.com.

IntechOpen Book Series

Environmental Sciences

Volume 21

Aims and Scope of the Series

Scientists have long researched to understand the environment and man's place in it. The search for this knowledge grows in importance as rapid increases in population and economic development intensify humans' stresses on ecosystems. Fortunately, rapid increases in multiple scientific areas are advancing our understanding of environmental sciences. Breakthroughs in computing, molecular biology, ecology, and sustainability science are enhancing our ability to utilize environmental sciences to address real-world problems.

The four topics of this book series - Pollution; Environmental Resilience and Management; Ecosystems and Biodiversity; and Water Science - will address important areas of advancement in the environmental sciences. They will represent an excellent initial grouping of published works on these critical topics.

Meet the Series Editor

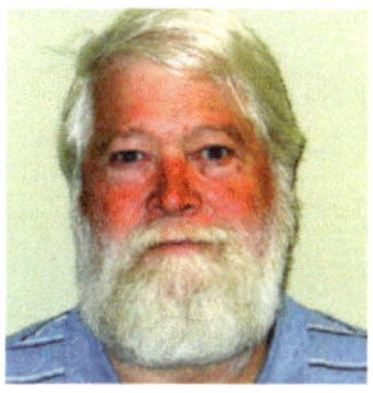

J. Kevin Summers is a Senior Research Ecologist at the Environmental Protection Agency's (EPA) Gulf Ecosystem Measurement and Modeling Division. He is currently working with colleagues in the Sustainable and Healthy Communities Program to develop an index of community resilience to natural hazards, an index of human well-being that can be linked to changes in the ecosystem, social and economic services, and a community sustainability tool for communities with populations under 40,000. He leads research efforts for indicator and indices development. Dr. Summers is a systems ecologist and began his career at the EPA in 1989 and has worked in various programs and capacities. This includes leading the National Coastal Assessment in collaboration with the Office of Water which culminated in the award-winning National Coastal Condition Report series (four volumes between 2001 and 2012), and which integrates water quality, sediment quality, habitat, and biological data to assess the ecosystem condition of the United States estuaries. He was acting National Program Director for Ecology for the EPA between 2004 and 2006. He has authored approximately 150 peer-reviewed journal articles, book chapters, and reports and has received many awards for technical accomplishments from the EPA and from outside of the agency. Dr. Summers holds a BA in Zoology and Psychology, an MA in Ecology, and Ph.D. in Systems Ecology/Biology.

Meet the Volume Editor

Prof. Mohamed Nageeb Rashed is a professor of analytical and environmental chemistry at the Faculty of Science, and the head of the unit of environmental studies at Aswan University, Egypt. His research interest has been analytical and environmental chemistry. He has published 95 scientific papers in peer-reviewed international journals. He has participated as an invited speaker at 35 international conferences worldwide. He was selected as an examiner for several Ph.D. theses in analytical chemistry from India, Azerbaijan, Malaysia, KSA and Botswana. Prof. Rashed acts as editor-in-chief of the Aswan University Journal of Environmental Studies as well as an editorial board member for several international journals. He received the Egyptian State Award for Environmental Research in 2001. He was recognized by Stanford University's list of the world's top 2% of scientists in 2020, 2021, 2022, and 2023.

Contents

Preface

Inland waters are permanent water bodies inland from areas and coastal zone regions whose features and uses are primarily influenced by the regular, irregular, or seasonal occurrence of flooding. Rivers, lakes, floodplains, and wetlands are examples of inland waters. Along with bogs, marshes, swamps, wetlands, and inland saline systems, the classification also includes ponds, streams, groundwater, reservoirs, springs, cave waters, and floodplains.

Aquatic ecosystems known as inland waters are contained inside the confines of land and offer crucial biological, environmental, hydrological, and socioeconomic benefits to both people and the environment.

Inland water bodies, which are essential for recreation and the provision of food, commerce, transportation, and human health, have come under threat from outside factors such as direct human activity and climate change. Since inland waterways are aquatic ecosystems surrounded by land, it is important to preserve them and their ecosystems.

This book gathers recent research by outstanding experts in the field related to inland waters' ecology, limnology, and environmental protection. It will sound to a wide readership from universities and industry. Also, the readers will obtain updated information on all aspects of inland waters, limnology, ecology, water monitoring, and water pollution treatment.

This book *Inland Waters – Ecology, Limnology, and Environmental Protection* deals with several aspects of inland waters. The book is divided into three sections and structured in 12 chapters. The first section includes water pollution and treatment and presents three chapters: (1) "Adsorptive Removal of Water Pollutants: Modeling and Consequences"; (2) "Plastic Pollution in Inland Waters – A Threat to Life"; and (3) "Removal of Heavy Metals and Purification of Surface Waters". The second section explains the hydromorphological quality of inland water and includes four chapters: (1) "Analysis of Heavy Metal Toxicity in the Surface and Bottom Waters of Lower Lake Bhopal, M.P. (India)"; (2) "Determination of the Physicochemical and Bacteriological Parameters of the Waters of Lake Sonfonia Commune of Ratoma (Republic of Guinea) 2021"; (3) "Hydropower Reservoirs as Arbiters of Climate Change"; and (4) "The European REFORM Project for Hydromorphological Quality in River Basin Management". The third section explains the ecology and limnology of inland water and presents five chapters: (1) "Experimental Study of a Fish Behavioral Barrier Based on Bubble Curtains for a River Water Intake"; (2) "Time Trends in Fish Tissue Methylmercury in Northern Watersheds: Implications of Phosphorus Loading and Eutrophication on Subsistence Fisheries "; (3) "Fish Fauna and Fishery in Ethiopia, Africa"; (4) "Timber and Trout: An Examination of the Logging Legacy and Restoration Efforts in Headwater Streams in New England (USA)"; and (5) "Wetlands and the Ecological Services that They Provide on Multiple Spatial Scales, from Landscape Down to Soil".

So, in this book, the readers will obtain updated information on all aspects of inland waters: ecology, limnology, and environment protection. Scientists from different scientific fields report their findings in this book. The book *Inland Waters – Ecology, Limnology, and Environmental Protection* offers an important information source for researchers and professionals working in water, ecology, and environmental fields.

I would like to express my sincere thanks to my colleagues from Aswan University, Egypt: Dr. Arwa Hassan, Faculty of Science; Prof. Abdel-Moamen Mohamed, Faculty of Energy Engineering; Dr. Hickmat Hossen, Faculty of Engineering; Dr. Abdallah Elshawadfy Elwakeel, Faculty of Agriculture and Natural Resources; and Dr. Ragaa Abdallah Ahmed, Faculty of Fish and Fisheries Technology. Also, thanks to my colleques from the National Institute of Oceanography and Fisheries, Egypt, Prof. Dr. Hala Elshahat Ghannam and Dr. Sally Salaah Eldin Elshalqamy, for their contribution in the scientific revision of the book chapters.

Also, I would like to express my thanks to the IntechOpen Publishing Process Manager, Ms. Iva Horvat. I'm also grateful to the book authors for their hard work and worthy contributions.

Dr. Mohamed Nageeb Rashed
Professor,
Faculty of Science,
Aswan University,
Aswan, Egypt

Section 1

Water Pollution and Treatment

Chapter 1

Adsorptive Removal of Water Pollutants: Modeling and Consequences

Lotfi Sellaoui and Fatma Dhaouadi

Abstract

In this chapter, the metal-organic framework (MOF) was employed to investigate the adsorption mechanism of different water pollutants such as dyes (Direct Blue 1: D1 and Direct Yellow-4: D2) and pharmaceuticals (amoxicillin: PHM1 and doripenem: PHM2) *via* physical approach in single and binary systems (SS and BS). Based on an experimental assessment, it was indicated that the adsorption capacities of dye and pharmaceutical pollutants (D1, D2, PHM1, and PHM2) were reduced when the second pollutant is present in solution. This investigation highlighted that an antagonistic impact was occurred during the adsorption process. The competition between D1 and D2, and PHM1 and PHM2 on the same MOF adsorbent site was interpreted *via* the physical model parameters. The application of models on D1, D2, PHM1, and PHM2 data indicated that an aggregation process was present with lower degree that is due to the lower interactions between the pollutants in the solution. A detailed analysis reflected that our adsorbent presents an excellent performance to remove D1, D2, PHM1, and PHM2 from environment compared to other materials. Overall, this chapter presents a deeper analysis of the adsorption process and its relevant impact to protect the environment from known water pollutants.

Keywords: water pollutants, adsorption, physical models, MOF, environment

1. Introduction

In recent years, the industrial landscape has undergone rapid evolution, catalyzing advancements that have significantly enhanced our society well-being and quality of life. This progress has been marked by innovations in various sectors, leading to the creation of new products and the adoption of cutting-edge technologies and processes [1–3]. However, amidst these commendable strides, an environmental pollution was totally identified. The unchecked growth of industrial activities has resulted in the generation of a diverse array of pollutants, which find their way into our waterways [4–7]. These pollutants stem from the manufacturing processes, waste disposal practices, and the widespread use of chemicals inherent in industrial operations. Moreover, the commercialization of new products often introduces novel substances into the environment whose long-term effects remain uncertain. Additionally, the adoption of advanced transformation technologies and processes, while improving efficiency and productivity, can inadvertently contribute to the pollution load. Contaminated water sources pose risks not only to aquatic ecosystems but also to human health, as water is a fundamental resource for drinking, agriculture, and sanitation. Furthermore, the impact extends beyond local ecosystems, with pollutants often accumulating and spreading through interconnected water systems, affecting regions far removed from their source [1–7].

In this context, the protection of the environment that is mainly related to the water contamination has become an important subject for society due to the increment population and growing of different industrial activities [1–7]. The minimization of concentrations of environmental pollutants is necessary and has a relevant role to retain the environment and human health. Note that the pollution source is due to the wide use of the organic and inorganic pollutants in the industrial fields. Among these water pollutants; the pharmaceuticals (PHMs) and dyes (Ds) compounds are posing potential risks on human health [8, 9]. As by-products in wastewater, the PHMs (amoxicillin and doripenem) and Ds (Direct Blue 1 and Direct Yellow-4) are discharged into diverse water bodies, posing considerable toxicity to aquatic life, plants, and humans. Despite the efforts regarding the wastewater remediation, these mentioned pollutants persist in environmental waters. Therefore, the exploration of diverse techniques for their complete remediation remains pertinent. In this context, various methods for removing these organic wastes have been applied, including coagulation, precipitation, reverse osmosis, advanced oxidation processes, filtration, and biodegradation [6–9].However, the implementation of these technologies in treating actual effluents faces significant constraints. For instance, they are costly and difficult in manipulation compared to other process.

It was proved that the water pollutants, mainly the PHMs and Ds, can be eliminated from wastewater, rivers, lakes, etc., *via* the adsorption process [10, 11] to minimize their dangers. The adsorption process stands out as a highly effective and versatile method for the removal of PHMs and Ds from contaminated effluents. Its excellent performance, simplicity in manipulation, and compatibility with existing treatment infrastructure make it a preferred choice for addressing water contamination challenges and safeguarding environmental and human health [12–14]. The adsorbent choice is the first key for obtaining good PHMs and Ds elimination. Hence, metal-organic frameworks (MOFs) have proved to be promising adsorbents than known other materials like activated carbons [15–24]. In general, this material is typified by excellent physicochemical properties such as high specific surface area. In addition, its surface contains different functional groups that can play a relevant role to remove both mentioned water pollutants. Therefore, the objective of this chapter is to provide advanced explanation of the PHMs and Ds adsorption mechanisms and explain the impact of the presence of a second pollutant in solution *via* physical models. In particular, an advanced vision was developed in this chapter *via* non-classical approach to select the best conditions to protect the environment.

2. Materials and methods: determination of PHMs and Ds adsorption data

The D1, D2, PHM1, and PHM2 adsorption data expressing the variation of the adsorption capacities as function of their concentrations were achieved at three different temperatures ranging from 30 to 50°C. The determination of the adsorption isotherms of D1, D2, PHM1, and PHM2 represents a necessary step to offer a comprehensive analysis concerning the adsorption mechanisms. In particular, these isotherms can provide ideas about the interactions between the D1, D2, PHM1, and PHM2 and MOF adsorbent, the saturation process, etc. To determine these data, a concentration of 250 mg/L was prepared that contains D1 or D2 and PHM1 or PHM2 with RO water. D1, D2, PHM1, and PHM2 adsorption investigations were conducted using Erlenmeyer flasks with 50 mL of dye and pharmaceutical solution and MOF quantities ranging from 0.01 to 0.1 g. Erlenmeyer flasks with the suspensions were later immersed in a controlled water bath to maintain a specific temperature. Subsequently, the water bath was agitated at a rotational speed of 200 rpm. Note that the equilibrium time is of 12 h. To separate MOF from the solution, a centrifugation was used. The initial and equilibrium concentrations of D1, D2, PHM1, and PHM2 were calculated by using a SHIMADZU UV/VIS-1700 PharmaSpec instrument operated at a wavelength of 565 nm for D1, 497 nm for D4, 628 nm for PHM1 and 448 nm for PHM2. The same method was used to determine binary adsorption data. All the determined data were fitted *via* advanced theoretical models that were developed by elements in statistical physics.

3. Results and discussion

3.1 Advantages of the models

The models that will be applied on D1, D2, PHM1, and PHM2 data have numerous advantages compared to the classical models. Note that they consider that the MOF adsorbent site (i.e., functional group) can receive a variable number of water pollutants. This consideration is not consistent with the Langmuir model hypotheses, which assumes that the main adsorbent site of all adsorbents can detect only one water pollutant. This assumption that is generalized by the Langmuir model can provide classical investigation of the adsorption mechanism and limit the space of the treatment of the adsorption process of water pollutants mainly the PHMs and Ds. Classically, the application of this model can offer simple and questionable investigation of the adsorption mechanism. For instance, the analysis of the parameter n of these models that represents the number of captured water pollutants per site provides additional insights and a plausible understanding of the adsorption mechanism. It can estimate the aggregation degree of organic pollutant under different experimental conditions. It is important to note that is possible to describe the adsorbates configuration of these used pollutants *via* the values of this parameter at different solution temperatures.

3.2 D1, D2, PHM1, and PHM2 isotherms and their modeling in single systems

The relationship between the D1, D2, PHM1, and PHM2 adsorption capacities and their equilibrium concentrations in single systems is summarized according to **Figure 1**. These D1, D2, PHM1, and PHM2 adsorption data were fitted *via* a physical model that suggests a monolayer process occurring at different temperatures. This model represents the general form of Langmuir model. Its expression is described as follows [25, 26]:

$$Q_e = \frac{n_p S}{1 + \left(\frac{C_{hs}}{C_e}\right)^{n_D}} \quad (1)$$

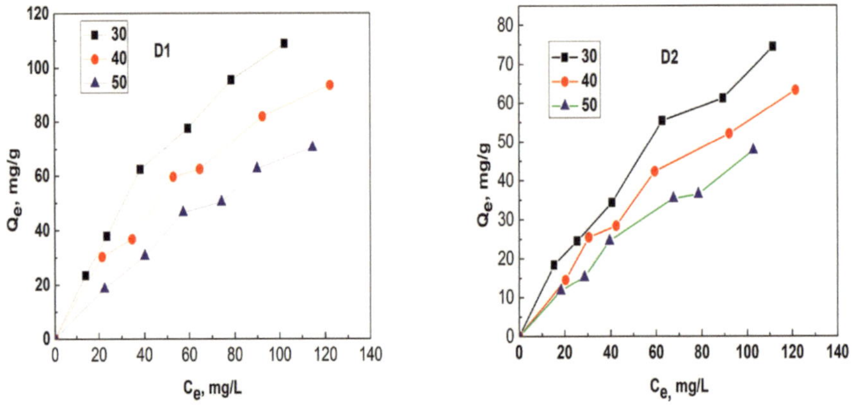

Figure 1.
D1 and D2 adsorption capacities as function of their equilibrium concentrations (T = 30, 40, and 50°C).

Based on this model, the D1, D2, PHM1, and PHM2 adsorption can be controlled *via* three parameters defined such as:

Parameter n_p: this parameter estimates the pollutant amount per MOF site (number of pollutant per adsorbent site of MOF),

Parameter S: this parameter represents the density of sites at saturation,

Parameter C_{hs}: the concentration at half-saturation of the pollutant layer.

3.3 D1, D2, PHM1, and PHM2 isotherms and their modeling in binary systems

The D1, D2, PHM1, and PHM2 adsorption quantities as function of their concentrations are represented in **Figure 2**. Comparatively, a significant reduction in adsorption capacity from binary to single systems was observed. Overall, the model of binary systems considers that these pollutants (D1 and D2, PHM1, and PHM2) were removed *via* the contribution of the same MOF adsorbent site to explain and understand the competition between these molecules. The model expression is defined according to the next expression [27, 28]:

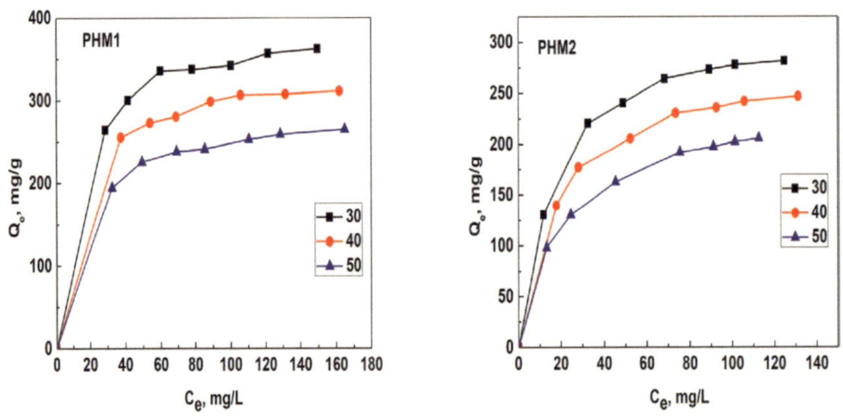

Figure 2.
PHM1 and PHM2 adsorption capacities as function of their equilibrium concentrations (T = 30, 40, and 50°C).

$$Q_{b_1} = \frac{n_{1b} D_b \left(\frac{C_e}{C_{1b}}\right)^{n_{1b}}}{1 + \left(\frac{C_e}{C_{1b}}\right)^{n_{1b}} + \left(\frac{C_e}{C_{2b}}\right)^{n_{2b}}} \tag{2}$$

$$Q_{b_2} = \frac{n_{2b} D_b \left(\frac{C_e}{C_{2b}}\right)^{n_{2_b}}}{1 + \left(\frac{C_e}{C_{1b}}\right)^{n_1} + \left(\frac{C_e}{C_{2b}}\right)^{n_2}} \tag{3}$$

Both relationships contain different parameters that can be defined such as: The n_{1b} and n_{2b} represent the quantities of D1, D2, PHM1, and PHM2 that are captured per MOF site, D_b can estimate the number of MOF sites that are filled by these adsorbates, and $C_{ib}(i = 1,2)$ is the concentrations at half-saturation(mg/L) (**Figures 3** and **4**).

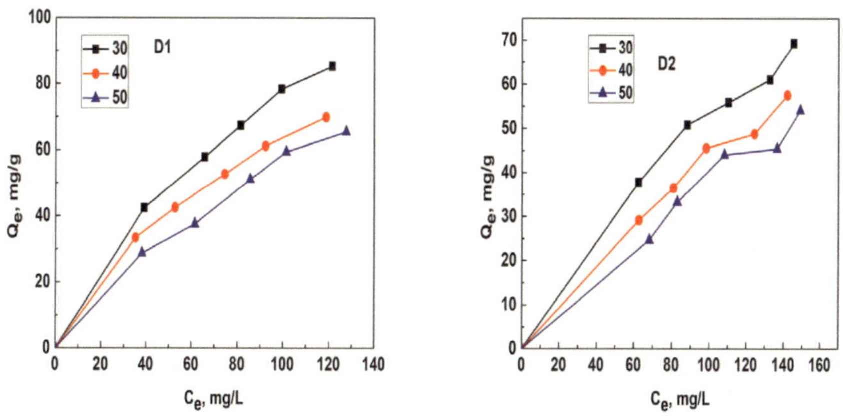

Figure 3.
D1 and D2 adsorption capacities as function of their concentrations in binary systems (T = 30, 40, and 50°C).

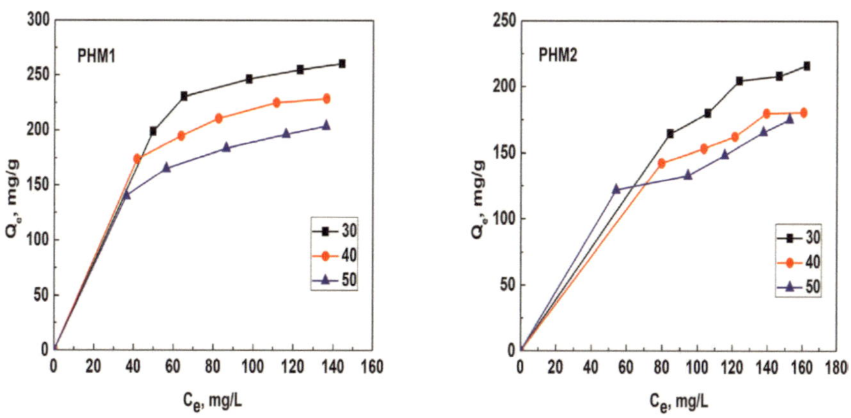

Figure 4.
PHM1 and PHM2 adsorption capacities as function of their concentrations in binary systems (T = 30, 40, and 50°C).

3.4 Interpretation of the model parameters

The objective of this section was to analyze the adsorption process of these pollutants in single and binary systems *via* both model parameters and to select the best working conditions to remove these tested water pollutants in terms of environment protection. In this direction, these two models were applied on all data to retrieve their parameters. The impact of temperature on all model parameters was described in the next sections.

3.4.1 Physical assessment of D1, D2, PHM1, and PHM2 adsorption mechanisms

3.4.1.1 Physicochemical analysis of the n_D, n_1, and n_2 parameters

The data fitting showed that the estimated values of the n_D parameter that is in relationship with single systems of dyes are close to the unity (**Figure 5**). The determined values are 1.17, 0.95, and 1.22 for D1 adsorption, while for D2 adsorption they are 0.92, 1.16, and 0.95. Based on these fitted values, we can conclude that D1 and D2 were adsorbed with the presence of a monomer process. This means that an

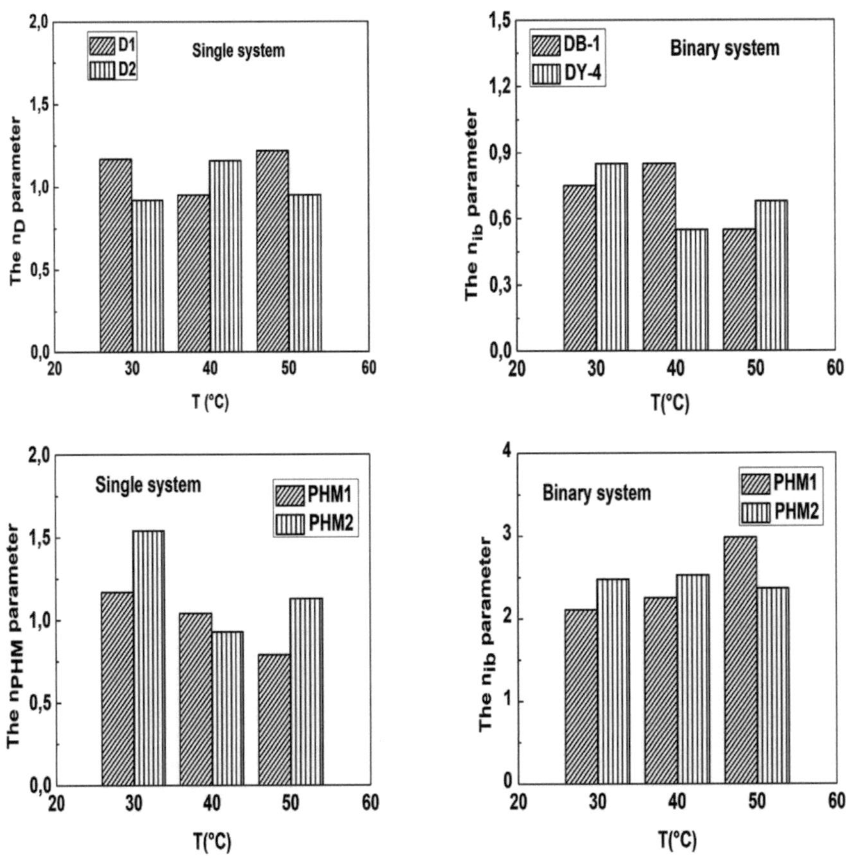

Figure 5.
Impact of the temperature on the n_D, n_{PHM} (n pollutant), and n_{ib} parameters.

aggregation process with lower degree [29, 30] was present. Concerning the pharmaceuticals adsorption, the values are 1.54, 0.93, and 1.13 and 1.17, 1.04, and 0.79 for PHM1 and PHM2, respectively. Most of these values are also close to the unity suggesting that PHM1 and PHM2 are also removed with an aggregation process with lower degree. The presence of this aggregation with lower degree is probably due to the lower interaction energies between the molecules in solutions (before adsorption). The presence of a second pollutant in the solution (D1 with D2, and PHM1 with PHM2) reduced the n_D values at different temperatures (30, 40, and 50°C). This reduction is mainly related to the competition that is established between D1 and D2, and PHM and PHM2 on the same MOF site. It is convenient to note that the presence of a second water pollutant can exclude the first one leading to this reduction of water pollutant amount per MOF site. Referring to this investigation, we can deduce that both pollutants share the MOF site during the adsorption process. In general, the temperature impact plays a relevant to analyze the adsorption process. This effect on the n_D, n_1, and n_2 parameters for D1, D2, PHM1, PHM2-MOF and D1 (D1 + D2), D2 (D1 + D2), PHM1 (PHM1 + PHM2), PHM2 (PHM1 + PHM2) systems is depicted in **Figure 5**. For instance, for dyes (D1 and D2) removal, it was observed that two opposite behaviors were identified in single and binary systems. In this context, for D1 and D2 adsorption systems, it was obtained that there is no linear trend showing that both D1 and D2 shared their MOF sites for adsorption. For D1(D1 + D2), D2 (D1 + D2) systems, an opposite trend was observed indicating that D1 partially excludes D2 or inversely.

3.4.1.2 Estimation and interpretation of the maximum adsorption capacity: Q_m and Q_{mbi} (i = 1,2)

The estimation of the maximum adsorption capacities in single and binary cases can describe the MOF performance (see **Figure 6**) and provides useful results concerning the protection of the environment. A simple comparison between the results in single systems indicated that this adsorbent is characterized by a good performance to eliminate D2 and PHM2 than D1 and PHM1 from environment [31–33]. For an excellent analysis of the presence of a second pollutant in solution, three cases can be discussed *via* the calculation of this report $r = Q_{ib}/Q_s$ (Q_{ib} and Qs represent the adsorption uptakes in binary and single systems, respectively) [34, 35]:

- If r = 1: This investigation indicates that the adsorption capacities in single and binary systems are the same. This means that the presence of D1 with D2, or PHM1 with PHM2 is totally negligible.

- If r < 1: This case corroborates that the presence of D1 with D2, or PHM1 with PHM2decreases the adsorption capacity.

- If r > 1: This behavior shows that there is an increment of binary adsorption capacity: Presence of synergetic effect.

A comparison was achieved indicating that that all maximum adsorption capacities in binary systems were inferior to single systems. This reduction corroborates that during the binary adsorption, an antagonistic impact was present between D1 and D2, and PHM and PHM2.

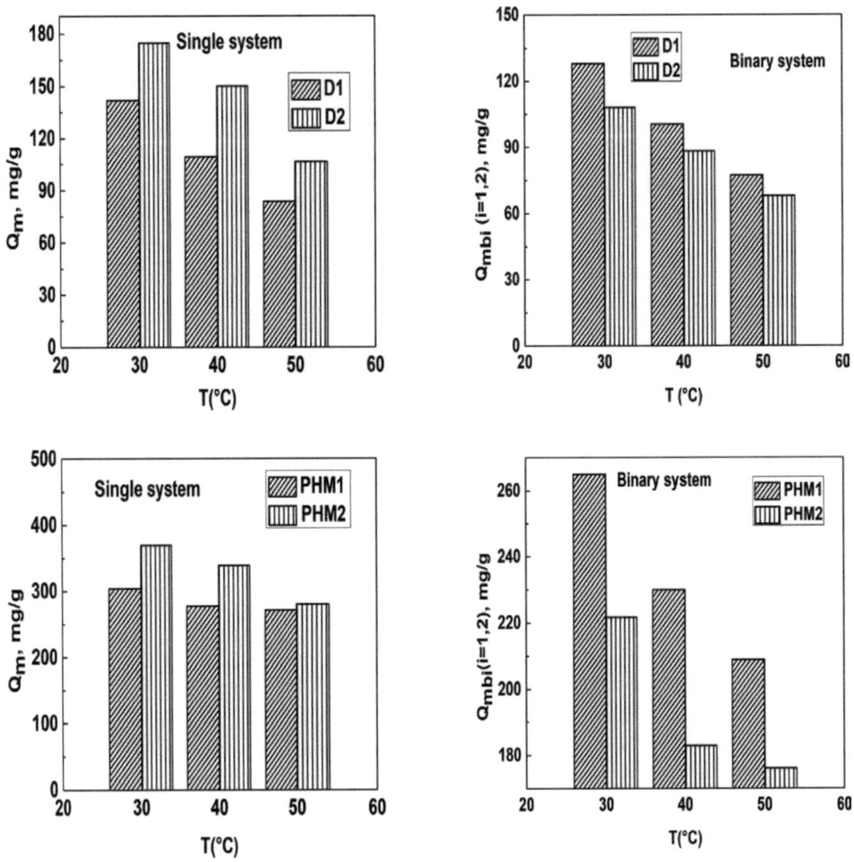

Figure 6.
Difference of adsorption capacity in single and binary systems.

3.4.1.3 Investigation of the parameter S and D_b

The analysis of the adsorption density based on the S and D_b parameters can complete the assessment of the D1, D2, PHM1, and PHM2 adsorption mechanisms. **Figure** 7 explains the impact of temperature on these parameters.

It is convenient to note that the temperature has an opposite impact on these parameters in single and binary systems. This opposite phenomenon was related to the role of temperature on the number of D1 and D2 (also for PHM1 and PHM2) that are removed by the main MOF site. For instance, for D1 and D2 systems, when the uptake of these dyes increased per MOF site, a reduction of its adsorption site density was identified and inversely. For D1 and D2 mixture solutions, an opposite behavior was obtained that is also due in particular due to the competitive effect.

3.4.1.4 Description of the adsorbent performance

In general, the removal of the liquid or gas pollutants can be performed *via* different adsorbents as discussed in introduction. According to the literature, there is

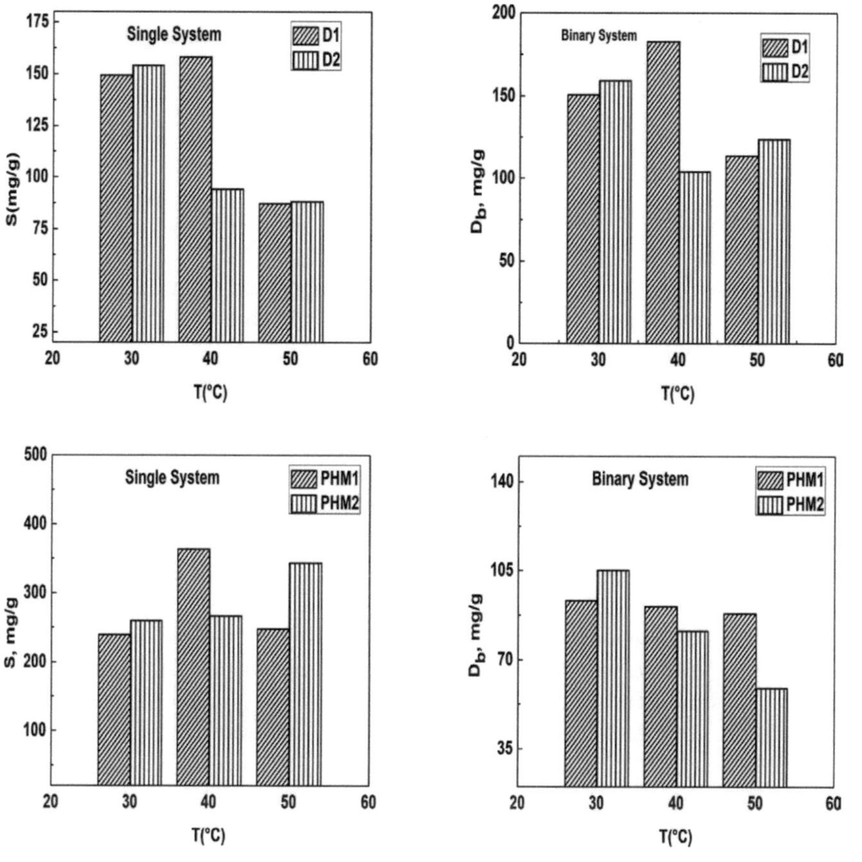

Figure 7.
Impact of the temperature on the S and D_b parameters.

a competition about the choice of the adsorbents to remove these pollutants particularly these dyes and pharmaceuticals. To select the best adsorbents for different applications, it is worth to compare their performances. Comparatively, it was proved that our adsorbent is an excellent material to remove various water pollutants mainly the Direct Blue 1, Direct Yellow-4, amoxicillin, and doripenem. In this direction, its maximum adsorption capacity for these pollutants was compared with other available adsorbents from literature. For instance, the maximum adsorption capacity of amoxicillin (PHM1) was about 300 mg/g (see **Figure 6**) in this study. The maximum adsorption capacities of this PHM1 are 26, 2.3, and 24.7 mg/g by using Kaolinite clay [36], Zr-MOFs nanoparticles [37], and ordered-mesoporous silica SBA-15 [38], respectively. This investigation indicated that our used adsorbent is relevant to protect the environment such as the water of rivers and lakes in single and mixture systems.

3.4.1.5 Energetic assessment

The models offer good option to provide additional insights for a better treatment of all adsorption systems of the used pollutants that is based on energetic investigation. Based on the estimated values of the concentrations at half-saturation, different adsorption energies were determined through the next expressions [39–45]:

$$\Delta E_s = \text{RTln}\left(\frac{C_s}{C_{\text{hs}}}\right) \tag{4}$$

$$\Delta E_{b2} = \text{RTln}\left(\frac{C_s}{C_{\text{ib}}}\right) \tag{5}$$

These expressions characterize the interactions between the pollutants and MOF surface in single and binary systems, respectively. Note that C_S is the water pollutant solubility and R is constant of idea gas. In single systems (D1-MOF, D2-MOF, PHM1-MOF, and PHM2-MOF), the determined values are lower than 23 kJ/mol. This evidence reflected that these water pollutants were eliminated *via* physics interactions. According to the chemical structures of these adsorbates, hydrogen bonding and Van der Walls interactions can mainly participate in the elimination of D1, D2, PHM1, and PHM2. Returning the binary systems (D1 + D2 and PHM1 + PHM2), the calculation indicated that a reduction in values of adsorption energies at different temperatures. They are around 16 kJ/mol. This significant reduction is explained by the competition between D1 and D1, and PHM1 and PHM2. Since the values of these energies are not strong, the inverse process is possible at different temperatures. This means that our adsorbent can be used for additional applications. It is clear that these systems were controlled by all single and binary model parameters. By investigation of the amount of D1, D2, PHM1, and PHM2 per MOF site, it was concluded that these water pollutants shared the main MOF site. This means that the presence of a second pollutant on MOF site excludes the first pollutant and inversely. This partial exclusion is the result to the reduction of the pollutant amount per MOF site in binary systems. This reduction plays a clear role concerning the reduction in D1, D2, PHM1, and PHM2 adsorption capacities in binary systems. In terms of comparison, it was also obtained that there is a reduction in adsorption energies from single to binary systems. Based on these scientific investigations, it is clear that this adsorption process of potential water pollutants was well governed by steric and energetic assessments contrary to the classical studies of these and other water pollutants.

4. Conclusion

In this chapter, an advanced analysis of the adsorption of two relevant pharmaceuticals (PHM1 and PHM2) and dyes (D1 and D2) was developed *via* physical models. Steric investigations were detailed *via* the model parameters. The impact of temperature was also studied on the adsorption data. This study indicated that both tested water pollutants were removed *via* lower aggregation degree. It was deduced that the presence of a second adsorbate in solution reduced the adsorption capacities and also the amounts of the pharmaceuticals and dyes per site. An energetic assessment was developed signaling that physical interactions contributed in the elimination of all pollutants. The experimental and theoretical studies indicated that our adsorbent is a promising material for water treatment from pharmaceuticals and dyes.

DOI: http://dx.doi.org/10.5772/intechopen.1005456

Author details

Lotfi Sellaoui[1,2]* and Fatma Dhaouadi[2]

1 CRMN, Centre for Research on Microelectronics and Nanotechnology of Sousse, NANOMISENE, LR16CRMN01, Sousse, Tunisia

2 Faculty of Sciences of Monastir, Laboratory of Quantum and Statistical Physics, LR18ES18, Monastir University, Monastir, Tunisia

*Address all correspondence to: sellaouilotfi@yahoo.fr

References

[1] Li C, Wang H, Liao X, Xiao R, Liu K, Bai J, et al. Heavy metal pollution in coastal wetlands: A systematic review of studies globally over the past three decades. Journal of Hazardous Materials. 2022;**424**:127312

[2] Jeong C, Ansari MZ, Anwer AH, Kim SH, Nasar A, Shoeb M, et al. A review on metal-organic frameworks for the removal of hazardous environmental contaminants. Separation and Purification Technology. 2023;**305**: 122416

[3] Salahshoori I, Nasirian D, Rashidi N, Hossain MK, Hatami A, Hassanzadeganroudsari M. The effect of silica nanoparticles on polysulfone–polyethylene glycol (PSF/PEG) composite membrane on gas separation and rheological properties of nanocomposites. Polymer Bulletin. 2021; **78**:3227-3258. DOI: 10.1007/s00289-020-03255-8

[4] Salahshoori I, Seyfaee A, Babapoor A. Recent advances in synthesis and applications of mixed matrix membranes. Synthesis and Sintering. 2021;**1**:1-27

[5] Daija L, Selberg A, Rikmann E, Zekker I, Tenno T, Tenno T. The influence of lower temperature, influent fluctuations and long retention time on the performance of an upflow mode laboratory-scale septic tank. Desalination and Water Treatment. 2016;**57**:18679-18687. DOI: 10.1080/19443994.2015.1094421

[6] Salahshoori I, Seyfaee A, Babapoor A, Cacciotti I. Recovery of manganese ions from aqueous solutions with Cyanex 272 using emulsion liquid membrane technique: A Design of Experiment Study. Journal of Sustainable Metallurgy. 2021;7:1074-1090. DOI: 10.1007/s40831-021-00396-6

[7] Salahshoori I, Hatami A, Seyfaee A. Investigation of experimental results and D-optimal design of hafnium ion extraction from aqueous system using emulsion liquid membrane technique. Journal of the Iranian Chemical Society. 2021;**18**:87-107. DOI: 10.1007/s13738-020-02007-9

[8] Salahshoori I, Montazeri N, Yazdanbakhsh A, Golriz M, Farhadniya R, Khonakdar HA. Insights into the adsorption properties of mixed matrix membranes (Pebax 1657- g -chitosan-PVDF-bovine serum albumin@ZIF-CO 3 -1) for the antiviral COVID-19 treatment drugs Remdesivir and Nirmatrelvir: An In Silico study. ACS Applied Materials & Interfaces. 2023;**15**:31185-31205. DOI: 10.1021/acsami.3c03943

[9] Jabeen S, Sufaid Khan M, Khattak R, Zekker I, Burlakovs J, Rubin SS, et al. Palladium-supported zirconia-based catalytic degradation of rhodamine-b dye from wastewater. Water. 2021;**13**: 1522. DOI: 10.3390/w13111522

[10] Garg VK, Gupta R, Yadav AB, Kumar R. Dye removal from aqueous solution by adsorption on treated sawdust. Bioresource Technology. 2003; **89**:121-124. DOI: 10.1016/S0960-8524(03)00058-0

[11] Bhatnagar A, Jain AK. A comparative adsorption study with different industrial wastes as adsorbents for the removal of cationic dyes from water. Journal of Colloid and Interface Science. 2005;**281**: 49-55. DOI: 10.1016/j.jcis.2004.08.076

[12] Ahmad A, Mohd-Setapar SH, Chuong CS, Khatoon A, Wani WA, Kumar R, et al. Recent advances in new

generation dye removal technologies: Novel search for approaches to reprocess wastewater. RSC Advances. 2015;**5**: 30801-30818. DOI: 10.1039/C4RA16959J

[13] Sellaoui L, Bouzidi M, Franco DS, Alshammari AS, Gandouzi M, Georgin J, et al. Exploitation of Bauhinia forficata residual fruit powder for the adsorption of cationic dyes. Chemical Engineering Journal. 2023;**456**:141033. DOI: 10.1016/j.cej.2022.141033

[14] Sellaoui L, Wang H, Badawi M, Bonilla-Petriciolet A, Chen Z. Synergistic adsorption of Pb2+ and CrO42− on an engineered biochar highlighted by statistical physical modeling. Journal of Molecular Liquids. 2020;**312**:113483. DOI: 10.1016/j.molliq.2020.113483

[15] Burtch NC, Jasuja H, Walton KS. Water stability and adsorption in metal–organic frameworks. Chemical Reviews. 2014;**114**:10575-10612. DOI: 10.1021/cr5002589

[16] Panella B, Hirscher M, Pütter H, Müller U. Hydrogen adsorption in metal–organic frameworks: Cu-MOFs and Zn-MOFs compared. Advanced Functional Materials. 2006;**16**:520-524. DOI: 10.1002/adfm.200500561

[17] Khan NA, Hasan Z, Jhung SH. Adsorptive removal of hazardous materials using metal-organic frameworks (MOFs): A review. Journal of Hazardous Materials. 2013;**244–245**: 444-456. DOI: 10.1016/j.jhazmat.2012.11.011

[18] Alhamami M, Doan H, Cheng CH. A review on breathing behaviors of metal-organic-frameworks (MOFs) for gas adsorption. Materials. 2014;**7**:3198-3250. DOI: 10.3390/ma7043198

[19] Du C, Zhang Z, Yu G, Wu H, Chen H, Zhou L, et al. A review of metal organic framework (MOFs)-based materials for antibiotics removal via adsorption and photocatalysis. Chemosphere. 2021;**272**:129501. DOI: 10.1016/j.chemosphere.2020.129501

[20] Nuhnen A, Janiak C. A practical guide to calculate the isosteric heat/enthalpy of adsorption via adsorption isotherms in metal–organic frameworks, MOFs. Dalton Transactions. 2020;**49**: 10295-10307. DOI: 10.1039/D0DT01784A

[21] Mansoorianfar M, Nabipour H, Pahlevani F, Zhao Y, Hussain Z, Hojjati-Najafabadi A, et al. Recent progress on adsorption of cadmium ions from water systems using metal-organic frameworks (MOFs) as an efficient class of porous materials. Environmental Research. 2022;**214**:114113. DOI: 10.1016/j.envres.2022.114113

[22] Karmakar A, Prabakaran V, Zhao D, Chua KJ. A review of metal-organic frameworks (MOFs) as energy-efficient desiccants for adsorption driven heat-transformation applications. Applied Energy. 2020;**269**:115070. DOI: 10.1016/j.apenergy.2020.115070

[23] Abdollahi N, Moussavi G, Giannakis S. A review of heavy metals' removal from aqueous matrices by metal-organic frameworks (MOFs): State-of-the art and recent advances. Journal of Environmental Chemical Engineering. 2022;**10**:107394. DOI: 10.1016/j.jece.2022.107394

[24] Zhang H, Hu X, Li T, Zhang Y, Xu H, Sun Y, et al. MIL series of metal organic frameworks (MOFs) as novel adsorbents for heavy metals in water: A review. Journal of Hazardous Materials. 2022;**429**:128271. DOI: 10.1016/j.jhazmat.2022.128271

[25] Sellaoui L, Gómez-Avilés A, Dhaouadi F, Bedia J, Bonilla-Petriciolet A, Rtimi S, et al. Adsorption of emerging pollutants on lignin-based activated carbon: Analysis of adsorption mechanism via characterization, kinetics and equilibrium studies. Chemical Engineering Journal. 2023;**452**:139399. DOI: 10.1016/j.cej.2022.139399

[26] Sellaoui L, Silva LFO, Badawi M, Ali J, Favarin N, Dotto GL, et al. Adsorption of ketoprofen and 2-nitrophenol on activated carbon prepared from winery wastes: A combined experimental and theoretical study. Journal of Molecular Liquids. 2021;**333**:115906. DOI: 10.1016/j.molliq.2021.115906

[27] Sellaoui L, Soetaredjo FE, Ismadji S, Lima ÉC, Dotto GL, Lamine AB, et al. New insights into single-compound and binary adsorption of copper and lead ions on a treated sea mango shell: Experimental and theoretical studies. Physical Chemistry Chemical Physics. 2017;**19**:25927-25937. DOI: 10.1039/C7CP03770H

[28] Sellaoui L, Soetaredjo FE, Ismadji S, Bonilla-Petriciolet A, Belver C, Bedia J, et al. Insights on the statistical physics modeling of the adsorption of Cd2+ and Pb2+ ions on bentonite-chitosan composite in single and binary systems. Chemical Engineering Journal. 2018;**354**: 569-576. DOI: 10.1016/j.cej.2018.08.073

[29] Dhaouadi F, Sellaoui L, Dotto GL, Bonilla-Petriciolet A, Erto A, Lamine AB. Adsorption of methylene blue on comminuted raw avocado seeds: Interpretation of the effect of salts via physical monolayer model. Journal of Molecular Liquids. 2020;**305**:112815. DOI: 10.1016/j.molliq.2020.112815

[30] Sellaoui L, Dhaouadi F, Reynel H, Mendoza-Castillo D, Bonilla-Petriciolet A, Trejo-Valencia R, et al. Physicochemical assessment of anionic dye adsorption on bone char using a multilayer statistical physics model. Environmental Science and Pollution Research. 2021;**28**:1-8. DOI: 10.1007/s11356-021-15264-9

[31] Huang L, Shen R, Shuai Q. Adsorptive removal of pharmaceuticals from water using metal-organic frameworks: A review. Journal of Environmental Management. 2021;**277**:111389. DOI: 10.1016/j.jenvman.2020.111389

[32] Zango ZU, Jumbri K, Sambudi NS, Abu Bakar HH, Garba ZN, Isiyaka HA, et al. Selective adsorption of dyes and pharmaceuticals from water by UiO metal–organic frameworks: A comprehensive review. Polyhedron. 2021;**210**:115515. DOI: 10.1016/j.poly.2021.115515

[33] Allahbakhshi M, Mahmoodi NM, Mosaferi M, Kazemian H, Aslani H. Synthesis of functionalized metal-organic framework metal-organic framework (MIL-53)/chitosan for removing dye and pharmaceuticals. Surfaces and Interfaces. 2022;**35**:102471. DOI: 10.1016/j.surfin.2022.102471

[34] Mohan D, Singh KP. Single- and multi-component adsorption of cadmium and zinc using activated carbon derived from bagasse – An agriculture waste. Water Resources. 2002;**36**:2304-2318. DOI: 10.1016/S0043-1354(01)00447-X

[35] Qin FB, Xie YN, Liu T, Zhang SZ, Khan SU. Mechanism of competitive adsorption of Pb, Cu and Cd on peat. Environmental Pollution. 2006;**144**: 669-680. DOI: 10.1016/j.envpol.2005.12.036

[36] El, Azzouzi L, El Aggadi S, Ennouhi M, Ennouari A, Kabbaj OK,

Zrineh A. Removal of the amoxicillin antibiotic from aqueous matrices by means of an adsorption process using kaolinite clay. Scientific African. 2022; **18**:01390. DOI: 10.1016/j.sciaf.2022.e01390

[37] Liu L, Cui W, Lu C, Zain A, Zhang W, Shen G, et al. Analyzing the adsorptive behavior of amoxicillin on four Zr-MOFs nanoparticles: Functional groups dependence of adsorption performance and mechanisms. Journal of Environmental Management. 2020;**268**: 110630. DOI: 10.1016/j.jenvman.2020.110630

[38] Pandey P, Shankar A, Biney M, Saini VK. Enhancement in amoxicillin adsorption and regeneration properties of SBA-15 after surface modification with polyaniline. Colloid and Interface Science Communications. 2021;**43**: 100432. DOI: 10.1016/j.colcom.2021.100432

[39] Sellaoui L, Dhaouadi F, Abdulaziz F, Alsehli A, Alsowayigh M, Dotto G, et al. Adsorption of perfluorooctanoic carboxylic and heptadecafluorooctane sulfonic acids via magnetic chitosan: Isotherms and modeling. Environmental Science and Pollution Research. 2023;**30**: 1-8. DOI: 10.1007/s11356-023-30600-x

[40] Atri A, Dhaouadi F, Mechi N, Sellaoui L, Echabaane M, Chaabane RB, et al. Removal of textile pollutants from aqueous medium using biosynthesized CuO nanoparticles: Theoretical comparative investigation via analytical model. Heliyon. 2024;**10**:26285. DOI: 10.1016/j.heliyon.2024.e26285

[41] Sellaoui L, Gerhardt R, Dhaouadi F, Chebaane S, Azhary A, Saedoon H, et al. Novel films prepared from spirulina and chitosan for textile pollutant removal: Experiments and theoretical study of adsorption equilibrium via an advanced theoretical approach. Separation and Purification Technology. 2023;**329**; 125158. DOI: 10.1016/j.seppur.2023.125158

[42] Diao Z, Zhang L, Li Q, Gao X, Gao X, Seliem MK, et al. Adsorption of food dyes from aqueous solution on a sweet potato residue-derived carbonaceous adsorbent: Analytical interpretation of adsorption mechanisms via adsorbent characterization and statistical physics modeling. Chemical Engineering Journal. 2024;**482**:148982. DOI: 10.1016/j.cej.2024.148982

[43] Sellaoui L, Dhaouadi F, Deghrigue M, Bouzidi M, Khmissi H, Dotto GL, et al. A multilayer adsorption of perfluorohexanesulfonic and perfluorobutanesulfonic acids on bio-based polyurethane/chitosan foam: Advanced interpretation of the adsorption mechanism. Chemical Engineering Journal. 2024;**489**:151173. DOI: 10.1016/j.cej.2024.151173

[44] Li Z, Sellaoui L, Dotto GL, Bonilla-Petriciolet A, Lamine AB. Understanding the adsorption mechanism of phenol and 2-nitrophenol on a biopolymer-based biochar in single and binary systems via advanced modeling analysis. Chemical Engineering Journal. 2019;**371**:1-6

[45] Cheknane B, Zermane F, Bouras O, Debord J, Harel M, Bollinger JC, et al. Modeling of binary and ternary batch adsorption systems via multidimensional logistic distribution and statistical physics. Journal of Environmental Chemical Engineering. 2021;**9**:105664. DOI: 10.1016/j.jece.2021.105664

Chapter 2

Plastic Pollution in Inland Waters – A Threat to Life

Marie Serena McConnell

Abstract

This chapter explores a comprehensive approach to mitigating plastic pollution in freshwater ecosystems, aligning recommendations with the United Nations Sustainable Development Goals (SDGs). Regulatory measures, including extended producer responsibility and waste management infrastructure, are essential to curb plastic production. Mitigation strategies emphasize technological innovations, nature-based solutions, and individual actions. Education and outreach activities, targeting schools, communities, and businesses, play a pivotal role in preventing future pollution. Challenges and research needs highlight the evolving nature of the issue, necessitating a deeper understanding of plastic sources, long-term effects, and effective monitoring techniques. The interdisciplinary approach presented integrates environmental, social, and economic dimensions, emphasizing the interconnectedness of sustainability efforts.

Keywords: plastic pollution, freshwater ecosystems, sustainability, sustainable development goals, challenges

1. Introduction

Inland waters refer to bodies of water that are not part of the ocean, including rivers, lakes, ponds, reservoirs, and wetlands [1]. These water systems play a crucial role in the Earth's ecosystem and are of significant importance to both the environment and human societies [2, 3].

Inland waters support a diverse range of aquatic species, including fish, amphibians, insects, and plants. These ecosystems contribute to global biodiversity and provide habitat for many unique and often specialized species. They serve as a vital source of freshwater for human consumption, agriculture, and industrial processes. Rivers and lakes are important reservoirs that store and supply water for various uses, sustaining human life and economic activities [4].

Inland waters provide a variety of ecosystem services, such as water purification, nutrient cycling, and flood control. Wetlands, for example, act as natural filters, improving water quality by trapping pollutants and sediments [5]. Lakes and rivers offer recreational opportunities such as boating, fishing, swimming, and wildlife observation. These activities contribute to the tourism industry and provide social and economic benefits to local communities [6].

Historically, rivers and lakes have been essential for transportation. Even today, many rivers are used for shipping goods, and lakes provide important navigation routes. Inland waters are crucial for agriculture, providing water for irrigation. Many civilizations throughout history have flourished near rivers and lakes due to the availability of water for farming. Rivers are often harnessed for hydropower generation, providing a renewable and relatively clean source of energy. Dams and reservoirs are constructed to regulate water flow and generate electricity [7].

Inland waters play a role in climate regulation by influencing local weather patterns and helping to moderate temperature extremes. They can also act as carbon sinks, sequestering carbon and mitigating climate change. They often hold cultural and spiritual significance for communities. Many societies have developed around rivers and lakes, and these water bodies are integral to their myths, traditions, and rituals. Inland waters are considered biodiversity hotspots, with high concentrations of unique species. Protecting these ecosystems is crucial for maintaining overall global biodiversity [5].

2. From miracle material to global menace: Plastics' rise and fall

The world is draped in plastic. From the clothes we wear to the toys our children play with, from the food we eat to the bottles we sip from, plastic has become omnipresent, infiltrating every facet of our lives. What began as a revolutionary invention in the early 1900s has morphed into a ubiquitous material fuelling a "use and throw" culture with dire environmental consequences [8, 9].

Plastics boast remarkable properties – lightweight, durable, and affordable. These advantages propel them into everything from packaging to furniture, medical equipment to cleaning products. They have revolutionized industries, streamlined transportation, and offered countless conveniences. Yet, the very qualities that make plastics so useful also contribute to their sinister side. Their resistance to degradation, once a marvel, translates to an alarming reality – they persist in the environment for centuries [10].

Landfills groan under the weight of discarded plastic, while oceans choke on an endless tide of plastic debris. Microplastics, insidious fragments born from larger plastic breakdown, infiltrate food chains, poisoning ecosystems and potentially endangering human health. The statistics are staggering: millions of tonnes of plastic waste dumped annually, plastic pollution threatening countless species, and a growing plastic island swirling in the Pacific Ocean [10, 11].

The convenience plastic offers comes at a steep price. Our addiction to this ubiquitous material has unleashed a global pollution crisis. But, amidst the devastation, glimmers of hope emerge. Innovative minds are developing biodegradable alternatives, exploring plastic recycling methods, and promoting responsible consumption habits. Governments are implementing stricter regulations to curb plastic production and encourage better waste management.

The challenge we face is monumental but not insurmountable. We must transition from a disposable mindset to one of responsible stewardship. Investing in sustainable alternatives, embracing reuse and repair, and actively seeking solutions to manage existing plastic waste are crucial steps on this path. By rethinking our relationship with this ubiquitous material, we can rewrite the narrative of plastic, transforming it from a global menace back into a force for good [12–14].

2.1 Beyond the blue bin: Unveiling the hidden sources of plastic pollution

2.1.1 Human reliance on single-use plastics

Single-use plastics like bottles, bags, straws, and food packaging offer ease and convenience, contributing to their popularity. The cheap production cost of these plastics makes them readily available and tempting to manufacturers and consumers. Many people underestimate the far-reaching consequences of discarding single-use plastics after a single use [15].

2.1.2 Inadequate waste management

Many countries lack proper waste collection and disposal systems, leading to plastic ending up in landfills, waterways, and the environment. Practices like open dumping and burning of plastic waste release harmful toxins and microplastics into the air and soil. Even in developed countries, recycling capacity for certain types of plastic is limited, and recycled content in new products remains low [15, 16].

2.1.3 Unregulated industrial practices

Microfibers from synthetic clothing, beads in personal care products like scrubs, and pre-production pellets all contribute to microplastic pollution. Production and transportation of plastic materials can lead to accidental spills and the release of microplastics. Lost or discarded fishing gear, primarily made of plastic, constitutes a significant source of water pollution [16].

2.1.4 Consumer choices

Our preference for readily available and disposable products fuels the production and use of single-use plastics. We prefer "convenient" over sustainability. Lack of awareness about the lifecycle of plastic and its environmental impact can lead to irresponsible disposal practices. Availability and affordability of sustainable alternatives to plastic products sometimes remain a challenge (**Figure 1**) [16, 17].

2.2 Unraveling the web of plastic pollution in inland waters

2.2.1 Invisible menace

The silent tide of plastic pollution has swept inland, infiltrating our rivers, lakes, and freshwater ecosystems. This insidious threat emerges from diverse sources, weaving a complex web of contamination [2, 18].

2.2.2 Micro mania

Microplastics, tiny warriors less than 5 mm in size, dominate the battlefield. Some emerge preformed, shed from cosmetics, textiles, and even fishing nets. Others are battle-hardened veterans, broken down from larger plastic warriors like bottles, their resilience a curse that allows them to linger for centuries. Research tells us these seasoned veterans, the secondary microplastics, outnumber their younger counterparts [19].

Figure 1.
Classification of plastic materials. Adapted from: https://www.plasticsforchange.org/blog/different-types-of-plastic.

2.2.3 Transformation's dance

The fate of these microplastic warriors is a complex ballet. They migrate from manufacturing to consumption, then discard, only to return in an intricate circle back to human society. Within inland waters, they undergo a five-act transformation: suspended pieces, settling sediments, resuspended nomads, buried soldiers, and finally, travelers to the marine world. This intricate dance with the environment allows them to entangle aquatic plants, be ingested by unsuspecting organisms, and ultimately impact us all [19–21].

2.2.4 Agricultural echoes

Beyond the direct battlefield, agriculture plays a subtle role. Millions of tonnes of sewage sludge containing microplastics are used to nourish land, while discarded agricultural films leave their plastic ghost behind. Even irrigation water, tainted with these tiny warriors, carries the pollution further, contaminating soil and eventually seeping into freshwater bodies [21].

2.2.5 Rain's downpour

From the heavens above, another enemy arrives. Rainfall-runoff and atmospheric deposition, like silent assassins, carry pollutants, including plastics, directly into vulnerable small water bodies. Mismanagement and negligence become accomplices, further bolstering the enemy's ranks [22].

DOI: http://dx.doi.org/10.5772/intechopen.1005201

2.2.6 Macroplastics and megaplastics

This web also ensnares larger warriors. Macroplastics, visible to the naked eye, litter riverbanks and lake beds, while megaplastics, giants from the seas and landfills, cast their long shadows. Mismanaged waste, individual carelessness, and industrial leaks all feed the ranks of these visible enemies. Wind, surface runoff, and the very flow of rivers become their chariots, spreading their dominion further. Despite the initial focus on plastic pollution being due to these types of plastic, recent research has not given them much attention [18, 22–24].

Understanding the multifaceted nature of plastic pollution in inland waters is crucial for crafting effective solutions. By recognizing the diverse sources, the dance of transformation, and the silent allies like rain and agriculture, we can begin to untangle this complex web and protect our precious freshwater ecosystems.

3. Freshwater ecosystems: choking on plastic's grip

The crystal-clear image of a serene lake belies a harsh reality – freshwater ecosystems are drowning in plastic. This global environmental crisis is not just confined to the oceans; it is plaguing rivers, lakes, and wetlands at an alarming rate [25–29].

3.1 Biodiversity under siege

The very fabric of freshwater life is unraveling under the onslaught of plastic. Studies reveal over 206 freshwater species, from tiny invertebrates to majestic mammals, bearing the scars of plastic ingestion [25]. Entangled limbs, choked intestines, and internal injuries – the plastic menace mimics its marine counterpart in inflicting a gruesome toll on freshwater fauna [26].

3.2 A toxic cocktail

Beyond physical harm, plastic acts as a Trojan horse, transporting a chemical cocktail into the bodies of aquatic life. These additives and waterborne pollutants can wreak havoc, triggering reproductive problems, growth abnormalities, and even altering organ function [28].

3.3 Ecosystems on edge

Plastic disrupts the delicate dance of life in freshwater ecosystems. It smothers habitats, alters water flow, and hinders natural processes, leaving these vital systems crippled in the face of climate change. Ultimately, the consequences ripple outwards, impacting millions of people's livelihoods, food security, and well-being [29].

3.4 Plankton in peril

Microplastics are not just an eyesore; they pose a grave threat to the ecosystem's silent workhorses – plankton. These microscopic organisms, crucial for carbon sequestration and ecosystem health, are being adversely affected by the plastic invasion, jeopardizing the planet's ability to combat climate change [27].

3.5 Cascading consequences

The boundaries between ecosystems are blurry. Plastic pollution in one system does not stay put; it cascades through interconnected webs, impacting terrestrial, aquatic, and marine realms alike [27]. From land-based sources, plastic contaminates rivers and wetlands, eventually finding its way to the depths of the ocean, leaving no corner untouched.

The plastic pandemic in freshwater ecosystems demands immediate and coordinated action. Robust recycling programs, responsible disposal habits, stringent legislation, regular inspections, and replacing synthetic polymers with eco-friendly alternatives are all crucial steps in this fight. We must not underestimate the severity of this crisis; the situation in freshwater may be as dire as in the oceans, yet it remains largely under-recognized [25].

Only by acknowledging the full scope of this plastic chokehold and taking decisive action can we hope to heal our freshwater ecosystems and restore the delicate balance of life they sustain.

4. Microplastics and nanoplastics: a tangled web of human health threats

Microplastics, minuscule invaders less than 5 millimeters in size, and their even tinier cousins, nanoplastics, are silently pervading our world. We breathe them in the air, drink them in our water, and eat them in our food. This pervasive presence ignites a crucial question: What are the potential impacts of these minuscule menaces on human health and society?

4.1 Ingestion pathways

Our food and water, the very sustenance of life, become unwitting Trojan horses for these plastic particles. Seafood, drinking water, and even the air we breathe are increasingly contaminated [30–36]. This alarming reality suggests a silent invasion, with microplastics insinuating themselves into our very bodies.

4.2 Tissue presence

The grim reality is not limited to external exposure. Recent studies have unearthed microplastics within human tissues, including the gastrointestinal tract, placenta and breast milk [37–40]. While the full implications remain unclear, this discovery sends shivers down the spine, raising potent questions about their potential to harm us from within.

4.3 Bioaccumulation concerns

Could these plastic infiltrators become unwanted tenants in our bodies? Evidence suggests the chilling possibility of bioaccumulation, where microplastics build up in the food chain, eventually reaching our plates. The extent and health effects of this phenomenon are still unfolding, but the potential implications are far-reaching [41–44].

4.4 Potential for physical harm

The mere size of these invaders raises concerns. Microplastics, like silent saboteurs, could cause physical harm due to their ability to accumulate in tissues. Studies hint at inflammation, oxidative stress, and even disruptions in nutrient absorption and the gut microbiome, a vital ecosystem within us [41–44].

4.5 Toxicological threats

Microplastics are not just inert trespassers; they can act as sponges, adsorbing and concentrating environmental pollutants. When ingested, these pollutants are released within our bodies, raising alarms about potential toxicological effects [42, 45]. This adds another layer of complexity to the already murky picture of their impact.

4.6 Inflammatory responses

Beyond physical harm, some research suggests that microplastics can trigger unwelcome guests in the form of inflammatory responses and immunological reactions [46–48]. While the long-term consequences remain obscured, the possibility of internal battles against these invaders is unsettling.

4.7 Unraveling the web of uncertainty

Our understanding of microplastics and nanoplastics and their impact on human health is like a half-woven tapestry. We see glimpses of potential harm – physical disruption, internal battles, and toxic cocktail delivery – but the full picture remains veiled. More research is urgently needed to untangle this web of uncertainty and assess the true scope of the threat these tiny trespassers pose. While answers remain elusive, one thing is clear: inaction is not an option. We must prioritize research, implement preventive measures to curb plastic pollution, and develop strategies to mitigate the potential risks. The health of future generations and the integrity of our ecosystems depend on it.

5. Microplastics and human health: future trends and research focus

5.1 Expanding research efforts

The field of microplastics and human health is anticipated to see a surge in research efforts aimed at understanding the sources, pathways, and potential health effects of microplastics [49–51]. With the rapid advancement in analytical techniques, our ability to detect and quantify microplastics in various biological samples is expected to improve. This will provide more accurate data, thereby enhancing our understanding of the extent and implications of microplastic pollution [52, 53].

5.2 Standardization of methods

There is a growing need for standardized methods to measure and characterize microplastics in human tissues. The development of consistent methodologies will enhance the comparability of studies and contribute to a more comprehensive understanding of exposure levels. This is crucial for establishing reliable baselines and tracking trends over time [52–54].

5.3 Long-term health studies

Future research may involve more longitudinal studies to assess the long-term health effects of microplastic exposure. This includes investigating chronic exposure scenarios and potential cumulative impacts. Such studies are critical for understanding the potential health risks associated with prolonged exposure to microplastics [41–53].

5.4 Regulatory considerations/actions

As awareness of the potential impacts of microplastic pollution on human health grows, there may be increased attention from regulatory bodies. This could lead to the development of guidelines or regulations to mitigate exposure. Such measures are essential for protecting public health and preventing further environmental degradation [55–57].

5.5 Technological innovations/advances

Advances in analytical techniques and technologies may enhance our ability to detect and characterize microplastics, including nanoplastics, in various matrices. This will allow for more accurate risk assessments and inform mitigation strategies [58–60]. The integration of artificial intelligence in microplastic research is also a promising trend, offering new possibilities for data analysis and interpretation [58].

In conclusion, the field of microplastics and human health is poised for significant advancements in the coming years. Through expanding research efforts, standardization of methods, long-term health studies, regulatory actions, and technological innovations, we can gain a deeper understanding of this global issue and develop effective strategies to mitigate its impacts.

6. The impact of micro and nanoplastics on aquatic life in freshwater ecosystems

This segment explores the impact of microplastic and nanoplastic on organisms residing in freshwater ecosystems, including plants, animals, and microbes. Ongoing research in these domains is essential, as much more investigation is required to fully comprehend the situation and understand how organisms respond to plastic pollution.

6.1 The growing threat to freshwater flora

The menace of plastic pollution poses an escalating threat to freshwater flora, impacting the lifeblood of aquatic ecosystems worldwide. The infiltration of microplastics and nanoplastics, derived from larger debris and everyday products, has become pervasive in lakes, rivers, and wetlands, posing multifaceted dangers to aquatic plants. Studies on duckweed and water hyacinths reveal the physical harm inflicted by sharp plastic fragments, causing lacerations and punctures on delicate photosynthetic tissues [61–63]. These injuries compromise plant vigor and productivity. Additionally, nanoparticles create a coating on biofilm surfaces, potentially impeding algal growth [64].

The obstruction of sunlight, essential for photosynthesis, emerges as a significant concern, with nanoplastics lingering in the water column, creating a haze that obstructs sunlight penetration [65]. This plastic-induced shade suppresses the growth of submerged plants, hindering their capacity to fix carbon and maintain biomass. Ingestion of microplastics further intensifies the threat, with studies reporting accumulation on and within plant tissues, raising concerns about enzyme inhibition, cell stress, and metabolic decline [66, 67].

Plastics' role as carriers of harmful chemicals and sponges for waterborne pollutants exacerbates the problem. Common additives like phthalates and flame retardants, confirmed endocrine disruptors, leach into the water from plastics, while organic contaminants like Polychlorinated Biphenyls (PCBs) and pesticides readily bind to plastic surfaces, heightening plant exposure to toxins [68]. Documented impacts include altered cell structures, disrupted pigment production, and weakened enzymatic activity [67].

In the genus *Lemna*, crucial as bioindicators and base feeders in aquatic food webs, exposure to microplastics leads to stunted growth, diminished pigmentation, and cellular membrane disruption [69]. Further concerns arise as research indicates microplastics may enhance metal bioaccumulation in water hyacinths, potentially promoting the spread of invasive plants in disturbed ecosystems [70, 71].

The comprehensive understanding of plastic's impact on freshwater flora is still unfolding, yet existing evidence paints a dire picture. From physical harm to chemical threats, aquatic plants globally demonstrate vulnerability to this pervasive pollution. Safeguarding freshwater ecosystems demands concerted efforts to curtail plastic releases, improve waste management, and advance biodegradable alternatives. Urgent research extending beyond lab studies is crucial to quantifying impacts on natural plant communities.

6.2 Menacing consequences for freshwater fauna

Freshwater ecosystems, vital conduits of life for numerous species, face a formidable adversary – plastic pollution. Microplastics and nanoplastics, minute fragments originating from larger debris and everyday items, now extensively taint rivers, lakes, and streams worldwide. This pervasive menace poses a myriad of challenges for our freshwater fauna.

Ingestion stands out as a major concern, as a diverse array of animals mistakenly consume microplastics, leading to false satiety, nutrient dilution, and potentially fatal blockages. Research, such as a study on Great Lakes fish, reveals that over 80% of them have microplastics in their guts [72, 73]. Ingestion not only hampers growth and reproduction across various species, from zooplankton to fish [74].

Entanglement introduces another layer of peril. Abandoned fishing gear, packaging bands, and other plastic debris ensnare and fatally harm turtles, waterfowl, fish, and more [75, 76]. Apart from causing drowning or strangulation, entanglement inflicts severe injuries, impeding mobility and escalating predation risk.

Plastic debris also acts as a Trojan horse for invasive species and toxins. Microplastics readily absorb pollutants and pathogens, subsequently delivering them to animal tissues upon ingestion [77]. Studies confirm the transfer of heavy metals, petroleum compounds, and other harmful chemicals from plastics to organisms [78]. Furthermore, floating plastic serves as a habitat for invasive species, facilitating their spread [79].

Beyond these documented effects, plastic disrupts natural behaviors like algal grazing and can alter endocrine systems and respiration [80, 81]. The full picture of chronic plastic exposure remains under investigation, but the combined physical and chemical threats are incontrovertible.

Specific instances illustrate a grim reality. Perch larvae exposed to microplastics experience a 40% mortality rate within days [82]. Over half of Asian carp in Illinois rivers harbor nanoplastics [83]. Plastic ingestion by freshwater birds mirrors the devastating impacts observed in their marine counterparts [84].

Even iconic species like river turtles are not immune. More than 50 freshwater turtle species have been documented ingesting or entangling themselves in plastic [76]. For these long-lived reptiles, plastic accumulation can lead to severe health decline over decades, compounding the threats to their already endangered status.

The sheer volume of plastic infiltrating freshwater systems poses a dire threat to biodiversity and ecosystem health. With plastic production expected to surge, immediate action is imperative. Stricter policies, enhanced waste management infrastructure, widespread education on proper disposal and alternatives, and ongoing research on plastic impacts and mitigation are essential (**Figure 2**) [85, 86].

6.3 Microbial battleground: plastic's covert assault on freshwater ecosystems

Freshwater ecosystems, the concealed engines of life, confront an imperceptible adversary: plastic pollution. Microplastics and nanoplastics, minute remnants from larger debris and daily items, now cover lake bottoms and riverbeds. Yet, this intrusion is not merely passive; it actively reshapes the unseen realm of freshwater microbes in intricate and unsettling ways.

Plastics create new habitats for certain bacteria, fostering adaptations like hydrocarbon degradation and specialized "grappling hooks." Genera such as Pseudomonas, Bacillus, and Vibrio dominate these "plastispheres," showcasing remarkable adjustments to their plastic environments [86, 87]. While the overall impact on ecosystem function remains uncertain, it is clear that plastics instigate novel microbial relationships.

However, this apparent symbiosis has a sinister side. Plastics function as carriers for disease-spreading, with pathogenic bacteria like *Vibrio cholera*, *Legionella*, and *Pseudomonas aeruginosa* exploiting these platforms for accelerated dispersal [88, 89]. This disturbing trend suggests that plastic pollution may be facilitating the swift spread of harmful microbes in freshwater systems.

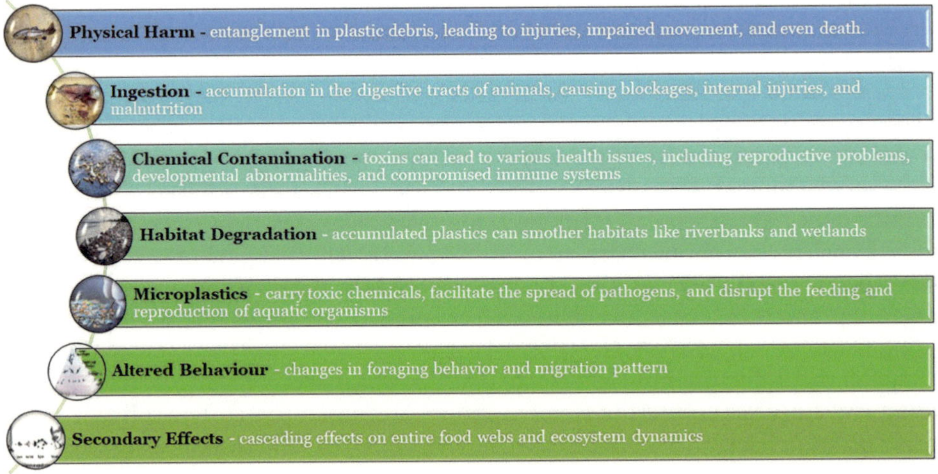

Figure 2.
Consequences of plastic pollution for aquatic life.

The narrative extends further. The intense competition on plastics initiates an evolutionary arms race among bacteria, promoting the emergence of more virulent strains through toxin and antibiotic release [90, 91]. Plastic additives like flame retardants and Bisphenol A (BPA) contribute to this selection pressure, potentially elevating mutation rates and genetic adaptation [92]. In essence, plastic pollution favors the development of hardier, more harmful microbes in freshwater.

Beyond targeting specific species, plastic disrupts microbial diversity at its core. Studies comparing pristine and plastic-laden sediments reveal significant declines in community richness and evenness [93, 94]. This imbalance could have cascading effects, disrupting the crucial roles played by sediment bacteria in nutrient cycling and aquifer health.

Moreover, plastics release a toxic blend of chemicals, including monomers, stabilizers, and pesticides, potent enough to hinder microbial growth at environmentally relevant concentrations [77]. UV-degraded plastics release even more hazardous derivatives, intensifying chemical warfare in an already hostile environment [95].

The threat transcends surface contamination. Vertical transport of microplastics into sediments endangers vital communities responsible for bioremediation and nutrient cycling. N-cycling bacteria and sulfur-oxidizing populations, crucial for water decontamination, are impacted by plastics [96, 97]. This unseen battleground beneath the surface underscores the profound impact plastics have on the foundational elements of freshwater ecosystems.

Adding to the complexity, invasive aquatic plants may benefit from plastic pollution, as debris offers new habitats and microplastics enhance biofilm formation, potentially providing protection from pollutants [79, 98]. This alliance between plastic and invaders may accelerate the spread of undesirable species in waterways.

Finally, plastic fragments infiltrate the sensitive hyporheic zone, the crucial meeting point of groundwater and surface water with unique microbial communities. Significant differences in microbial biodiversity are observed between plastic-contaminated and pristine hyporheic zones, raising concerns about plastic's influence on these critical life support systems.

In conclusion, plastic pollution poses a disconcerting threat to freshwater microbes. While some may flourish on plastic islands, the larger ecosystem bears the brunt of the impact. Reduced diversity, toxic leaching, disrupted biogeochemical services, and the rise of potentially dangerous bacterial strains are alarming consequences.

The future of our freshwater systems, and ultimately our own well-being, hinges on safeguarding these microscopic life forms. Neglecting the hidden world of microbes would be akin to ignoring the very roots of a tree. Let us act now to ensure these vital communities thrive, preserving the health and resilience of our freshwater ecosystems for generations to come.

7. Regulatory and mitigatory measures: a comprehensive approach to plastic pollution

This section is not just a laundry list of Sustainable Development Goals (SDGs) – it is a battle cry. For over 30 years, we have danced around the fringes of reducing, reusing, and recycling plastics. The Earth is at a breaking point, and we cannot waltz our way out anymore. We need to link these actions directly to the SDGs, transforming them into potent weapons against plastic pollution. The time for talk is over; the time for strategic action is now.

Curbing plastic pollution demands a three-pronged attack: stricter regulations, innovative mitigation strategies, and empowered individuals. Linking these efforts to the UN's Sustainable Development Goals (SDGs) offers a roadmap for lasting change. Governments and NGOs are already leading the charge with awareness campaigns and initiatives to curb plastic use. By anchoring the fight within the SDG framework, we gain a shared understanding and a springboard for effective action [99–107].

7.1 Regulatory and policy measures

Our insatiable appetite for convenience has fuelled a global plastic crisis. We churn through 300 million tonnes of plastic annually, half of it disposable, creating a staggering waste stream roughly equivalent to the weight of humanity itself. This plastic tide pollutes ecosystems, endangers marine life, and even infiltrates our food chain. Recognizing the urgency, governments worldwide are stepping up with regulatory measures linked to the UN Sustainable Development Goals (SDGs):

7.1.1 SDG 14: life below water – regulating plastic production

Countries like Rwanda and Kenya are leading the charge with bans on specific single-use items like bags and straws. The EU has followed suit, and India recently enacted a comprehensive ban targeting various disposable plastics. This aligns with SDG 14's call to "conserve and sustainably use the oceans and marine resources" [99, 102].

7.1.2 SDG 12: responsible consumption and production – extended producer responsibility (EPR)

France's heavy levies on plastic packaging producers exemplify EPR policies aimed at shifting responsibility to manufacturers, encouraging them to design for reusability and recyclability. This echoes SDG 12's focus on "promoting sustainable production and consumption patterns" [108, 109].

7.1.3 SDG 15: life on land – protecting ecosystems

Establishing and enforcing protected areas, as Bangladesh is doing with international support, safeguards ecosystems from plastic pollution. This aligns with SDG 15's goal of "protecting, restoring and sustainably managing terrestrial ecosystems" [109].

7.1.4 SDG 6: clean water and sanitation – water quality standards

Setting stringent water quality standards and implementing monitoring systems, as proposed by the UN Environment Program, are crucial steps toward SDG 6's target of "ensuring access to clean water and sanitation for all" [55, 110].

7.2 Mitigation strategies

7.2.1 SDG 13: climate action – reducing the carbon footprint

The fight against plastic goes hand-in-hand with mitigating climate change. Initiatives like Kenya's bamboo straw movement, replacing plastic with a

DOI: http://dx.doi.org/10.5772/intechopen.1005201

fast-growing, naturally compostable alternative, are not just reducing plastic waste but also sequestering carbon [111]. In India, companies like Ecoware are pioneering edible seaweed-based cutlery, further slashing the carbon footprint associated with plastic production [112]. These stories exemplify SDG 13's call to "take urgent action to combat climate change and its impacts."

7.2.2 SDG 11: sustainable cities and communities – waste management infrastructure

Cities stand at the frontline of the plastic battle. Bogota, Colombia, has transformed its landfill into a sprawling eco-park, diverting tonnes of plastic waste from landfills [113]. Meanwhile, Amsterdam is piloting "smart bins" that monitor fill levels and optimize collection routes, boosting efficiency and reducing plastic leakage [114]. Such initiatives embody SDG 11's goal of "making cities and human settlements inclusive, safe, resilient and sustainable."

7.2.3 SDG 9: industry, innovation, and infrastructure – research and innovation

Innovation holds the key to breaking free from our plastic dependence. Researchers around the world are harnessing fungi that degrade plastic, offering a bioremediation solution with immense potential [115, 116]. Meanwhile, start-ups in the United States are developing enzyme cocktails that can break down various types of plastic into recyclable components [117–119]. These cutting-edge solutions align with SDG 9's focus on "promoting sustainable industrial development" and "fostering innovation."

7.3 Individual and community actions

7.3.1 SDG 3: good health and well-being – consumer awareness

Knowledge is power in the fight against plastic. Stories like Adithya Mukerjiee, the 16-year-old activist from India who raised awareness about the dangers of plastic straws, exemplify the potential of individual action [120]. Empowering individuals with information about the health risks associated with microplastic ingestion through campaigns and educational platforms like The Story of Stuff Project's Plastic Free July initiative encourages informed choices toward sustainable alternatives [121]. This resonates with SDG 3's goal of "ensuring healthy lives and promoting well-being at all ages."

7.3.2 SDG 4: quality education – educational campaigns

Educating the next generation is crucial for breaking the cycle of plastic dependence. Initiatives like the Plastic Pollution Coalition's educational resource kits for schools or the Green Schools program in Mexico, which integrates environmental education into the curriculum, are sowing the seeds of change [122]. These efforts embody SDG 4's focus on "ensuring inclusive and equitable quality education and promoting lifelong learning opportunities for all."

7.3.3 SDG 17: partnerships for the goals – community engagement

Collective action holds the key to tackling the plastic crisis head-on. Local campaigns like the Great Pacific Garbage Patch Cleanup, with its dedicated volunteers tackling plastic debris in the ocean [123], or the "Adopt-a-River" initiatives in many countries, demonstrate the power of community engagement [124]. These partnerships between governments, businesses, NGOs, and communities align with SDG 17's call to "strengthen the means of implementation and revitalize the global partnership for sustainable development."

7.4 Everyday actions

The scale of the problem can feel overwhelming, but amidst the plastic tide, a wave of individual action is rising, driven by the spirit of the UN Sustainable Development Goals (SDGs). These everyday choices, big and small, have the power to break our dependence on plastic and build a healthier future for our planet.

7.4.1 SDG 12: responsible consumption and production – plastic-free living

Our everyday choices can create a ripple effect of change. Opting for products with minimal packaging or choosing to refill your own containers at stores like Package Free Shop in London are simple yet impactful steps [125]. Carrying a reusable water bottle and coffee mug, like those offered by companies like Hydro Flask, can dramatically reduce your reliance on single-use plastics [126, 127]. By embracing a "plastic-free living" mantra, we align with SDG 12's call to "promote sustainable production and consumption patterns."

7.4.2 SDG 14: life below water – cleanup initiatives

From organizing local beach cleanups like those coordinated by Surfrider Foundation to joining riverbank restoration projects, individuals can directly combat existing plastic pollution [128–131]. Responsible disposal of collected plastics ensures they do not re-enter the environment, further supporting SDG 14's goal of "conserve and sustainably use the oceans and marine resources."

7.4.3 SDG 6: clean water and sanitation – mindful water use

Conserving water goes hand-in-hand with reducing plastic waste. Every liter saved means fewer plastic bottles are produced and consumed. Additionally, choosing sustainable personal care products and properly disposing of them helps prevent microplastics from entering our water bodies, contributing to SDG 6's mission of "ensuring access to clean water and sanitation for all" [55, 56, 110].

The plastic pollution crisis demands urgent and collaborative action at all levels to create transformative change. While individual actions like recycling and using reusable products are invaluable, we must also harness our collective power to advocate for policy changes, invest in infrastructure, and drive widespread education and awareness. The key is scaling up pilot programs, fostering partnerships to accelerate innovation, empowering communities, and demanding stronger regulations worldwide. By embracing the collective spirit of the SDGs, we can phase out single-use plastics, develop alternative materials and technology, and usher in a new culture of sustainability. The road ahead is long, but each small act creates ripples that can swell into waves of change. If we work together, combining individual action with systemic transformation, we can stem the tide of plastic pollution and secure a healthy planet for generations to come.

8. Future challenges and research needs

The study of plastic pollution in freshwater bodies presents ongoing challenges and areas for future research. As the field evolves, researchers continue to uncover new dimensions of the issue. This section lists out the challenges and areas where research advancement is needed [16, 115, 132–137].

8.1 Comprehensive understanding: challenge

Achieving a comprehensive understanding of the sources, pathways, and fate of plastic pollution in freshwater bodies. *Research Needs*: Conduct detailed studies to identify and quantify various sources of plastic pollution, including primary sources (e.g., mismanaged waste) and secondary sources (e.g., fragmentation of larger plastics).

8.2 Microplastics and nanoplastics: challenge

Understanding the full extent of the impacts of microplastics and, especially, nanoplastics on freshwater ecosystems and human health. *Research Needs*: Investigate the mechanisms of nanoplastic formation, their behavior in aquatic environments, and potential toxicological effects on aquatic organisms and humans.

8.3 Long-term effects: challenge

Assessing the long-term ecological and human health consequences of plastic pollution in freshwater ecosystems. *Research Needs*: Implement longitudinal studies to monitor the persistence of plastics in the environment, their interactions with ecosystems over time, and the potential accumulation of plastics in organisms.

8.4 Remote sensing techniques: challenge

Developing effective remote sensing techniques for monitoring and mapping plastic pollution in large and remote freshwater bodies. *Research Needs*: Explore the use of satellite imagery, drones, and other remote sensing technologies to assess the distribution and dynamics of plastic pollution in lakes, rivers, and other freshwater environments.

8.5 Ecological impacts: challenge

Evaluating the ecological impacts of plastic pollution on different components of freshwater ecosystems, including flora, fauna, and microbial communities. *Research Needs*: Conduct in-depth studies on the effects of plastic pollution on biodiversity, habitat structure, nutrient cycling, and other ecological processes in both aquatic and riparian ecosystems.

8.6 Emerging contaminants: challenge

Identifying and understanding the role of emerging contaminants associated with plastics in freshwater systems. *Research Needs*: Investigate the potential release of chemicals from plastics into the environment and their effects on aquatic organisms and ecosystem functioning.

8.7 Social and economic dimensions: challenge

Incorporating social and economic dimensions into plastic pollution research to address the human dimensions of the issue. *Research Needs*: Explore the social and economic impacts of plastic pollution on communities dependent on freshwater resources. Investigate the effectiveness of different policy measures and community engagement strategies.

8.8 Mitigation and remediation strategies: challenge

Developing effective and scalable mitigation and remediation strategies for reducing plastic pollution and removing existing plastics from freshwater bodies. *Research Needs*: Evaluate the performance of different cleanup technologies, such as floating barriers, skimmers, and innovative materials for plastic capture. Explore nature-based solutions and the potential for natural processes to assist in plastic degradation.

8.9 Stakeholder engagement: challenge

Enhancing stakeholder engagement and collaboration to address plastic pollution at various levels, including government, industry, and local communities. *Research Needs*: Investigate the social dynamics and factors influencing the adoption of sustainable practices. Evaluate the effectiveness of education and outreach programs in changing behaviors related to plastic use and disposal.

8.10 Policy effectiveness: challenge

Assessing the effectiveness of existing policies and regulations in mitigating plastic pollution in freshwater bodies. *Research Needs*: Evaluate the enforcement, compliance, and overall impact of current regulations. Identify gaps and opportunities for strengthening policy frameworks at local, national, and international levels.

As researchers continue to address these challenges and research needs, advancements in the understanding of plastic pollution in freshwater ecosystems will contribute to more effective management strategies and policy interventions. Additionally, interdisciplinary collaboration and the integration of diverse perspectives will be crucial in addressing the complexity of this global environmental issue.

9. Education and outreach: powerful tools in the fight against plastic pollution

The power of education and outreach offers a potent antidote. By informing individuals, communities, and organizations, we can inspire behavioral changes, promote responsible consumption, and foster a sense of environmental stewardship [103, 138–145].

9.1 Engaging minds, shaping habits

At the heart of this revolution lies school curriculum integration. Weaving lessons on plastic pollution and sustainable practices into various subjects equips students with knowledge and empowers them to become responsible citizens. Field trips to recycling centres or polluted environments further solidify learning through visceral experiences.

DOI: http://dx.doi.org/10.5772/intechopen.1005201

9.2 Amplifying the message

But education alone is not enough. Public awareness campaigns utilizing media, celebrities, and influencers ignite a shared urgency. Imagine hard-hitting visuals paired with the voice of a beloved actor urging viewers to ditch single-use plastics. Such campaigns pierce through apathy and spark individual action.

9.3 Mobilizing communities

Community engagement takes education a step further. Interactive workshops empower residents with practical skills for reducing plastic use and proper waste disposal. Cleanup drives and eco-fairs transform awareness into action, fostering a sense of collective responsibility and environmental pride.

9.4 Collaboration and amplification

No one stands alone in this fight. Partnerships with NGOs and environmental organizations leverage their expertise and networks to magnify outreach efforts. Joint initiatives like awareness campaigns or community projects ensure diverse voices and resources reach a wider audience.

9.5 Online resources and action

In an age of digital ubiquity, web platforms and online courses democratize access to knowledge. Imagine an interactive website where users learn about the plastic life cycle, explore sustainable alternatives, and track their own plastic footprint. Such tools empower individuals to become agents of change, regardless of location.

9.6 Corporate responsibility and engagement

Businesses are crucial partners in this journey. Employee training programs on plastic reduction and eco-friendly practices empower workforces to become ambassadors of sustainability within their communities. Collaborations with corporations can promote sustainable packaging, reduce single-use plastics, and amplify awareness campaigns through their extensive reach.

9.7 Creativity and artistic expression

Art speaks where words falter. Artistic installations, sculptures, and performances deliver powerful messages about plastic pollution in a captivating way. Imagine a child's drawing of a choked dolphin going viral, sparking conversations, and prompting action at dinner tables across the globe.

9.8 Citizen science and data collection

Empowering citizens in data collection fosters a sense of ownership and responsibility. Imagine a mobile app where individuals report plastic pollution hotspots, creating a real-time map of the crisis that galvanizes collective action.

9.9 Culturally tailored messages

One size does not fit all. Reaching diverse audiences requires targeted messaging. Culturally aware materials and languages ensure inclusivity and resonate with specific communities, ensuring no voice is left behind.

9.10 Continuous monitoring and improvement

Education is a dynamic process. Feedback mechanisms like surveys and focus groups enable us to assess the effectiveness of outreach programs and adapt based on emerging trends and audience needs.

9.11 Learning in action: education & outreach impact stories

Educational Initiatives Educational initiatives play a crucial role in raising awareness about plastic pollution and its impacts. For instance, Unilever partnered with The Economist Educational Foundation to create a workbook about plastics and packaging [141]. This workbook encourages young people to think about the role of plastic in our society and what more could be done to keep plastic out of the environment.

Another example is the Ocean Purpose Project created by Mathilda D'Silva. After developing three autoimmune diseases due to prolonged contact with polluted water, D'Silva sought to gain a scientific understanding of how plastic molecules work and their impact on health and the environment. Her project aims to transform water polluted by plastic microorganisms into hydrogen, thereby creating energy [144].

Outreach Programs are another effective way to combat plastic pollution. The Zero Waste Challenge, supported by UNESCO Green Citizens, is an example of an outreach program that has made a significant impact [146]. The challenge includes a range of public awareness activities, such as a monthly zero-waste newsletter, a Facebook page for promoting zero-waste events, and an online resource showcasing best practice projects [139].

GreenKayak, co-founded by Oke Carstensen, is another innovative project committed to the fight against plastic pollution. The concept behind GreenKayak is a "win-win": having a good time by renting a kayak free of charge, taking a stroll along the river, and picking up any garbage you come across. By giving people a chance to get active and act collectively, Carstensen's project educates people by offering them an enjoyable, sustainable tourism experience [145, 147].

The Role of Schools: Schools play a vital role in educating students about plastic pollution [139, 142, 147]. The Ocean Plastics Academy, for example, helps students develop a deep understanding of the plastics challenge [139, 145]. The Plastic Clever Schools initiative provides real-world experiences where students can apply new knowledge, practice skills, and build confidence [148, 149].

10. Conclusion

To combat plastic pollution in freshwater ecosystems comprehensively, a multifaceted strategy is necessary, addressing regulatory, mitigation, and individual aspects. The United Nations' Sustainable Development Goals (SDGs) serve as a guiding framework for equitable solutions. Regulatory measures, crucial for tackling the issue at its source, involve enforcing production and disposal regulations, promoting

DOI: http://dx.doi.org/10.5772/intechopen.1005201

sustainable alternatives, and implementing Extended Producer Responsibility (EPR) policies. Mitigation efforts focus on efficient waste management infrastructure, technological innovations, and research into alternative materials to reduce environmental impacts. Nature-based solutions and ecosystem resilience enhancement are also valuable. Individual actions play a pivotal role, with education and outreach initiatives driving behavioral change through awareness campaigns and empowerment. Aligning recommendations with specific SDGs, such as SDG 14 (Life Below Water) and SDG 12 (Responsible Consumption and Production), emphasizes the interconnectedness of environmental and societal goals. Ongoing challenges highlight the evolving nature of plastic pollution, emphasizing the need for continued research, including understanding sources, pathways, and long-term effects. Education and outreach activities emerge as powerful tools for prevention, targeting diverse audiences and utilizing creative approaches. Ultimately, collaboration, innovation, and a commitment to SDGs are essential for a sustainable and resilient future amid the challenges of plastic pollution in freshwater ecosystems.

Acknowledgements

The author acknowledges the use of the free version of Grammarly and Word Tune for the language polishing of the manuscript.

Author details

Marie Serena McConnell
Lady Doak College, Madurai, Tamil Nadu, India

*Address all correspondence to: marieserena@ldc.edu.in

References

[1] Wetzel RG. Limnology: Lake and River Ecosystems. 3rd ed. San Diego: Academic Press; 2001. ISBN 978-0127447605. OCLC 46393244

[2] Grizzetti B, Poikane S. Chapter 2. The importance of inland waters. In: Jones ID, Smol JP, editors. Wetzel's Limnology. 4th ed. Academic Press; 2024. pp. 7-13. ISBN 9780128227015. Available from: https://www.sciencedirect.com/science/article/pii/B9780128227015000021

[3] Spyrakos E, O'Donnell R, Hunter PD, Miller C, Scott M, Simis SGH, et al. Optical types of inland and coastal waters. Limnology and Oceanography. 2018;**63**(2):846-870. DOI: 10.1002/lno.10674

[4] Aylward B, Bandyopadhyay J, Belausteguigotia JC, Borkey P, Cassar AZ, Meadors L, et al. Freshwater ecosystem services. Ecosystems and Human Well-Being: Policy Responses. 2005;**3**:213-256. Available from: https://www.scirp.org/reference/referencespapers?referenceid=2254220

[5] Keddy PA. Wetland Ecology: Principles And Conservation. 2nd ed. Vol. 1 Archived 2013-04-11 at the Wayback Machine. New York: Cambridge University Press; 2010. ISBN 978-0521519403

[6] Dodds WK, Perkin JS, Gerken JE. Human impact on freshwater ecosystem services: A global perspective. Environmental Science & Technology. 2013;**47**(16):9061-9068. DOI: 10.1021/es4021052

[7] Carpenter SR, Stanley EH, Vander Zanden MJ. State of the world's freshwater ecosystems: Physical, chemical, and biological changes. Annual Review of Environment and Resources. 2011;**36**(1):75-99. DOI: 10.1146/annurev-environ-021810-094524

[8] Andrady AL, Neal MA. Applications and societal benefits of plastics. Philosophical Transactions of the Royal Society of London. Series B, Biological Sciences. 2009;**364**(1526):1977-1984. DOI: 10.1098/rstb.2008.0304

[9] Thompson RC, Moore CJ, Vom Saal FS, Swan SH. Plastics, the environment and human health: Current consensus and future trends. Philosophical Transactions of the Royal Society of London. Series B, Biological Sciences. 2009;**364**(1526):2153-2166. DOI: 10.1098/rstb.2009.0053

[10] Chamas A, Moon H, Zheng J, Qiu Y, Tabassum T, Jang JH, et al. Degradation rates of plastics in the environment. ACS Sustainable Chemistry & Engineering. 2020;**8**(9):3494-3511. DOI: 10.1021/acssuschemeng.9b06635

[11] Aljaibachi R, Callaghan A. Impact of polystyrene microplastics on Daphnia magna mortality and reproduction in relation to food availability. PeerJ. 2018;**6**:e4601. DOI: 10.7717/peerj.4601

[12] Earn A, Bucci K, Rochman CM. A systematic review of the literature on plastic pollution in the Laurentian Great Lakes and its effects on freshwater biota. Journal of Great Lakes Research. 2021;**47**(1):120-133. DOI: 10.1016/j.jglr.2020.11.001

[13] Wagner M, Scherer C, Alvarez-Muñoz D, Brennholt N, Bourrain X, Buchinger S, et al. Microplastics in freshwater ecosystems: What we know and what we need to know. Environmental Sciences Europe.

2014;**26**(1):12. DOI: 10.1186/s12302-014-0012-7

[14] Barnes DK, Galgani F, Thompson RC, Barlaz M. Accumulation and fragmentation of plastic debris in global environments. Philosophical Transactions of the Royal Society of London. Series B, Biological Sciences. 2009;**364**(1526):1985-1998. DOI: 10.1098/rstb.2008.0205

[15] Cause of Plastic Pollution in the Environment. Plastic Soup Foundation [Internet]. Plastic Soup Foundation; 2022. Available from: https://www.plasticsoupfoundation.org/en/plastic-problem/plastic-soup/cause-plastic-pollution/.

[16] Iroegbu AOC, Ray SS, Mbarane V, Bordado JC, Sardinha JP. Plastic pollution: A perspective on matters arising: Challenges and opportunities. ACS Omega. 2021;**6**(30):19343-19355. DOI: 10.1021/acsomega.1c02760

[17] Obebe SB, Adamu AA. Plastic pollution: Causes, effects and preventions. International Journal of Applied Science and Technology. 10 May 2020. pp. 85-95. DOI: 10.33564/ijeast.2020.v04i12.011

[18] Cai Y, Li C, Zhao Y. A review of the migration and transformation of microplastics in inland water systems. International Journal of Environmental Research and Public Health. 2021;**19**(1):148. DOI: 10.3390/ijerph19010148

[19] Cheang CC, Ma Y, Fok L. Occurrence and composition of microplastics in the seabed sediments of the coral communities in proximity of a metropolitan area. International Journal of Environmental Research and Public Health. 2018;**15**(10):2270. DOI: 10.3390/ijerph15102270

[20] Corinaldesi C, Canensi S, Dell'Anno A, Tangherlini M, Di Capua I, Varrella S, et al. Multiple impacts of microplastics can threaten marine habitat-forming species. Communications Biology. 2021;**4**(1):431. DOI: 10.1038/s42003-021-01961-1

[21] Koutnik VS, Leonard J, Alkidim S, DePrima FJ, Ravi S, Hoek EMV, et al. Distribution of microplastics in soil and freshwater environments: Global analysis and framework for transport modeling. Environmental Pollution. 2021;**274**:116552. DOI: 10.1016/j.envpol.2021.116552

[22] Issac MN, Kandasubramanian B. Effect of microplastics in water and aquatic systems. Environmental Science and Pollution Research International. 2021;**28**(16):19544-19562. DOI: 10.1007/s11356-021-13184-2

[23] Cera A, Gallitelli L, Scalici M. Macroplastics in lakes: An underrepresented ecological problem? Water. 2022;**15**(1):60. DOI: 10.3390/w15010060

[24] van Emmerik T, Mellink Y, Hauk R, Waldschläger K, Schreyers L. Rivers as plastic reservoirs. Frontiers in Water. 2022;**3**:786936. DOI: 10.3389/frwa.2021.786936

[25] Schwarz AE, Ligthart TN, Boukris E, Van Harmelen T. Sources, transport, and accumulation of different types of plastic litter in aquatic environments: A review study. Marine Pollution Bulletin. 2019;**143**:92-100. DOI: 10.1016/j.marpolbul.2019.04.029

[26] Azevedo-Santos VM, Brito MFG, Manoel PS, Perroca JF, Rodrigues-Filho JL, Paschoal LRP, et al. Plastic pollution: A focus on freshwater biodiversity. Ambio. 2021;**50**(7):1313-1324. DOI: 10.1007/s13280-020-01496-5

[27] Ding R, Tong L, Zhang W. Microplastics in freshwater environments: Sources, fates and toxicity. Water, Air, and Soil Pollution. 2021;**232**(5):181. DOI: 10.1007/s11270-021-05081-8

[28] Khare D. What is the Effect of Plastic on Ecosystems? 2021. Available from: http://www.mongabay.com [Accessed: January 1, 2024]

[29] Bellasi A, Binda G, Pozzi A, Galafassi S, Volta P, Bettinetti R. Microplastic contamination in freshwater environments: A review, focusing on interactions with sediments and benthic organisms. Environments. 2020;**7**(4):30. DOI: 10.3390/environments7040030

[30] Wikipedia Contributors. Plastic Pollution [Internet]. Wikipedia. Wikipedia; 2024. Available from: https://en.wikipedia.org/wiki/Plastic_pollution

[31] Oliveri Conti GO, Ferrante M, Banni M, Favara C, Nicolosi I, Cristaldi A, et al. Micro- and nano-plastics in edible fruit and vegetables. The first diet risks assessment for the general population. Environmental Research. 2020;**187**:109677. ISSN 0013-9351. DOI: 10.1016/j.envres.2020.109677

[32] Zhang Q, Xu EG, Li J, Chen Q, Ma L, Zeng EY, et al. A review of microplastics in table salt, drinking water, and air: Direct human exposure. Environmental Science & Technology. 2020;**54**(7):3740-3751. DOI: 10.1021/acs.est.9b04535

[33] Cox KD, Covernton GA, Davies HL, Dower JF, Juanes F, Dudas SE. Human consumption of microplastics. Environmental Science & Technology. 2019;**53**(12):7068-7074. DOI: 10.1021/acs.est.9b01517

[34] Barboza LGA, Vethaak AD, Lavorante BRBO, Lundebye AK, Guilhermino L, Debris MM. Marine microplastic debris: An emerging issue for food security, food safety and human health. Marine Pollution Bulletin. 2018;**133**:336-348. DOI: 10.1016/j.marpolbul.2018.05.047

[35] Mason SA, Welch VG, Neratko J. Synthetic polymer contamination in bottled water. Frontiers in Chemistry. 2018;**6**:407. DOI: 10.3389/fchem.2018.00407

[36] Wright SL, Kelly FJ. Plastic and human health: A micro issue? Environmental Science & Technology. 2017;**51**(12):6634-6647. DOI: 10.1021/acs.est.7b00423

[37] Sykes EA, Dai Q, Tsoi KM, Hwang DM, Chan WC. Nanoparticle exposure in animals can be visualized in the skin and analysed via skin biopsy. Nature Communications. 2014;**5**:3796. DOI: 10.1038/ncomms4796

[38] Jenner LC, Rotchell JM, Bennett RT, Cowen M, Tentzeris V, Sadofsky LR. Detection of microplastics in human lung tissue using μFTIR spectroscopy. Science of the Total Environment. 2022;**831**:154907. ISSN 0048-9697. DOI: 10.1016/j.scitotenv.2022.154907

[39] Leslie HA, van Velzen MJM, Brandsma SH, Vethaak AD, Garcia-Vallejo JJ, Lamoree MH. Discovery and quantification of plastic particle pollution in human blood. Environment International. 2022;**163**:107199. ISSN 0160-4120. DOI: 10.1016/j.envint.2022.107199

[40] Ragusa A, Notarstefano V, Svelato A, Belloni A, Gioacchini G, Blondeel C, et al. Raman microspectroscopy detection and characterisation of microplastics in human breastmilk. Polymers (Basel). 2022;**14**(13):2700. DOI: 10.3390/polym14132700

[41] Ragusa A, Svelato A, Santacroce C, Catalano P, Notarstefano V, Carnevali O, et al. Plasticenta: First evidence of microplastics in human placenta. Environment International. 2021;**146**:106274. ISSN 0160-4120. DOI: 10.1016/j.envint.2020.106274

[42] Li Y, Tao L, Wang Q, Wang F, Li G, Song M. Potential health impact of microplastics: A review of environmental distribution, human exposure, and toxic effects. Environmental Health. 2023;**1**(4):249-257. DOI: 10.1021/envhealth.3c00052

[43] Campanale C, Massarelli C, Savino I, Locaputo V, Uricchio VF. A detailed review study on potential effects of microplastics and additives of concern on human health. International Journal of Environmental Research and Public Health. 2020;**17**(4):1212. DOI: 10.3390/ijerph17041212

[44] van Raamsdonk LWD, van der Zande M, Koelmans AA, Hoogenboom RLAP, Peters RJB, Groot MJ, et al. Current insights into monitoring, bioaccumulation, and potential health effects of microplastics present in the food chain. Food. 2020;**9**(1):72. DOI: 10.3390/foods9010072

[45] Yee MSL, Hii LW, Looi CK, Lim WM, Wong SF, Kok YY, et al. Impact of microplastics and nanoplastics on human health. Nanomaterials (Basel). 2021;**11**(2):496. DOI: 10.3390/nano11020496

[46] Sun A, Wang W. Human exposure to microplastics and its associated health risks. Environmental Health. 2023;**1**(3):139-149. DOI: 10.1021/envhealth.3c00053

[47] Yang W, Jannatun N, Zeng Y, Liu T, Zhang G, Chen C, et al. Impacts of microplastics on immunity. Frontiers in Toxicology. 2022;**4**:956885. DOI: 10.3389/ftox.2022.956885

[48] Amobonye A, Bhagwat P, Raveendran S, Singh S, Pillai S. Environmental impacts of microplastics and nanoplastics: A current overview. Frontiers in Microbiology. 2021;**12**:768297. DOI: 10.3389/fmicb.2021.768297

[49] Gigault J, El Hadri H, Nguyen B, Grassl B, Rowenczyk L, Tufenkji N, et al. Nanoplastics are neither microplastics nor engineered nanoparticles. Nature Nanotechnology. 2021;**16**(5):501-507. DOI: 10.1038/s41565-021-00886-4

[50] Wang Y, Liu G, Wang Y, Mu H, Shi X, Wang C, et al. The global trend of microplastic research in freshwater ecosystems. Toxics. 2023;**11**(6):539. DOI: 10.3390/toxics11060539

[51] Jenkins T, Persaud BD, Cowger W, Szigeti K, Roche DG, Clary E, et al. Current state of microplastic pollution research data: Trends in availability and sources of open data. Frontiers in Environmental Science. 2022;**10**:912107. DOI: 10.3389/fenvs.2022.912107

[52] Ya-di Z, Tian-jie S, Yan-hua W, Rui-yuan W. Review and future trends of soil microplastics research: Visual analysis based on citespace. Environmental Sciences Europe. 2022;**34**(1):122. DOI: 10.1186/s12302-022-00703-2

[53] Huang Z, Hu B, Wang H. Analytical methods for microplastics in the environment: A review. Environmental Chemistry Letters. 2023;**21**(1):383-401. DOI: 10.1007/s10311-022-01525-7

[54] Potter PM. Microplastics: Emerging trends and research gaps. In: ACS Spring 2021 National Meeting; 05-16 April 2021.

Cincinnati: United States Environmental Protection Agency. Available from: https://cfpub.epa.gov/si/si_public_record_Report.cfm?dirEntryId=351748&Lab=CESER

[55] Al-Azzawi MSM, Kefer S, Weißer J, Reichel J, Schwaller C, Glas K, et al. Validation of sample preparation methods for microplastic analysis in wastewater matrices—Reproducibility and standardization. Water. 2020;**12**(9):2445. DOI: 10.3390/w12092445

[56] European Commission. Directorate-General for Environment, EU Action against Microplastics. Publications Office of the European Union; 2023. Available from: https://data.europa.eu/doi/10.2779/917472

[57] Li Y. Legislation and policy on pollution prevention and the control of marine microplastics. Water. 2022;**14**(18):2790. DOI: 10.3390/w14182790

[58] Gago J, Booth AM, Tiller R, Maes T, Larreta J. Microplastics pollution and regulation. In: Rocha-Santos T, Costa M, Mouneyrac C, editors. Handbook of Microplastics in the Environment. Cham: Springer; 2020. DOI: 10.1007/978-3-030-10618-8_52-1

[59] Zhang Y, Zhang D, Zhang Z. A critical review on artificial intelligence—Based microplastics imaging technology: Recent advances, hotspots and challenges. International Journal of Environmental Research and Public Health. 2023;**20**(2):1150. DOI: 10.3390/ijerph20021150

[60] Lv M, Jiang B, Xing Y, Ya H, Zhang T, Wang X. Recent advances in the breakdown of microplastics: Strategies and future prospectives. Environmental Science and Pollution Research International. 2022;**29**(44):65887-65903. DOI: 10.1007/s11356-022-22004-0

[61] Hollerová A, Hodkovicová N, Blahová J, Faldyna M, Franc A, Pavlokova S, et al. Polystyrene microparticles can affect the health status of freshwater fish – Threat of oral microplastics intake. Science of the Total Environment. 2023;**858**(3):159976. DOI: 10.1016/j.scitotenv.2022.159976

[62] Jia L, Liu L, Zhang Y, Fu W, Liu X, Wang Q, et al. Microplastic stress in plants: Effects on plant growth and their remediations. Frontiers in Plant Science. 2023;**14**:1226484. DOI: 10.3389/fpls.2023.1226484

[63] Ge J, Li H, Liu P, Zhang Z, Ouyang Z, Guo X. Review of the toxic effect of microplastics on terrestrial and aquatic plants. Science of the Total Environment. 2021;**791**:148333. DOI: 10.1016/j.scitotenv.2021.148333

[64] Rehse S, Kloas W, Zarfl C. Short-term exposure with high concentrations of pristine microplastic particles leads to immobilisation of Daphnia magna. Chemosphere. 2016;**153**:91-99. DOI: 10.1016/j.chemosphere.2016.02.133

[65] Yang L, Wang X, Fu W, Xia X, Liu C, Min J, et al. Interactions between nano/micro plastics and suspended sediment in water: Implications on aggregation and settling. Water Research [Internet]. 2019;**161**:486-495. DOI: 10.1016/j.watres.2019.06.018

[66] Chen G, Li Y, Wang J. Occurrence and ecological impact of microplastics in aquaculture ecosystems. Chemosphere. 2021;**274**:129989. DOI: 10.1016/j.chemosphere.2021.129989

[67] Osman DM, Yuan W, Shabaka S, Nyaga MP, Geng J, Yu Y, et al. The threat of micro/nanoplastic to aquatic plants:

Current knowledge, gaps, and future perspectives. Aquatic Toxicology. 2023;**265**:106771. DOI: 10.1016/j.aquatox.2023.106771

[68] Wang W, Gao H, Jin S, Li R, Na G. The ecotoxicological effects of microplastics on aquatic food web, from primary producer to human: A review. Ecotoxicology and Environmental Safety. 2019;**173**:110-117. DOI: 10.1016/j.ecoenv.2019.01.113

[69] Rozman U, Kalčíková G. The response of duckweed Lemna minor to microplastics and its potential use as a bioindicator of microplastic pollution. Plants (Basel). 2022;**11**(21):2953. DOI: 10.3390/plants11212953

[70] Churko EE, Nhamo L, Chitakira M. Phytoremediation capacity of water hyacinth (Eichhornia crassipes) as a nature-based solution for contaminants and physicochemical characterization of lake water. Water. 2023;**15**(14):2540. DOI: 10.3390/w15142540

[71] Ali S, Abbas Z, Rizwan M, Zaheer IE, Yavaş İ, Ünay A, et al. Application of floating aquatic plants in phytoremediation of heavy metals polluted water: A review. Sustainability. 2020;**12**(5):1927. DOI: 10.3390/su12051927

[72] Baldwin AK, Corsi SR, Mason SA. Plastic pollution in Great Lakes tributaries: A watershed-scale, citizen science study. Journal of Great Lakes Research. 2016;**42**:1042-1049

[73] Welden NA, Cowie PR. Environment and gut morphology influence microplastic retention in langoustine, Nephrops norvegicus. Environmental Pollution. 2016;**219**:890-897

[74] Burns EE, Boxall ABA. Microplastics in the aquatic environment: Evidence for or against adverse impacts and major knowledge gaps. Environmental Toxicology and Chemistry. 2019;**38**(11):2776-2796

[75] Sigler M. The effects of plastic pollution on aquatic wildlife: Current situations and future solutions. Water, Air, and Soil Pollution. 2013;**224**:2184

[76] Duncan EM, Broderick AC, Fuller WJ, Galloway TS, Godfrey MH, Hamann M, et al. Microplastic ingestion ubiquitous in marine turtles. Global Change Biology. 2019;**25**(2):744-752. DOI: 10.1111/gcb.14519

[77] Prokić MD, Radovanović TB, Gavrić JP, Faggio C. Ecotoxicological effects of microplastics: Examination of biomarkers, current state and future perspectives. TrAC Trends in Analytical Chemistry. 2019;**111**:37-46. DOI: 10.1016/j.trac.2018.12.001

[78] Avio CG, Gorbi S, Milan M, Benedetti M, Fattorini D, d'Errico G, et al. Pollutants bioavailability and toxicological risk from microplastics to marine mussels. Environmental Pollution. 2015;**198**:211-222. DOI: 10.1016/j.envpol.2014.12.021

[79] Rech S, Macaya-Caquilpán V, Pantoja JF, Rivadeneira MM, Jofre Madariaga D, Thiel M. Rivers as a source of marine litter – A study from the SE Pacific. Marine Pollution Bulletin. 2014;**82**(1-2):66-75. DOI: 10.1016/j.marpolbul.2014.03.019

[80] Stachurski P, Świątkowski W, Ciszewski A, Sarna-Boś K, Michalak A. A short review of the toxicity of dentifrices—Zebrafish model as a useful tool in ecotoxicological studies. International Journal of Molecular Sciences. 2023;**24**(18):14339. DOI: 10.3390/ijms241814339

[81] Mattsson K, Ekvall MT, Hansson LA, Linse S, Malmendal A, Cedervall T. Altered behavior, physiology, and metabolism in fish exposed to polystyrene nanoparticles. Environmental Science & Technology. 2015;**49**(1):553-561. DOI: 10.1021/es5053655

[82] Geremia E, Muscari Tomajoli MTM, Murano C, Petito A, Fasciolo G. The impact of micro- and nanoplastics on aquatic organisms: Mechanisms of oxidative stress and implications for human health—A review. Environments. 2023;**10**(9):161. DOI: 10.3390/environments10090161

[83] Battaglin WA, Duncker JJ, Terrio PJ, Bradley PM, Barber LB, DeCicco L. Evaluating the potential role of bioactive chemicals on the distribution of invasive Asian carp upstream and downstream from river mile 278 in the Illinois waterway. Science of the Total Environment. 2020;**735**:139458. DOI: 10.1016/j.scitotenv.2020.139458

[84] Martins I, Almeida CMR, Ramos S. Plastic pollution in aquatic ecosystems: From research to public awareness. In: Leal Filho W, Azul AM, Brandli L, Lange Salvia A, Wall T, editors. Life Below Water. Encyclopedia of the UN Sustainable Development Goals. Cham: Springer; 2022. pp. 822-833. DOI: 10.1007/978-3-319-98536-7_141

[85] Dikareva N, Simon KS. Microplastic pollution in streams spanning an urbanisation gradient. Environmental Pollution. 2019;**250**:292-299. DOI: 10.1016/j.envpol.2019.03.105. Epub 2019 Apr 2. PMID: 31003141

[86] McCormick AR, Hoellein TJ, Mason SA, Schluep J, Kelly JJ. Microplastic is an abundant and distinct microbial habitat in an urban river. Environmental Science & Technology. 2014;**48**(20):11863-11871. DOI: 10.1021/es503610r

[87] Kumar R, Sharma P, Manna C, Jain M. Abundance, interaction, ingestion, ecological concerns, and mitigation policies of microplastic pollution in riverine ecosystem: A review. Science of the Total Environment. 2021;**782**:146695. DOI: 10.1016/j.scitotenv.2021.146695

[88] Pedrotti ML, de Figueiredo Lacerda AL, Petit S, Ghiglione JF, Gorsky G. Vibrio spp and other potential pathogenic bacteria associated to microfibers in the North-Western Mediterranean Sea. PLoS One. 2022;**17**(11):e0275284. DOI: 10.1371/journal.pone.0275284

[89] Arias-Andres M, Klümper U, Rojas-Jimenez K, Grossart HP. Microplastic pollution increases gene exchange in aquatic ecosystems. Environmental Pollution. 2018;**237**: 253-261. DOI: 10.1016/j.envpol.2018.02.058

[90] Marathe NP, Bank MS. The microplastic-antibiotic resistance connection. In: Environmental Contamination Remediation and Management. Switzerland AG: Springer Nature; 2022. pp. 311-322. DOI: 10.1007/978-3-030-78627-4_9

[91] Zhang Q, Fan D, Pang X, Zhu W, Zhao J, Xu J. Effects of polyethylene microplastics on the fate of antibiotic resistance genes and microbial communities in anaerobic digestion of dairy wastes. Journal of Cleaner Production. 2021;**292**:125909. DOI: 10.1016/j.jclepro.2021.125909

[92] Wang F, Guan W, Xu L, Ding Z, Ma H, Ma A, et al. Effects of nanoparticles on algae: Adsorption, distribution, ecotoxicity and fate.

Applied Sciences. 2019;**9**(8):1534. DOI: 10.3390/app9081534

[93] Grbić J, Helm PA, Athey SN, Rochman CM. Microplastics entering northwestern Lake Ontario are diverse and linked to urban sources. Water Research. 2020;**174**:115623. DOI: 10.1016/j.watres.2020.115623

[94] Mani T, Primpke S, Lorenz C, Gerdts G, Burkhardt-Holm P. Microplastic pollution in benthic midstream sediments of the Rhine River. Environmental Science & Technology. 2019;**53**(10):6053-6062. DOI: 10.1021/acs.est.9b01363

[95] Rauchschwalbe MT, Höss S, Haegerbaeumer A, Traunspurger W. Long-term exposure of a free-living freshwater micro- and meiobenthos community to microplastic mixtures in microcosms. Science of the Total Environment. 2022;**827**:154207. DOI: 10.1016/j.scitotenv.2022.154207

[96] Shen M, Song B, Zhou C, Almatrafi E, Hu T, Zeng G, et al. Recent advances in impacts of microplastics on nitrogen cycling in the environment: A review. Science of the Total Environment. 2022;**815**:152740. DOI: 10.1016/j.scitotenv.2021.152740

[97] Triebskorn R, Braunbeck T, Grummt T, Hanslik L, Huppertsberg S, Jekel M, et al. Relevance of nano- and microplastics for freshwater ecosystems: A critical review. TrAC Trends in Analytical Chemistry [Internet]. 2019;**110**:375-392. DOI: 10.1016/j.trac.2018.11.023

[98] Zhang Y, Lü J, Wu J, Wang J, Luo Y. Potential risks of microplastics combined with superbugs: Enrichment of antibiotic resistant bacteria on the surface of microplastics in mariculture system. Ecotoxicology and Environmental Safety [Internet]. 2020;**187**:109852. DOI: 10.1016/j.ecoenv.2019.109852

[99] Academies of Sciences, Engineering, and Medicine. Emerging technologies to advance research and decisions on the environmental health effects of microplastics. In: Brief. Washington. Proceedings of the Workshop. DC: The National Academies Press; 2020. DOI: 10.17226/25862

[100] United Nations. Community Action against Plastic Waste (CAPWs); 2022. Available from: https://www.un.org/en/civil-society/community-action-against-plastic-waste-capws

[101] World Economic Forum. Circular Economy, How Empowering Local Communities Can Help Solve Global Plastic Wastes. 2022. Available from: https://www.weforum.org/agenda/2022/01/empowering-local-communities-help-solve-global-plastic-waste/

[102] Diana Z, Vegh T, Karasik R, Bering J, Juan D, Pickle A, et al. The evolving global plastics policy landscape: An inventory and effectiveness review. Environmental Science & Policy. 2022;**134**:34-45. ISSN 1462-9011. DOI: 10.1016/j.envsci.2022.03.028

[103] Ministry of Environment Forest and Climate Change. Government of India; August 13 2021, Gazette Notification: Guidelines for Extended Producer Responsibility Given Legal Force. Available from: http://pib.gov.in/PressReleaseIframePage.aspx?PRID=1745433

[104] Maureen V. Public Awareness [Internet]. Plastic Smart Cities; 2023. Available from: https://plasticsmartcities.org/public-awareness/

[105] EARTHDAY.ORG. The 6 Rs of Plastic Pollution - Earth Day [Internet].

Earth Day; 2023. Available from: https://www.earthday.org/actions/act-on-plastic-pollution/

[106] Deena R. 15 Ways to Reduce Your Plastic Waste. 2021. Available from: https://earth.org/ways-to-reduce-your-plastic-waste

[107] Engler S. 10 Ways to Reduce Plastic Pollution [Internet]. 2016. Available from: https://www.nrdc.org/stories/10-ways-reduce-plastic-pollution

[108] Cho R. What You Can Do to Fight Plastic Pollution, State of the Planet. Columbia Climate School; 2018. Available from: https://news.climate.columbia.edu/2018/05/11/can-fight-plastic-pollution/

[109] De Clercq G. France Bans Plastic Packaging for Fruit and Vegetables. 2021. Available from: https://www.reuters.com/business/environment/france-bans-plastic-packaging-fruit-vegetables-2021-10-11/.

[110] Chakma P. Laws and Policies on Plastic Waste Management. 2023. Available from: https://www.thedailystar.net/law-our-rights/news/laws-and-policies-plastic-waste-management-3374806

[111] UNEP – UN Environment Programme. Tackling Plastic Pollution: Legislative Guide for the Regulation of Single-Use Plastic Products. 2021. Available from: https://www.unep.org/resources/toolkits-manuals-and-guides

[112] Cstraight Media - http://www.cstraight.com. Launch of the Green Belt Movement's Community-Based Bamboo Craft Center|The Green Belt Movement [Internet]. Available from: https://www.greenbeltmovement.org/node/911

[113] Khakhariya N. In Gujarat, Seaweed Food Packs to Fight Use of Plastic. The Times of India [Internet]; 2023. Available from: https://timesofindia.indiatimes.com/city/rajkot/in-gujarat-seaweed-food-packs-to-fight-use-of-plastic/articleshow/103547386.cms

[114] Circular Economy Initiatives in Colombia [Internet]. CIL; 2023. Available from: https://www.circularinnovationlab.com/post/circular-economy-initiatives-in-colombia

[115] Cai Z, Li M, Zhu Z, Wang X, Huang Y, Li TJ, et al. Biological degradation of plastics and microplastics: A recent perspective on associated mechanisms and influencing factors. Microorganisms. 2023;**11**(7):1661. DOI: 10.3390/microorganisms11071661

[116] Ekanayaka AH, Tibpromma S, Dai DQ, Xu R, Suwannarach N, Stephenson SL, et al. A review of the fungi that degrade plastic. Journal of Fungi. 2022;**8**(8):772. DOI: 10.3390/jof8080772

[117] Colletta R. Mushroom Magic: Fantastic Fungi Fight Plastic Waste [Internet]. Earth Day; 2023. Available from: https://www.earthday.org/mushroom-magic-fantastic-fungi-fight-plastic-waste/

[118] Ballerstedt H, Tiso T, Wierckx N, Wei R, Avérous L, Bornscheuer UT, et al. MIXed plastics biodegradation and upcycling using microbial communities: EU Horizon 2020 project MIX-UP started January 2020. Environmental Sciences Europe. 2021;**33**(1):99. DOI: 10.1186/s12302-021-00536-5

[119] Knott BC, Erickson E, Allen MD, Gado JE, Graham R, Kearns FL, et al. Characterization and engineering of a two-enzyme system for plastics depolymerization. Proceedings of the National Academy of Sciences of the United States of America.

2020;**117**(41):25476-25485. DOI: 10.1073/pnas.2006753117

[120] Ong SK. The Living Things that Feast on Plastic. Knowable Magazine; Annual reviews. 24 Aug 2023. DOI: 10.1146/knowable-082423-1. Available from: https://knowablemagazine.org/content/article/food-environment/2023/how-to-recycle-plastic-with-enzymes

[121] Balaji R. Meet the 16-Year-Old Who's Made it his Mission to Stop the Use of Plastic Straws [Internet]. YourStory.com; 2020. Available from: https://yourstory.com/socialstory/2020/09/aditya-plastic-straw-waste-awareness-campaign

[122] Fight Plastic Pollution – Story of Stuff [Internet]. Story of Stuff; 2023. Available from: https://www.storyofstuff.org/plastic/

[123] Plastic Pollution Coalition. Plastic-Free School Resources [Internet]. Plastic Pollution Coalition; 2024. Available from: https://www.plasticpollutioncoalition.org/plastic-free-school-resources

[124] The Ocean Cleanup. The Great Pacific Garbage Patch • The Ocean Cleanup [Internet]. The Ocean Cleanup; 2024. Available from: https://theoceancleanup.com/great-pacific-garbage-patch/

[125] Community Action for Fresh Water [Internet]. UNEP – UN Environment Programme. Available from: https://www.unep.org/topics/fresh-water/nature-based-solutions/working-nature/community-action-fresh-water#:~:text=The%20earlier%20Rotary%20International%20and,the%20local%20level%20under%20the%20%22

[126] Bell J. 17 Zero Waste Shops in London for Sustainable Living in the Big Smoke [Internet]. Sustainable Jungle; 2024. Available from: https://www.sustainablejungle.com/zero-waste/zero-waste-shops-london/

[127] AGI Glaspac. 7 Reasons Why You Should Carry a Reusable Water Bottle [Internet]. AGI Glaspac; 2021. Available from: https://www.agi-glaspac.com/7-reasons-why-you-should-carry-a-reusable-water-bottle/

[128] 6 Reusable Water Bottles You Can Carry With You [Internet]. MyBorosil. Available from: https://myborosil.com/blogs/borosil-blog/6-reusable-water-bottles-you-can-carry-with-you

[129] Beach Cleanups [Internet]. Surfrider Foundation. Available from: https://cleanups.surfrider.org/about/beach-cleanups/

[130] A Beach Clean-Up Like No Other in Mumbai [Internet]. WWF-India. Available from: https://www.wwfindia.org/news_facts/feature_stories/a_beach_clean_up_like_no_other_in_mumbai/

[131] Farm Advisory Scotland. Riverbank Restoration|Helping Farmers in Scotland. Farm Advisory Service [Internet]. FAS; 2022. Available from: https://www.fas.scot/environment/water-management/water-margins-2/riverbank-restoration/

[132] Seidl R, Stauffacher M. Evaluation of river restoration by local residents. Water Resources Research. 2013;**49**(10):7077-7087. DOI: 10.1002/2013WR013988

[133] Kibria MG, Masuk NI, Safayet R, Nguyen HQ, Mourshed M. Plastic waste: Challenges and opportunities to mitigate pollution and effective management. International Journal of Environmental Research. 2023;**17**(1):20. DOI: 10.1007/s41742-023-00507-z

[134] Tiwari RK, Azad N, Dutta D, Yadav BR, Kumar S. A critical review and

future perspective of plastic waste recycling. Science of the Total Environment. 2023;**881**:163433. DOI: 10.1016/j.scitotenv.2023.163433

[135] Diana ZT, Virdin J, Valiyaveettil S, Li H-X, Rittschof D. Editorial: Emerging challenges and solutions for plastic pollution. Frontiers in Marine Science. 2023;**10**:1162680. DOI: 10.3389/fmars.2023.1162680

[136] Grande-Tovar CD, Carrascal-Hernández DC, Trilleras J, Mora K, Arana VA. Microplastics' and nanoplastics' interactions with microorganisms: A bibliometric study. Sustainability. 2022;**14**(22):14761. DOI: 10.3390/su142214761

[137] Fava MF. Ocean Plastic Pollution an Overview: Data and Statistics. Ocean Literacy Portal; 2022. Available from: https://oceanliteracy.unesco.org/plastic-pollution-ocean/

[138] Detecting Plastic Pollution|CMEMS [Internet]. Available from: https://marine.copernicus.eu/explainers/phenomena-threats/plastic-pollution/detecting-plastic-pollution

[139] Educate and Inspire Young People to Take Action on Plastic. Unilever [Internet]. Unilever; 2023. Available from: https://www.unilever.com/planet-and-society/take-action/initiative/educate-and-inspire-young-people-to-take-action-on-plastic/

[140] Liu J, Hu Z, Du F, Tang W, Zheng S, Lu S, et al. Environment education: A first step in solving plastic pollution. Frontiers in Environmental Science. 2023;**11**:1130463. DOI: 10.3389/fenvs.2023.1130463

[141] Chauhan NS, Punia A. Role of education and society in dealing plastic pollution in the future. In: Ahamad A, Singh P, editors. Plastic and microplastic in the environment: Management and health risks, Dhanesh Tiwary Online. Hoboken, NJ, USA: John Wiley & Sons Ltd; © 2022. ISBN: 9781119800897, Print ISBN: 9781119800781

[142] Plastic Pollution: Students and Leaders Share Solutions in an UNESCO. Campus [Internet]. United Nations Educational, Scientific and Cultural Organization; 2023. Available from: https://www.unesco.org/en/articles/plastic-pollution-students-and-leaders-share-solutions-unesco-campus

[143] Cristian Aedo JSSKS, Abhas JSSKS, Aedo C, Abhas JSSKS, Aedo C, Abhas JSSKS, et al. Tackling Plastic Pollution and Climate Change through Education [Internet]. World Bank blogs; 2023. Available from: https://blogs.worldbank.org/endpovertyinsouthasia/tackling-plastic-pollution-and-climate-change-through-education

[144] Ocean Purpose Project. Available from: https://www.oceanpurposeproject.com/about-1-1

[145] Mironenko O, Mironenko E. Education against plastic pollution: Current approaches and best practices. In: Stock F, Reifferscheid G, Brennholt N, Kostianaia E, editors. Plastics in the Aquatic Environment - Part II. The Handbook of Environmental Chemistry. Vol. 112. Cham: Springer; 2020. pp. 67-93. DOI: 10.1007/698_2020_486

[146] UNESCO: Building Peace through Education, Science and Culture, communication and information. n.d. Available from: https://www.unesco.org/en

[147] GreenKayak. Paddle for the Oceans! [Internet]. Available from: https://www.greenkayak.org/

[148] Seas C. Ocean Plastics Academy is Now Accessible to Schools across Greece [Internet]. Common Seas; 2021. Available from: https://commonseas.com/news/ocean-plastics-academy-is-now-accessible-to-schools-across-greece

[149] Plastic Clever Schools. Reducing Single-Use Plastics [Internet]. Available from: https://plasticcleverschools.co.uk/

Chapter 3

Removal of Heavy Metals and Purification of Surface Waters

Mario Mariglia and Edineldo Lans-Ceballos

Abstract

In many parts of the world, the removal of heavy metals and turbidity in surface waters for potabilization is a focus of study, as it has become a public health problem. Many researchers develop methodologies for the removal of heavy metals; however, these are generally difficult to implement in rural areas due to the danger associated with the use of chemical reagents by individuals with little or no education. That is why this chapter aims to conduct a comprehensive literature review, including an original research project developed by our working group, to identify affordable methods for the potabilization of surface waters in small communities with minimal technology. Additionally, it aims to provide data for better understanding. These economical, sustainable, and efficient methods will help improve the treatment of surface waters for human consumption, using plants and their extracts for the removal of turbidity and various heavy metals. The methods proposed in this chapter for the potabilization of natural water are a contribution to green analytical chemistry. The objective of this chapter is to present metals, removal techniques, and evaluate the efficiency of removing Fe and Mn during phytoremediation processes. To carry out this work, three stages were considered. In the first stage, the quantity of plant material required for the process is selected. In the second stage, the metal removal capacity is determined with the selected mass, and in the third stage, the aim is to understand the removal levels of extracts from *Opuntia ficus-indica* and *Hylocereus triangularis* to compare it with *Hydrilla verticillata* (L.f.) Royle. The determination of metals was performed using atomic absorption spectroscopy with a previously validated method. The *Hydrilla* exhibited a 100% removal of Fe and Mn, as well as a reduction in turbidity and color of 92% and 94%, respectively. The *Hydrilla verticillata* (L.f.) Royle and *Opuntia ficus-indica* are plants that can be used for the removal of Fe, Mn, turbidity, and color in natural waters. The *Hylocereus triangularis* did not show efficient results in removing these metals in natural waters. Similar to *Hylocereus* and *Hydrilla, Opuntia* can be used for the removal of turbidity from natural waters, which is useful for rural communities.

Keywords: plant material, manganese, iron, water, removal

1. Introduction

Aquatic ecosystems essential for human consumption are riddled with numerous contaminants, including heavy metals, which constitute a major issue in developing countries due to the significant pollution they introduce to surface waters [1]. This level of contamination is a result of anthropogenic activities. While various water treatment methods exist, many of them are effective but costly [2]. Hence, there is a need to employ appropriate and low-cost technologies for the removal of such contaminants. Considering that rural areas in developing countries lack the capacity for expensive water treatments, it is essential to highlight the different methods used for contaminant removal. Furthermore, it is of utmost importance to showcase effective and economical methods that enable the removal of heavy metals.

There is interest in a wide variety of heavy metals as they pose a health problem for humans. Among them, we can find zinc (Zn), copper (Cu), cadmium (Cd), lead (Pb), arsenic (As), iron (Fe), manganese (Mn), and chromium (Cr) [2, 3]. These metals can cause harmful effects on the environment and the living organisms inhabiting it [2]. Hence, it is important to monitor water quality concerning the various contaminants that may be present in surface waters.

To assess water quality, the Water Quality Index (WQI) [4] is used, which indicates the condition of a water source. If the WQI falls between 91 and 100, the water quality is considered good; 71–90 indicates acceptable water quality; 51–70 suggests regular water quality; 26–50 denotes poor water quality, and 0–25 implies the lowest level of water quality. This index varies based on the geography and water usage of the country [5]. Water quality regarding heavy metals is determined by two indices: the Heavy Metal Pollution Index (HPI), which shows the combined effect of metals in surface waters [6], and the Heavy Metal Evaluation Index (HEI), which indicates the state of the water body. The state of the water body can be affected by various sources of pollution, including mining, agriculture, volcanoes, forest fires, and urbanization, among others.

One crucial aspect to consider regarding heavy metals is their toxicity, as they jeopardize the natural balance and human health [7]. Prolonged exposure to these metals is associated with various diseases such as Alzheimer's, neurological issues, osteoporosis [2], and can even lead to death. For these and many more reasons, this text aims to conduct a thorough review of different methods for heavy metal removal. We will focus on those methods that are cost-effective and also involve the use of plants or plant extracts for the decontamination of surface waters, such as coagulation/flocculation and phytoremediation. These methods are highly beneficial as they enable the removal of contaminants in a natural way without the use of chemicals.

The *B. papyrifera* and *K. paniculata* have been used for manganese removal in soil [8], *Duckweed* and *tape grass* for copper removal in water [9], Water Jacinta (*Eichhornia crassipes*), Water Lettuce (*Pistia stratiotes*) [10, 11] used for water remediation, Guasimo (*Guazuma ulmifolia*), prickly pear cactus (*Opuntia ficus-indica*), and dragon fruit (*Hylocereus triangularis*) [12]. These studies demonstrate that they are an environmentally friendly alternative.

For contaminant removal, it employs various types of hyperaccumulating plants known as macrophytes, which are beneficial in removing metals from water through mechanisms such as phytoextraction, rhizofiltration, phytovolatilization, and phytostabilization [13]. Chelation is the mechanism used by plants to retain heavy metals [14].

Phytoremediation has its advantages and disadvantages. Among its disadvantages is the difficulty of disposing of biomass with adsorbed contaminants. However, there are different disposal mechanisms such as incineration treatments, pyrolysis, gasification, and microbial extraction, among others [13, 15].

Hydrilla Verticillata is an aquatic plant belonging to the *Hydrocharitaceae* family. It is considered an aggressive and invasive species that has been used in studies for the removal of red dye 120 from simulated wastewater [16, 17] and phosphates in domestic wastewater [18]. The interest in studying this plant lies in its abundance in our aquatic ecosystems, the lack of contaminant removal through conventional treatments, and its potential use in removing Fe and Mn, which contribute to a brown color in the water, giving it an unhygienic appearance.

2. Heavy metals: Their sources and toxicity

2.1 Heavy metals

Heavy metals are considered those whose density is greater than (4 g/cm3), and they are toxic at low concentrations [19, 20]. Among these, we can find (Hg), (Zn), (Cu), (Cd), (Pb), (As), (Fe), (Mn), (Cr), (Ni), and (Co), among others [3, 20].

Mercury (Hg) is a silver-colored metal, liquid at room temperature, with oxidation states (+4, +2, +1). It accumulates in most living organisms, being extremely toxic and causing diseases such as digestive syndromes (nausea, bad breath, vomiting, diarrhea, etc.), neurological syndromes, ophthalmological syndromes, dermatitis, rhinitis, and hypersensitivity, among others [21].

Zinc (Zn) is an element with oxidation states (−2, 0, +1, +2), appearing as a solid with a silver-gray appearance. Although essential for humans, excessive intake increases the likelihood of prostate cancer, elevated cholesterol levels, increased testosterone, and can cause immune dysfunction [22].

Copper (Cu) with oxidation states (+1 and + 2), this reddish solid can cause stress to aquatic biota [23]. Additionally, in large quantities, it can exert toxic effects, physiological alterations, and acute toxicity are also consequences of exposure to Cu [24, 25].

Cadmium (Cd) is a metal characterized by a bluish-white solid appearance and oxidation states +1 and +2. This metal possesses the four most dangerous characteristics that can be attributed to a toxic contaminant: bioaccumulation, ease of transport through water and air, toxicity to humans, and persistence in the environment. Due to its bioaccumulative nature, it is a causative factor for conditions in the lungs, bones, and kidneys [26].

Lead (Pb) is a dark-gray element with oxidation states (+2, +4), harmful to the central nervous system and affecting balance, as well as the direction and range of muscle movement [27].

Arsenic (As) with oxidation states (+3, +5), this metal can be found in gray, yellow, and black colors due to its three allotropic states. It is extremely toxic, causing poisonings, as well as a predisposition to cancer in the lungs, skin, and bladder [28].

Iron (Fe) is a gray-silver metal with oxidation states (+2, +3). It can generate free radicals, leading to oxidative stress, and can cause poisoning due to its accumulation in the liver. Iron intoxication affects the central nervous, gastrointestinal, cardiovascular systems, and metabolic functions [29].

Chromium (Cr) with a bright whitish-gray color and oxidation states (+2, +3, +4, +6), can cause damage to the skin and eyes, and it is also carcinogenic. Its intake may lead to kidney failure, bronchitis, hemorrhages, and gastrointestinal damage [30].

Manganese (Mn) is a whitish-gray transition metal, with the most common oxidation states being (+2, +3, +4, +6, +7). It causes respiratory problems, blocks the neurotransmitter system, and also leads to a syndrome called manganism, where the patient exhibits symptoms similar to Parkinson's [31].

Niquel (Ni) Elemento solido de color blanco ligeramente amarillento con estados de oxidación (+3, +2, 0). Su exposición prolongada u ingesta puede provocar embolia pulmonar, fallos respiratorios, mareos, bronquitis crónica, erupciones cutáneas y cáncer de pulmón entre otras [32].

Nickel (Ni) is a solid element with a slightly yellowish-white color and oxidation states (+3, +2, 0). Prolonged exposure or ingestion can lead to pulmonary embolism, respiratory failure, dizziness, chronic bronchitis, skin rashes, and lung cancer, among other effects [32].

Cobalto (Co) is a metal pesado de color blanco azulado con estados de oxidación (+5, +4, +3, +2, +1, −1). Su ingesta puede ocasionar alteraciones en epitelios de la mucosa bucal [33], alteraciones visuales, hipotiroidismo, cardiomiopatía, erupciones en la piel [34].

Cobalt (Co) is a heavy metal with a bluish-white color and oxidation states (+5, +4, +3, +2, +1, −1). Its ingestion can cause alterations in the epithelium of the oral mucosa [33], visual disturbances, hypothyroidism, cardiomyopathy, and skin rashes [34].

2.2 Sources

According to the World Health Organization (WHO), contaminated waters are those whose composition has been altered in such a way that it no longer maintains optimal conditions for human consumption. Surface water sources receive a wide variety of contaminants from numerous natural and anthropogenic events, including mining, forest fires, agroindustry, urbanization, and volcanoes [2]. Each of these events releases a significant amount of heavy metals into the environment, posing a threat to human health and ecosystems (**Figure 1**).

Primarily due to anthropogenic activities, surface waters can become contaminated with a range of heavy metals, through pesticides, insecticides, and mining, among others. Specifically, for water used for human consumption, the presence of

Figure 1.
Heavy metal sources. Water pollution by heavy metals.

high concentrations of heavy metals in the body can lead to failures in the organisms that are exposed to it. In (**Table 1**), the heavy metals, their sources, ailments caused by these metals, and the maximum allowable amount in water are shown.

Metals	Sources	Suffering	Maximum allowed in drinking water (WHO)	Ref.
Cr	Corrosion	Kidney failure, bronchitis, bleeding, and gastrointestinal damage	0.05 mg/l	[3, 30, 35–37], [WHO]
Pb	Mining, corrosion, agriculture, and cattle farming	Loss of sense of balance and direction and range of muscle movement	0,01 mg/l	[3, 27, 35–38]
Zn	Mining, corrosion, agriculture, and cattle farming	Prostate cancer, increased cholesterol, testosterone, and immune dysfunction	3 mg/l	[3, 22, 35–37]
Ni	Mining, corrosion	Pulmonary embolism, respiratory failure, dizziness, bronchitis, skin rashes, and lung cancer	0,02 mg/l	[3, 32, 35–37]
Cu	Mining, corrosion, agriculture, and cattle farming	Physiological alterations and acute toxicity	2 mg/l	[3, 24, 25, 35–37]
Cd	Mining, corrosion	Conditions in the lungs, bones, and kidneys.	0,003 mg/l	[3, 26, 35–37]
As	Mining, agriculture, and cattle farming	Poisoning, carcinogenic prevalence in lungs, skin, and bladder	0.01 mg/l	[3, 28, 35–38].
Fe	Corrosion	Effects on the central nervous system, gastrointestinal, cardiovascular	NGL**	[3, 29, 35, 37, 39]
Co	Corrosion	Alterations in epithelia of the oral mucosa, visual, hypothyroidism, cardiomyopathy, skin rashes	NM	[3, 33–35, 37, 39]
Mn	Agriculture and cattle farming	Respiratory problems, blockage of the neurotransmitter system, and manganism syndrome	0,5 mg/l	[3, 31, 35, 37]
Hg	Mining	Nausea, vomiting, diarrhea, neurological syndromes, ophthalmological syndrome, dermatitis, rhinitis, and hypersensitivity	0,001 mg/l	[21, 36, 38]

*NM = Not mentioned; NGL = No Guideline, because it occurs in drinking water at concentrations well below those at which toxic effects may occur**No Guideline, because it is not of health concern at concentrations normally observed in drinking water, but may affect the acceptability of water at concentration above 0,3 mg/l.*

Table 1.
Major heavy metals present in surface waters and their anthropogenic sources.

2.3 Toxicity of heavy metals

Heavy metals such as chromium (VI), beryllium, arsenic, nickel, and cadmium have been classified by the International Agency for Research on Cancer (IARC) as human carcinogens, along with aluminum production and iron and steel foundries [38].

Some divalent metal cations are structurally similar, as is the case with Co^{2+}, Fe^{2+}, Zn^{2+}, Cu^{2+}, Ni^{2+}, and Mn^{2+}. It is likely that they may compete with each other, leading to complications in physiological functions within the cell [40]. An example of this is Ca, which is found in membranes and can be displaced by other metals, causing functional disorders [41].

The ingestion and/or prolonged exposure to heavy metals can lead to multiple diseases such as cancer, neurological syndromes, ophthalmological syndromes, dermatitis, rhinitis, hypersensitivity [21], visual impairments, hypothyroidism, cardiomyopathy, skin rashes [34], elevated cholesterol levels, increased testosterone, and immune dysfunction [22], conditions in lungs, bones, and kidneys [26], kidney failure, gastrointestinal damage [30], a syndrome called manganism [31], poisonings, among others [28]. This emphasizes the importance of researching and improving various methods for removing contaminants from waters intended for human consumption (**Figure 2**).

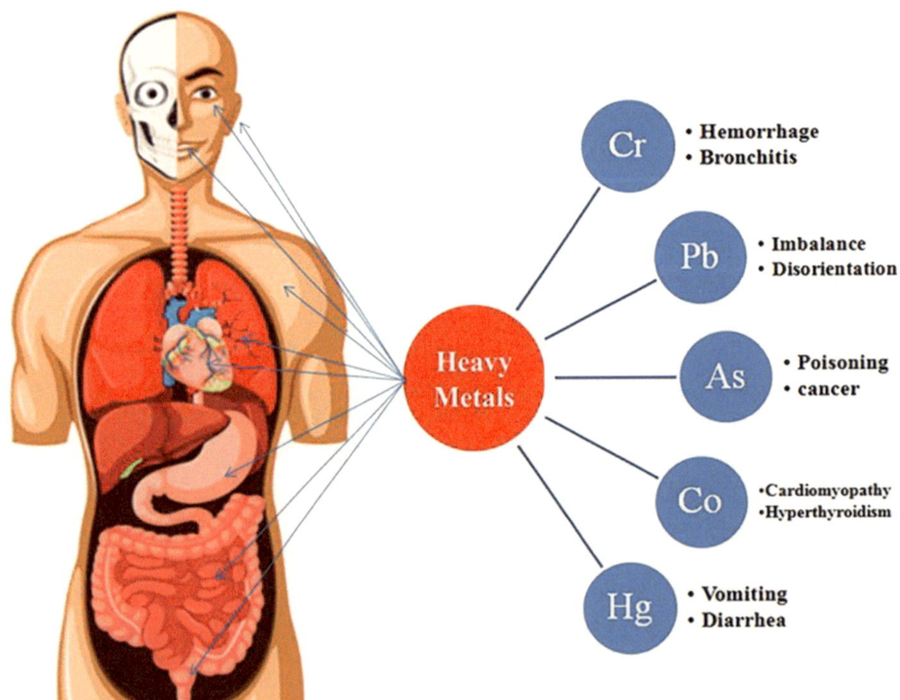

Figure 2.
Alterations and toxicity in humans caused by ingestion of heavy metals.

3. Methods for removal of heavy metals

There are various methods for treating surface waters, including chemical, physicochemical, and biological methods. Biological and physicochemical methods can involve the use of plants for water decontamination (phytoremediation) [7] (coagulation/flocculation). These two methods are significant as they can be employed in rural areas where large infrastructure for the treatment of surface waters for human consumption may be lacking.

3.1 Coagulation/flocculation

The method that uses plants or plant extracts with the ability to absorb contaminants, followed by a coagulation/flocculation process that allows for the easy separation of contaminants. However, this method generates a secondary pollution [42].

3.2 Phytoremediation

The method that uses plants to remove contaminants from water is environmentally friendly though it is a slow process and requires large land areas [43].

3.3 Photocatalysis

Photocatalysis este método posee la capacidad de degradar complejos de metales pesados, dispone de un alto poder oxidativo, descompone complejos orgánicos y además no produce lodos, su limitación es el alto costo a causa del equipamiento necesario para su funcionamiento [44].

3.4 Chemical precipitation

It is an economical and straightforward method that removes a significant amount of heavy metals from water, although it has some limitations such as the use of chemicals, generation of sludge, as well as handling expenses [45].

3.5 Electrochemical treatment

The method that minimizes the use of chemicals and is also an effective technique for recycling heavy metals. The limitations of this method include its lack of sensitivity, stability, and efficiency, as well as its high cost [46].

3.6 Reverse osmosis

The method that can be complemented with other methods, no use of chemicals is estimated, and it is simple to operate. Among the limitations, we can find reduced water permeability, the membrane may become contaminated, large amounts of energy are required, and there are high operation and equipment costs [47].

3.7 Flotation

Method for water treatment where physical phenomena are predominant, which can be low-cost when purchasing equipment, although maintenance and operational costs are some of its limitations [48].

3.8 AOPs

This method involves minimal chemical consumption and does not generate sludge. However, one of the major limitations of this method is its inapplicability on a large scale [49].

3.9 Ion-exchange

The method that uses a resin that maintains good regeneration and is also cost-effective. Its limitations are based on the adsorption of organic substances and difficulty in handling [50].

3.10 Membrane filtration

This method has good selectivity and provides efficient separation under low-pressure conditions. Its limitation lies in its expensive operability and difficult handling after use [51].

3.11 Adsortion

An efficient separation method, easy to use, with low costs, and applicable over a wide pH range. The limitation of this method is the need to regenerate the adsorbent [52].

As can be observed, all the methods shown above have their associated limitations. However, in this text, we want to highlight the method in which plants or plant extracts are used (phytoremediation). This method can be reproducible in rural areas where there are limited resources for the treatment of surface waters.

4. Methodology

4.1 Experimental design

To carry out this work, three stages were considered, for which glass experimental units with a capacity of 10 L were used, (2) for each evaluated factor. To assess the removal efficiency of Fe and Mn in natural water, *Hydrilla verticillata* (L.f.) Royle plants, and extracts from cactus (*Opuntia ficus-indica*) and dragon fruit (*Hylocereus triangularis*) were employed.

4.1.1 Stage 1

The amount of plant material, *Hydrilla verticillata* (L.f.) Royle, to be used in the subsequent stages of the work was determined. Four quantities of the plant were considered: 20,000, 80,000, 120,000, and 160,000 mg per 10 L of water.

The plant was extracted from a natural pond with all its roots and fresh foliar system, immediately weighed on a scale (Pioneer, TM Ohaus), and introduced into the experimental unit.

Initially, the levels of Fe and Mn present in the raw water were determined by AAS using a Thermoelectronatomic absorption system model S4AA System (Thermofisher Scientific, MA, USA).

200 mL of metals at a concentration of 2 ppm were added to each experimental unit. Every two hours after the start of the process (9 am, 11 am, 1 pm, and 3 pm), water samples (1 L) were taken from each experimental unit at a depth of 0.5 cm from the surface to determine the levels of Fe and Mn. This allows understanding of the influence of mass and determining the time range during which solar radiation has the greatest impact on removal.

4.1.2 Stage 2

Four experimental units with a capacity of 10 L each were taken, and (200 mL) of a concentration of 1.5, 2.0, 3.0, and 4.0 mg metal/L were added. The mixtures were homogenized through rapid mixing at 100 rpm for 1 minute and slow mixing at 30 rpm for 60 minutes.

The selected mass was distributed among the experimental units and remained for 18 days. Three samples were collected at 3 pm every six days, and the levels of total Fe and Mn, as well as other physicochemical parameters, were determined at the beginning, during, and at the end of the experiment. This was done to understand the influence of time and the maximum amount of plant absorption.

4.1.3 Stage 3

Obtaining the extracts 500 g (wet weight) of Pitahaya (*Hylocereus triangularis*) and Cactus (*Opuntia ficus-indica*) were weighed. They were manually grated and filtered three times under pressure with surgical gauze. Due to the viscosity and instability of the concentrated extracts, 10% and 20% (w/v) solutions were prepared for the cactus and dragon fruit, respectively. These solutions were then stored at 4°C [12]. To determine the appropriate dose for each extract, a jar test was conducted (*Standard Practice for Coagulation-Flocculation Jar Test of Water*).

To find and select the appropriate dose of the extracts, each 1 L container was filled with raw water, and increasing doses of 0.008, 0.01, 0.012, 0.015, and 0.020 L of the solutions previously prepared with the extracts under study were added. It underwent rapid and slow mixing at 100 rpm for 1 minute and 30 rpm for 60 minutes, respectively. The mixture was allowed to settle for 60 minutes.

5. Statistical analysis and analytical control

To determine the mass of plant material, a completely randomized design (CRD) was used for each of its masses, selecting the mass with the best performance through the *Tukey HDS* multiple range test. Assumptions of normality and homogeneity of variance of errors in the proposed designs were tested using *Shapiro-Wilk* and *Cochran* tests, respectively. When it was not possible to ensure them, the non-parametric *Kruskal-Wallis* test was used. All tests were carried out using the statistical package *Statgraphics Centurion* 16 and Excel 2016.

	Fe	Mn
Linear range	0.050–4000 mg/L	0.050–4000 mg/L
LDM	0.018 (+/−0.003) mg/L	0.021(+/−0.002) mg/L
LCM	0.050 (+/−0.004) mg/L	0.050 (+/−0.004) mg/L
Precision	1.11%	2.98%
Accuracy	99.69%	97.15%
Uncertainty	1.500(+/−0.119) mg/L	1.600(+/−0.088) mg/L

To ensure that there were no changes in the calibration curve during measurements, r^2, $m \pm s$, $b \pm s$ of the line, and residuals were checked every 12 measurements using a standard.

Table 2.
Method validation parameters.

For metal determinations, the direct flame atomic absorption method was used, following digestion with sulfuric-nitric acid for Fe and Mn. The method validation parameters are shown in the following **Table 2**.

6. Results and discussion

See **Table 3**.

The results indicate that the mass of plant material influences the removal of these metals, with the best results observed for the mass of 160,000 mg in both cases, at different times of the day, with an average of 78.84% for Fe and 66.56% for Mn, respectively.

It is observed that the lowest removal occurred in the morning hours for both metals at all masses. This is due to the short exposure time of the material to the metals dissolved in natural water.

In contrast, for water with longer exposure times to the plant, a higher percentage of metal removal is observed at all masses, as seen at midday (4 h) and in the afternoon (6 h). However, no significant differences in removal are observed between daytime hours in both cases (**Table 4**).

After six days of the experiment, experimental unit 1 shows a removal of 100% for both metals. Additionally, it is observed that as the concentration of metals

% removal Fe	20,000 mg	80,000 mg	120,000 mg	160,000 mg
Morning (2 h)	54.92 (±0.5)	59.58 (±0.3)	69.16 (±0.3)	75.19 (±0.4)
Half day (4 h)	65.32 (±0.7)	69.98 (±0.1)	73.54 (±0.4)	79.29 (±0.6)
Afternoon (6 h)	73.27 (±0.4)	75.46 (±0.7)	78.47 (±0.5)	82.04 (±0.6)
% removal Mn				
Morning (2 h)	59.81 (±0.1)	60.46 (±0.2)	61.93 (±0.5)	65.52 (±0.3)
Half day (4 h)	60.79 (±0.2)	62.26 (±0.4)	63.24 (±0.7)	66.67 (±0.2)
Afternoon (6 h)	62.26 (±0.2)	64.05 (±0.5)	65.85 (±0.9)	67.48 (±0.1)

Table 3.
Percentage removal of Fe and Mn (Hydrilla verticillata) as a function of solar radiation and mass.

	Expe Unit 1		Expe Unit 2		Expe Unit 3		Expe Unit 4	
	mg. Initial	mg. Final	mg. Initial	mg. Final	mg. Initial	mg. Final	mg. Initial	mg. Final
Fe	17.34	0	17.44	0.73	17.64	1.54	17.84	3.07
Mn	0.92	0	1.02	0.24	1.22	0.35	1.42	0.43
Turbidity NTU**	41.02	3.11	41.02	3.96				
%removal Fe	100		95.82		91.29		81.67	
%removal Mn	100		76.27		71.57		69.52	
% Turbidity NTU	92.42		90.35					

***Nephelometric turbidity units.*

Table 4.
Percentage of removal of Fe and Mn from each experimental unit after six days of exposure.

increases, the removal decreases, suggesting that there is a maximum concentration that the plant can tolerate. In experimental units 2, 3, and 4, the plants show signs of deterioration, and some died, possibly due to the surrounding climate (temperature, sunlight) and environmental conditions (sodium chloride, nutrients, and pH) [13], as these experiments were conducted in situ, where the plants were extracted.

The experimental units were set up under artisanal roofs. On days 12 and 18, it was not possible to carry out the experiment for the reasons mentioned above.

In experimental unit 1 and 2, after six days, a turbidity removal of 92.42% and 90.35%, respectively, was observed, indicating the usefulness of Hydrilla in the natural removal of turbidity. It can be used for this purpose in rural communities to avoid the use of chemicals and the risks associated with them.

In studies conducted by Lans et al. [12], turbidity removal percentages of 98.97% and 95% were observed for natural extracts of *Opuntia ficus-indica*, *Hylocereus triangularis* and *Guazuma ulmifolia*, respectively.

Figures 3 and **4** show the appearance of the water before and after treatment with the plant material. As can be seen, there is a drastic reduction in turbidity, as well as a decrease in Fe and Mn. *Hydrilla* is sensitive to high temperatures, so it must be controlled, and 323 K is the appropriate temperature [17]. Hence, it changes in appearance after being exposed to direct sunlight and high temperatures. This allowed us to avoid direct radiation on the plant to continue with the experiment (**Figure 5**).

Hydrilla verticillata has also been used to assess the removal efficiency of BOD, COD, and phosphates in freshwater and domestic wastewater, achieving removal efficiencies of up to 84% for BOD, up to 63% for COD, and up to 87% for phosphates [18].

Table 5 shows an efficient reduction in Fe levels by *Opuntia and Hydrilla*, 98.04% and 86.6% respectively, and 95.12% and 85.37% reduction in Mn for *Opuntia* and *Hydrilla* respectively. The Pitahaya extract showed a very poor reduction in these metals. *Opuntia* demonstrates a higher recovery level than *Hydrilla*; however, it is essential to consider that *Opuntia* requires an extract, involving prior work. For this reason, *Hydrilla* has an advantage, as it is only exposed to the water being treated. It is worth mentioning that this process is slower, as a specific waiting time is required

Figure 3.
Untreated raw water.

Figure 4.
Water condition after six days of contact with 160 g of Hydrilla mass.

Figure 5.
Change in the appearance of the plant after 1 day with direct sunlight and temperatures exceeding 32°C.

for removal. On the other hand, with *Hydrilla*, there is a risk that the plant may die without completing its work if Fe and Mn levels are too high.

Particularly, it was observed that the processes of turbidity and color removal by *Hydrilla* are slow compared to the *Opuntia* extract, where the result is almost immediate. Therefore, the plant material was left for a longer period (24 hours) after the process started. The removal of Fe and Mn is possibly due to the adsorption of these metals through an endothermic process [17].

	Ext. *Opuntia*		***Hydrilla***		**Ext. *Hylocereus***	
	[ ]. Initial	**[ ]. Final**	**[ ]. Initial**	**[ ]. Final**	**[ ]. initial**	**[ ]. Final**
Fe	1.704 ± 0.43	0.033	1.704 ± 0.43	0.228 ± 0.030	1.704 ± 0.43	1.557 ± 0.031
Mn	0.062 ± 0.02	0.003	0.062 ± 0.02	0.009 ± 0.004	0.062 ± 0.02	0.057 ± 0.010
% removal Fe	98.04		86.6		8.43	
% removal Mn	95.12		85.37		7.32	

[ ]= Concentration

Table 5.
Percentage removal of Fe and Mn by Hydrilla verticillata vs. extracts of Opuntia ficus and Hylocereus triangularis.

The *Opuntia* extract showed a removal different from *Hydrilla* and achieved better short-term results, both in turbidity and color, with removal values of 97% and 92%, respectively.

7. Conclusions

- The *Hydrilla verticillata* (l.f.) Royle, *Opuntia ficus indica* proved to be efficient in the removal of Fe, Mn, turbidity, and color in natural waters.
- The extracts of *Hylocereus triangularis* were found to be inefficient in the removal of the studied metals in natural waters.
- Similarly to *Opuntia* and *Hylocereus, Hydrilla* can be used for the removal of turbidity from natural waters, making it useful for rural communities.

Author details

Mario Mariglia[1*] and Edineldo Lans-Ceballos[2]

1 Instituto de Química, Pontificia Universidad Católica de Valparaíso, Valparaíso, Chile

2 Facultad de Ciencias Básicas Grupo de Investigaciones en Aguas Pesticidas y Metales Pesados (GIAMP), Universidad de Córdoba, Montería, Colombia

*Address all correspondence to: marsiglialans1987@gmail.com

References

[1] Zeng Y, Cai Y, Tan Q, Dai C. An integrated modeling approach for identifying cost-effective strategies in controlling water pollution of urban watersheds. Journal of Hydrology. 2020;**581**:124373

[2] Aziz K, HH, Mustafa FS, Omer KM, Hama S, Hamarawf RF, Rahmana KO. Heavy metal pollution in the aquatic environment: Efficient and low-cost removal approaches to eliminate their toxicity: A review. Royal Society of Chemistry Advances. 2023;**13**(26):17595-17610. DOI: 10.1039/d3ra00723e

[3] Lans-Ceballos E, Marsiglia M, Lans-Cuesta ES, Forero-Doria O, Guzman L. Water quality and heavy metal levels in the Sinú River, a drinking water source in the Colombian Caribbean. Universitas Scientiarum. 2023;**28**(2):161-182. DOI: 10.11144/Javeriana.SC282.wqah

[4] Sutadian AD, Muttil N, Yilmaz AG, Perera BJC. Development of a water quality index for rivers in West Java Province, Indonesia. Ecological Indicators. 2018;**85**(966):982. DOI: 10.1016/j.ecolind.2017.11.049

[5] Rojas O. Índices de Calidad del agua en Fuente de Captación. Seminario Internacional Sobre Calidad Del Agua Para Consumo. 1991

[6] Sheykhi V, Moore F. Geochemical characterization of kor river water quality, fars province, Southwest Iran. Water Quality Exposure and Health. 2012;**4**(1):25-38

[7] De La Cruz L, Antonio L, Méndez G, Blanca CA, Martha A, Portales L, et al. Suelos agrícolas y metales pesados, una relación tóxica que se puede remediar. Epistemus (Sonora). 2022;**16**(33):93-98

[8] Nong H, Liu J, Chen J, Zhao Y, Liang W, Tang Y, et al. Woody plants have the advantages in the phytoremediation process of manganese ore with the help of microorganisms. Science of the Total Environment. 2023;**863**:160995. DOI: 10.1016/j.scitotenv.2022.160995

[9] Enochs B, Meindl G, Shidemantle G, Wuerthner V, Akerele D, Bartholomew A, et al. Short and long-term phytoremediation capacity of aquatic plants in Cu-polluted environments. Heliyon. 2023;**9**. DOI: 10.1016/j.heliyon.2023.e12805

[10] Kristanti RA, Hadibarata T. Phytoremediation of contaminated water using aquatic plants, its mechanism and enhancement. Current Opinion in Environmental Science & Health. 2023;**32**:100451. DOI: 10.1016/j.coesh.2023.100451

[11] Wibowo YG, Nugraha AT, Rohman A. Phytoremediation of several wastewater sources using Pistia stratiotes and Eichhornia crassipes in Indonesia. Environmental Nanotechnology, Monitoring & Management. 2023;**20**:100781. DOI: 10.1016/j.enmm.2023.100781

[12] Lans-Ceballos E, Marsiglia M, Lans-Cuesta ES, Forero-Doria O, Guzman L. Improvement to quality of surface waters from rural communities: Vegetable extracts as a sustainable and cheap alternative. Environment, Development and Sustainability. 2020;**28**(2). DOI: 10.11144/Javeriana.SC282.wqah

[13] Tan HW, Pang YL, Lim S, Chong WC. A state-of-the-art of phytoremediation approach for sustainable management of heavy metals

recovery. Environmental Technology & Innovation. 2023;**30**. DOI: 10.1016/j.eti.2023.103043

[14] Mishra B, Chandra M. Evaluation of phytoremediation potential of aromatic plants: A systematic review. Journal of Applied Research on Medicinal and Aromatic Plants. 2022;**31**. DOI: 10.1016/j.jarmap.2022.100405

[15] Liu Z, Tran K-Q. A review on disposal and utilization of phytoremediation plants containing heavy metals. Ecotoxicology and Environmental Safety. 2021;**226**:112821. DOI: 10.1016/j.ecoenv.2021.112821

[16] Sousa WTZ. Hydrilla verticillata (Hydrocharitaceae), a recent invader threatening Brazil's freshwater environments: A review of the extent of the problem. Hydrobiologia. 2011;**669**: 1-20. DOI: 10.1007/s10750-011-0696-2

[17] Vij RK, Aruna Janani V, Subramanian D, Mistry CR, Devaraj G, Pandian S. Equilibrium, kinetic and thermodynamic studies for the removal of reactive red dye 120 using Hydrilla verticillata biomass: A batch and column study. Environmental Technology & Innovation. 2021;**24**:102009. DOI: 10.1016/j.eti.2021.102009

[18] Sangwijit C, Phoosuwan N, Sangwijit P, Thakong W, Saikhao L. Assessment of growth ability and removal efficiency of the Hydrilla (Hydrilla verticillata) in wastewater. The Egyptian Journal of Aquatic Research. 2022;**48**:199-204

[19] Londoño Franco LF, Londoño Muñoz PT, Muñoz Garcia FG. Los riesgos de los metales pesados en la salud humana y animal. Biotecnoloía En El Sector Agropecuario y Agroindustrial. 2016;**14**(2):145. DOI: 10.18684/bsaa(14)145-153

[20] Pabón SE, Benítez R, Sarria RA, Gallo JA. Contaminación del agua por metales pesados, métodos de análisis y tecnologías de remoción. Una revisión. Entre Ciencia e Ingeniería. 2020;**14**(27):9-18

[21] Ramírez AV. Intoxicación ocupacional por mercurio. Anales de la Facultad de Medicina. 2008;**69**(1):46-51

[22] Rubio C, Gonzales Waller D, et al. Zinc: an essential oligoelement. Nutrición Hospitalaria. 2007;**22**(1):101-107

[23] Hernández-Zamora M, Rodríguez-MiguelA,Martínez-JerónimoL, Martínez-Jerónimo F. Combined toxicity of glyphosate (Faena®) and copper to the American Cladoceran Daphnia exilis—A two-generation analysis. Water (Switzerland). 2023;**15**(11):2018. DOI: 10.3390/w15112018

[24] Griffitt RJ, Weil R, Hyndman KA, Denslow ND, Powers K, Taylor D, et al. Exposure to copper nanoparticles causes gill injury and acute lethality in zebrafish (Danio rerio). Environmental Science & Technology. 2007;**41**:8178-8186

[25] Xiao Y, Vijver MG, Chen G, Peijnenburg WJGM. Toxicity and accumulation of Cu and ZnO nanoparticles in Daphnia magna. Environmental Science & Technology. 2015;**49**:4657-4664

[26] Rofner F, Nelino. Review on maximum limits of cadmium in cocoa (Theobroma cacao L.). Revista de Ciencias de la Vida. 2021;**34**(2):113-126 http://doi.org/10.17163/lgr.n34.2021.08

[27] Ramírez AV. El cuadro clínico de la intoxicación ocupacional por plomo. Anales de la Facultad de Medicina. 2005;**66**(1):57-70

[28] WHO/CED/PHE/EPE/19.4.1. Organización Mundial de la Salud (OMS), ed. (febrero de 2018)

[29] Jover D, Francisco AG, Lucio RG, et al. Intoxicación aguda por hierro en un caso. Revista Médica de Chile. 2001;**129**(6):660-662

[30] Rojas Romero JE, Rincón Ramírez JE, Marín Leal JC, Ortega Fuenmayor PC, Buonocore Tovar R, Colina M, et al. Toxicidad y bioacumulación de Cromo (Cr +6) en la almeja Polymesoda solida del sistema estuarino Lago de Maracaibo. Boletín Del Centro De Investigaciones Biológicas. 2016;**49**(1):5-25

[31] Huante KEB, Velázquez KC. Intoxicación por manganeso: presentación de caso clínico: Manganese toxicity: case report. South Florida Journal of Health. 2022;**3**(1):2-6. DOI: 10.46981/sfjhv3n1-001

[32] Heredia R, Dunia. Intoxicación ocupacional por metales pesados. MEDISAN. 2017;**21**(12):3372-3385

[33] Felippini AL et al. Efectos del Cloruro de Cobalto sobre el Epitelio de Unión y Epitelio Reducido del Esmalte del Primer Molar Superior de Rata. International Journal of Morphology. 2009;**27**(4):1081-1088

[34] Pelayo-de Tomás JM, Novoa-Parra C, Gómez-Barbero P. Toxicidad por cobalto después de la revisión a una artroplastia total de cadera posterior a fractura de cabeza cerámica. Revista Española de Cirugía Ortopédica y Traumatología. 2017;**61**(3):203-207

[35] Vink R, Behrendt H, Salomons W. Development of the heavy metal pollution trends in severalEuropean rivers: An analysis of point and diffuse sources. Water Science and Technology. 1999;**39**(12):215-223

[36] Nieder R et al. Role of Potentially Toxic Elements in Soils en Soil Components and Human Health. 1st ed. Dordrecht: Springer Science & Business Media; 2018. pp. 375-450. DOI: 10.1007/978-94-024-1222-2

[37] Domenech X, Peral J. Química ambiental de los sistemas terrestres. Editorial Reverté, Páginas 256, Edición 1 Fecha publicación 01-01-2006 2006

[38] Ferrer A. Intoxicación por metales. Anales del Sistema Sanitario de Navarra. 2003;**26**(Supl. 1):141-153

[39] Mebrahtu G, Zerabruk S. Concentration and health implication of heavy metals in drinking water from urban areas of Tigray region, northern Ethiopia. Momona Ethiopian Journal of Science. 2011;**3**(1):105-121. DOI: 10.4314/mejs.v3i1.63689

[40] Navarro-Aviñó JP, Aguilar Alonso LMJR. Aspectos bioquímicos y genéticos de la tolerancia y acumulación de metales pesados en plantas. Ecosistemas: Revista Cietifica y Tecnica de Ecologia y Medio Ambiente. 2007;**16**(2):10-25

[41] Nies DH. Microbial heavy-metal resistance. Applied Microbiology and Biotechnology. 1999;**51**(6):730-750. DOI: 10.1007/s002530051457

[42] Sakhi D, Rakhila Y, Elmchaouri A, Abouri M, Souabi S, Jada A. Advanced intelligent systems for sustainable development (AI2SD´2018). In: Mostafa Ezziyyani, editor. Optimization of Coagulation Flocculation Process for the Removal of Heavy Metals from Real Textile Wastewater. Switzerland: Springer Nature; 2018. pp. 257-266. DOI: 10.1007/978-3-030-11881-5

[43] Bhat SA, Bashir O, Haq SAU, Amin T, Rafiq A, Ali M, et al. Phytoremediation

ofheavy metals in soil and water: An eco-friendly, sustainable and multidisciplinary approach. Chemosphere. 2022;**303**:134788

[44] Tang L, Deng Y, Wang J, Wang J, Zeng G. Nanohybrid and Nanoporous Materials for Aquatic Pollution Control. Changsha, P.R. China: Elsevier; 2019. pp. 125-153. DOI: 10.1016/B978-0-12-814154-0.00005-0

[45] Crini G, Lichtfouse E. Advantages and disadvantages of techniques used for wastewater treatment. Environmental Chemistry Letters. 2019;**17**:145-155

[46] Maarof HI, Daud WMAW, Aroua MK. Recent trends in removal and recovery of heavy metals from wastewater by electrochemical technologies. Reviews in Chemical Engineering. 2017;**33**:359-386

[47] Abdullah N, Yusof N, Lau W, Jaafar J, Ismail A, J. Recent trends of heavy metal removal from water/wastewater by membrane technologies. Industrial & Engineering Chemistry. 2019;**76**:17-38

[48] Sun Y, Zhou S, Zheng H, et al. Performance evaluation and optimization of flocculation process for removing heavy metal. Chemical Engineering Journal. 2020;**385**:123911

[49] Ji J, Yuan X, Zhao Y, Jiang L, Wang H. Mechanistic insights of removing pollutan in adsortion and advanced oxidation processes by sludge biochar. Journal of Hazard. 2022;**430**:128375. DOI: 10.1016/j.jhazmat.2022.128375

[50] Kansara N, Bhati L, Narang M, Vaishnavi R. Wastewater treatment by ion exchange method: A review of past and recent researches. Indian Journal of Environmental Sciences. 2016;**12**:143-150

[51] Imdad S, Dohare R, K. A review on heavy metals removal using ionic liquid membranes from the industrial wastewater. Chemical Engineering and Processing. 2022;**173**:108812. DOI: 10.1016/j.cep.2022.108812

[52] Chakraborty R, Asthana A, Singh A, et al. Adsorption of heavy metal ions by various low-cost adsorbents: A review. International Journal of Environmental Analytical Chemistry. 2022;**102**:342-379

Section 2

Hydromorphological Quality

Chapter 4

Analysis of Heavy Metal Toxicity in the Surface and Bottom Waters of Lower Lake Bhopal, M.P. (India)

Aarefa Jan, Suchitra Banerjee, Rajendra Chouhan, Subrata Pani and Saima Syed

Abstract

The present study was focused on the assessment of heavy metals in Lower Lake of Bhopal. With reference to toxic metal contamination, water samples were collected quaterly from four stations mentioned for a period of two years (January 2020 to December 2021). Heavy metals, i.e., iron, zinc, chromium, copper, and nickel were determined in surface and bottom waters taken from the Lower Lake, Bhopal, using atomic absorption spectroscopy (AAS) according to the Standard Methods of American Public Health Association (APHA). The range values of these metals were compared with the tolerance limits as laid down by the Central Pollution Control Board (CPCB), India. Results of this analysis revealed that the concentration of these metals was below the permissible limits both in surface and bottom waters except Fe, which was alarming. It was concluded that the metals (Fe, Zn, Cr, Cu, and Ni) were present in water, and the contamination was supposed to be due to a high degree of anthropogenic stress including idol immersion activity. The water quality of the Lake reveals that although the situation is not too bad, it is alarming. Proper conservation and management plans and strategies have to be formulated and implemented for the restoration, conservation, and management of these water bodies at the government and public level.

Keywords: Lower Lake, toxic metals, AAS, contamination, anthropogenic activities

1. Introduction

One of the main issues facing modern human society is environmental degradation [1]. Environmental pollution has been rising steadily over the last few decades because of rapidly expanding industry, rising energy consumption, and thoughtless loss of natural resources [2]. The soil and aquatic ecosystems are continually exposed to a variety of harmful organic and inorganic chemicals from various natural and anthropogenic sources. Heavy metals are one of them, and they contribute significantly to environmental pollution both because they are hazardous and because they have the capacity to bioaccumulate in the food chain [3]. Water quality assessment of water bodies like Lakes has become an important issue due to increased water pollution caused by human interference [4]. Lakes have been continuously contaminated by the encroachment of lake areas due to human settlements, release of industrial waste, sewage, dumping of garbage, etc. By assessing the water quality of lakes, pollution load caused by human activities can be monitored regularly, and water pollution can be controlled. The water quality index tells the water quality status of a water source and has been applied for both surface and groundwater quality assessment [5]. WQI reduces a large amount of water quality data into easily understandable information to the public and to the concerned authorities and policymakers [6]. WQI detects and evaluates the level of contamination of any water body and helps in water quality management [7]. Water purity is considered as a major concern for humanity since it is the main resource for sustaining life [8]. Due to the concomitant rise of human activity adjacent to the rivers over the past few decades, harmful chemical compounds are contaminating water supplies at an accelerated pace [2, 3]. Among the most significant pollutants are toxic metals, which pose a global problem due to their protracted retention in waterways and soils, rising geo-ecological dangers, and disruption of normal biochemical processes [9, 10]. Numerous toxic metals are created and discharged into our aquatic system because of human endeavors like agriculture, mining, industry, or urbanization. These metals then accumulate in soils and are biomagnified through the food chain cycle [11, 12].

In order to assess the quality of the water, physiochemical factors are crucial. The relationship between toxic metals and physicochemical characteristics may be seen in the way that a higher concentration of toxic metals is linked to a decrease in oxygen demand in water. Among the most harmful types of water contamination is toxic metals [13]. Cu and Zn are among those that are necessary for living things, whereas other elements like Pb, Cd, and Al are harmful to them [14]. Toxic metals are essential elements for drinking water guidelines in determining the purity of water for consumption [15]. The ecotoxicological impacts of toxic substances on living things include both necessary and non-essential elements [16]. In order to reduce sewage disposal, various preventive strategies and nature-based approaches are being used to limit toxic metal inflow into our water systems; however, toxic metal accumulation persists within aquatic habitats [17–21].

As a result, water contamination, evaluation, and monitoring were essential due to their immediate impacts on aquatic life as well as human health [22]. Because they were also non-degradable and hazardous even at small doses, lead and mercury can accumulate via food chain. On the contrary, important micronutrients, such as copper, zinc, and iron, have adverse impacts on the biology of living things when in large quantities [23, 24]. Heavy metals concentration in the groundwater could be due to various anthropogenic activities that release various types of contaminants [25]. The potential health risks of heavy metal contamination in groundwater sources of southwestern Punjab showed that the Hg, Pb, As, and Se concentrations are above the guideline values of the World Health Organization [26]. Contamination of freshwater sources can be caused by both anthropogenic and natural processes. According to the Central Pollution Control Board, Maharashtra along with two other states, contributes 80% of hazardous waste generated in India, including heavy metal pollution [27].

The present study has been carried out to evaluate the heavy metal toxicity in Lower Lake of Bhopal.

2. Methodology, results, and discussion

2.1 Description of sampling site

The Lower Lake, also known as Chhota Talaab, is a lake situated in Bhopal, the capital of the Indian province of Madhya Pradesh (**Figure 1**) (coordinates: 23016′0″N 77025′0″E). The Lake has a 1.29 km^2 (0.50 sq. mi) contact area, a maximal depth of 10.7 m, and normal depths of about 6.2 m (20 ft) (35 ft). In addition, the Lake's coverage area is 9.6 km^2 (3.7 sq. mi). It makes up the wetlands of Bhoj along with Bhojtal, or Lake Superior. The Chhota Talaab is plagued by contamination brought on by emptying nalllahs laden with sewage, a shortage of freshwater sources, and industrial laundry. The whole Lake was nutrient-rich, so the water was unfit for human consumption. Among all of these, tourists can find a great place to sit and enjoy the beauty of the city's surroundings at the Lake.

2.1.1 Study area

Lower Lake, which is a part of Bhoj Wetland (together known as Upper and Lower Lakes), was created by constructing an earthen dam (Pul Pukhta) in the seventeenth century. The Lake water was never used as potable water, even though it was not required because the other lake, i.e., upper lake water, was sufficient to meet the demand for potable water. The Lower Lake water was generally used for other purposes like irrigation, gardening, horticulture, washing clothes, bathing, and recreational activities.

2.2 Sampling technique

The current study was conducted from January 2020 to December 2021 to assess the water quality of Lower Lake from four stations. The sampling sites have been selected at various locations along Lower Lake Bhopal. For toxicological studies, water samples were collected during different seasons (quarterly), viz. January to March, April to June, July to September, and October to December. The descriptions of the sampling stations are described in **Table 1**.

Figure 1.
Lower Lake Bhopal (source: Google earth).

Sampling station	Latitude	Longitude	MASL
Ginnori (St-1)	23°01′09.81″N	77°23′37″E	437 m
Bhoipura (St-2)	23°07′09.99″N	77°35′40.23″E	411 m
Khatlapura (St-3)	23°11′55.82″N	77°39′01.63″E	405 m
Center (St-4)	23°17′01.74″N	77°41′31.56″E	399 m

Table 1.
Sampling stations along with global positioning system.

During the period of investigation quarterly samples were collected from four identified sampling stations. Water samples were collected in sterile glass bottles, jerry cans from each station following the standard methods [28].

2.3 Detailed description of the stations

Detailed description of the stations of both (**Table 1**) and (**Figure 2**) are given below:

2.3.1 Ginnori (St-1)

This station is close to Kamla Park at the Lake's northernmost end (Killole Park). The area of the Lake is among the most contaminated regions and has consistently shown exceptionally high levels of PO_4, NO_3, Ca, Mg, and other minerals. The water has an extremely high concentration of inorganic components since this area is a hub for

Figure 2.
Sampling stations of Lower Lake (Google source).

washermen's work. As a result, Microcystis aeruginosa outbreaks at high densities and were consistently seen over the seasons in all water columns. The epilimnion region is also confined to a few millimeters, which restricts the development of subsurface macrophytes and speeds up breakdown processes. Elevated biochemical oxygen demand (BOD) and chemical oxygen demand (COD) levels were also detected in the vicinity, which is consistent with washermen's activity.

2.3.2 Bhoipura (St-2)

This station lies close to Bhoipura. The wastewater that enters this site through the nearby inlets, which carries household pollutants from the numerous residential areas, has an effect on the water. This location is a highly impacted site for idol immersion practices as well as being vulnerable to human activities (swimming, bathing, etc.).

2.3.3 Khatlapura (St-3)

This station is among the main locations for idol immersion events and is close to Khatlapura Mandir. In addition to all this, local devotees who visit the temples for devotion also toss the puja items—flowers, incense sticks, coconuts, etc.—into the Lake. These are extremely degradable materials that are to blame for the water's rising biological oxygen requirement.

2.3.4 Center (St-4)

The sample station is located between MLB Hostel and MLB Campus (**Figure 2**). Sampling site 4, which is located in the Lower Lake's bottom zones, is among the Lake's highest contaminated regions. The purity of the water in this location, though, began to improve.

2.4 Determination of heavy metals

To evaluate the presence of heavy metals in water samples, saturated HCl was used to extract them, and the specimens were stored in a refrigerator until analysis. The metals assessed included Fe, Zn, Cr, Cu, and Ni. The analysis of heavy metals was carried out using an Atomic Absorption Spectrophotometer (Parkin Elmer Analyst

Parameter	Concentration (μg/l) in present investigation	Permissible limit (inland surface water in mg/l)	Remarks
Fe	2.036	3.0	Within permissible limits but alarming
Zn	1.236	5.0	Within permissible limit
Cr	0.501	2.0	Within permissible limit
Cu	0.009	3.0	Below permissible limit
Ni	0.064	3.0	Within permissible limit

***These standards shall be applicable for industries, operations, or processes other than those industries, operations or process for which standards have been specified in Schedule of the Environment Protection Rules, 1986.*

Table 2.
*General standards** for discharge of environmental pollutants [29].*

AA100) and a UV Visible Spectrophotometer (HACH DREL 4000), following the procedure outlined in the Hach Manual (2010) (**Table 2**).

3. Results

3.1 Iron

3.1.1 Surface water

Variation in iron in surface water during various months of 2020–2021 at different stations of Lower Lake is depicted in **Figure 3**.

During the period of investigation, iron at the surface was observed within the range of 0.144–2.036 μg/liter. The minimum value of iron was recorded at station 1, and the maximum value of iron was recorded at station 2. During this period, the maximum value of iron was observed in June 2020. Higher values of iron were recorded at station 2 and station 3 compared to other stations.

3.1.2 Bottom water

Variation in iron in bottom water during various months of 2020–2021 at different stations of Lower Lake is depicted in **Figure 4**.

During the period of investigation, iron at bottom water was observed within the range of 0.586 to 1.988 μg/liter. The minimum value of iron was recorded at station 2, and the maximum value of iron was recorded at station 3. During this period, the maximum value of iron was observed in November 2020. Higher values of iron were recorded at station 3 and station 1 compared to other stations.

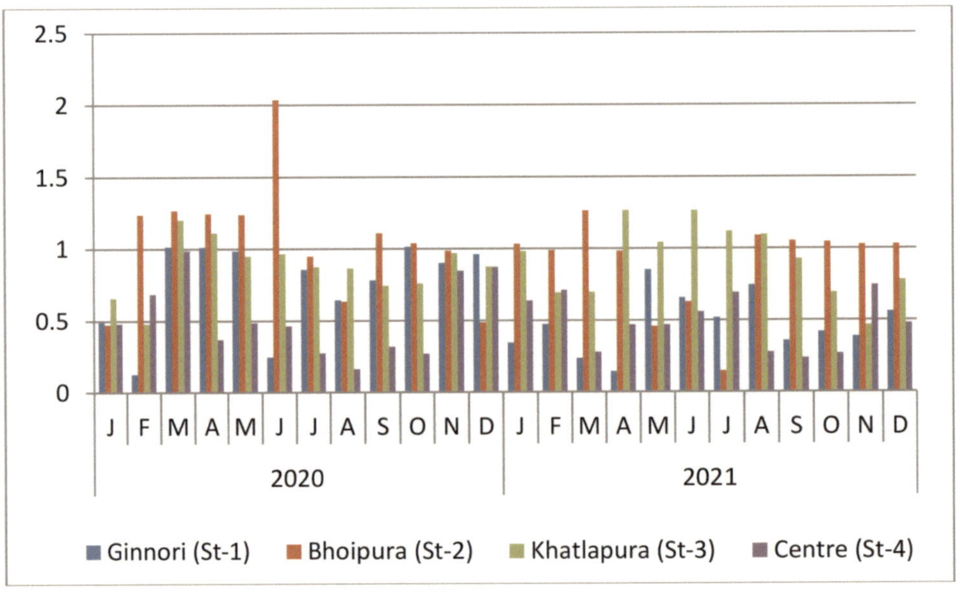

Figure 3.
Heavy metals in surface water of lower Lake: iron (μg/liter).

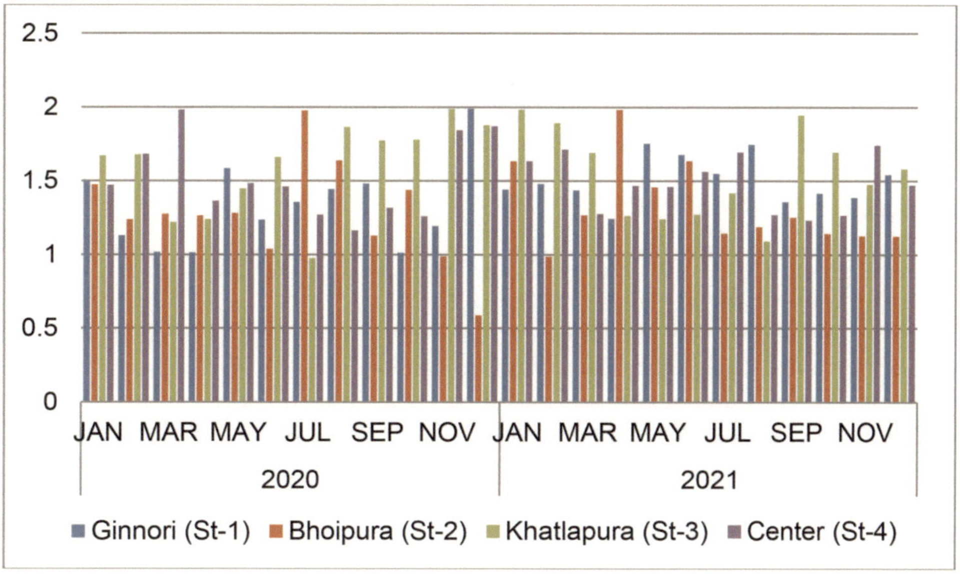

Figure 4.
Heavy metals in bottom water of lower Lake: iron (μg/liter) during 2020–2021.

3.2 Zinc

3.2.1 Surface water

Variation in zinc in surface water during various months of 2020–2021 at different stations of Lower Lake is depicted in **Figure 5**.

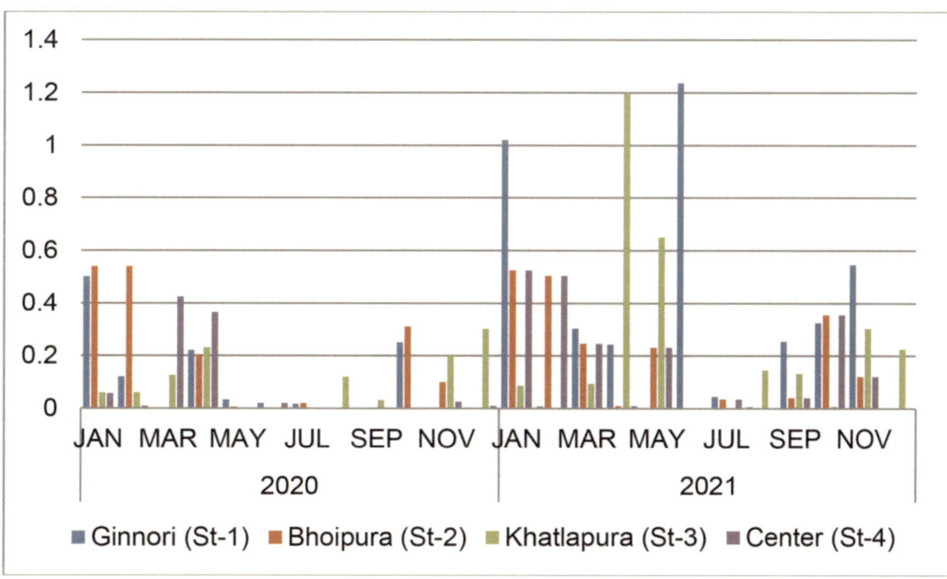

Figure 5.
Heavy metals in surface water of lower Lake: zinc (μg/liter) during 2020–2021.

During the period of investigation, zinc in surface water was observed within the range of nil to 1.236 μg/liter. The minimum value of zinc was recorded at many stations, and the maximum value of zinc was recorded at station 1. During this period, the maximum value of zinc was observed in June 2021. Higher values of zinc were recorded at station 1 and station 3 compared to other stations.

3.2.2 Bottom water

Variation in zinc in bottom water during various months of 2020–2021 at different stations of Lower Lake is depicted in **Figure 6**.

During the period of investigation, zinc at bottom water was observed within the range of 0.021–0.604 μg/liter. The minimum value of zinc was recorded at station 2, and the maximum value of zinc was also recorded at station 2. During this period, the maximum value of zinc was observed in September 2021, followed by January 2021. Higher values of zinc were recorded at station 2 and station 3 compared to other stations.

3.3 Chromium

3.3.1 Surface water

Variation in chromium in surface water during various months of 2020–2021 at different stations of Lower Lake is depicted in **Figure 7**.

During the period of investigation, chromium at surface water was observed within the range of nil to 0.5 μg/liter. The minimum value of chromium was recorded at station 1 and the maximum value of chromium was recorded at the station 3. During this period, the maximum value of chromium was observed in November 2021. Higher values of chromium were recorded at station 3 and station 2 compared to other stations.

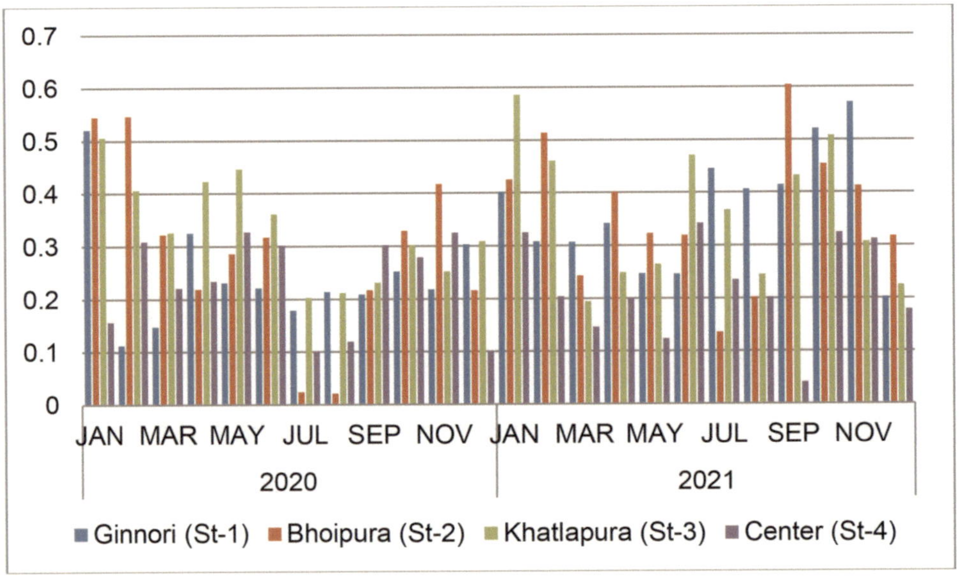

Figure 6.
Heavy metals in bottom water of lower Lake: zinc (μg/liter) during 2020–2021.

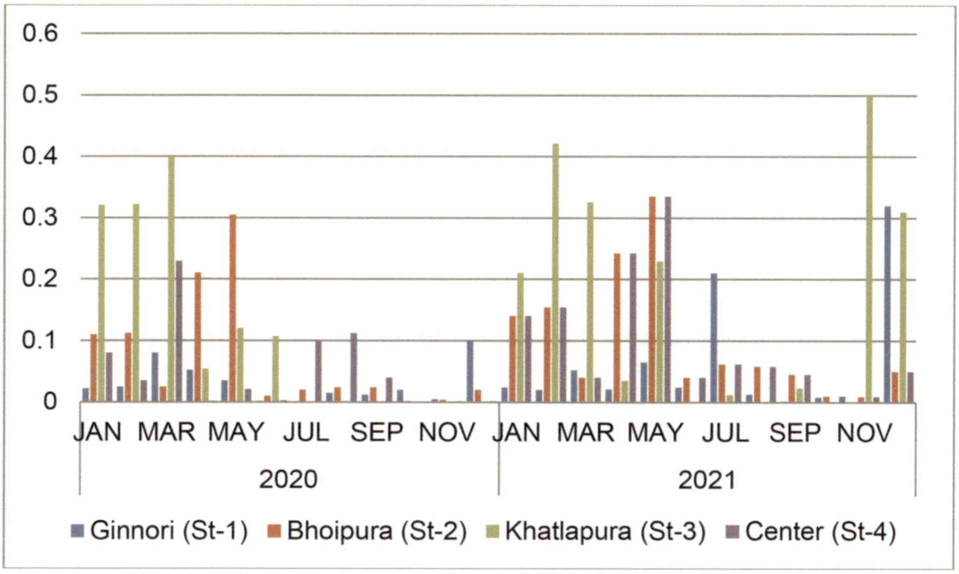

Figure 7.
Heavy metals in surface water of lower Lake: chromium (µg/liter) during 2020–2021.

3.3.2 Bottom water

Variation in chromium in bottom water during various months of 2020–2021 at different stations of Lower Lake is depicted in **Figure 8**.

During the period of investigation, chromium at bottom was observed within the range of 0.013–0.423 µg/liter. The minimum value of chromium was recorded at

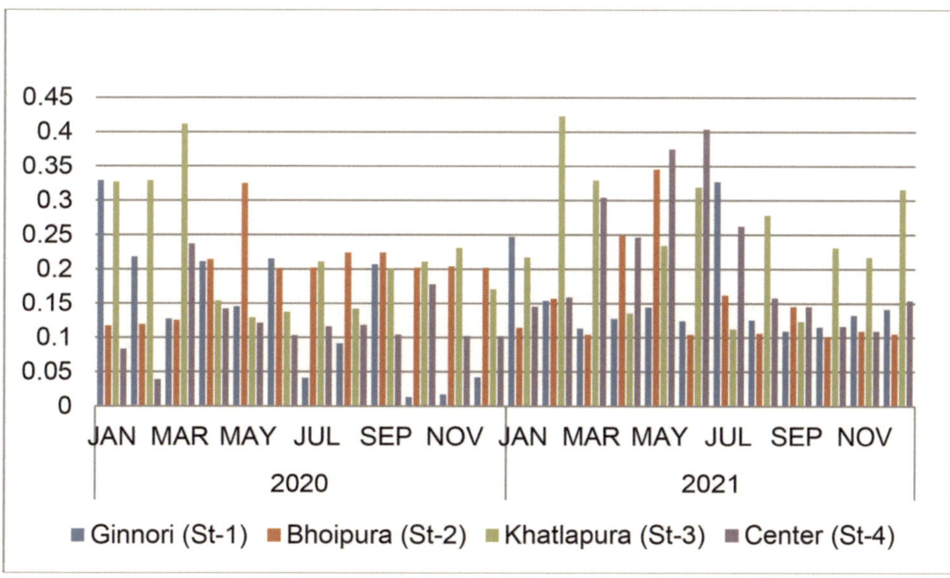

Figure 8.
Heavy metals in bottom water of lower Lake: chromium (µg/liter) during 2020–2021.

station 1 and the maximum value was recorded at station 3. During this period, the maximum value of chromium was observed in February 2021, followed by March 2020. Higher values of chromium were recorded at station 3 compared to other stations.

3.4 Copper

3.4.1 Surface water

Variation in copper in surface water during various months of 2020–2021 at different stations of Lower Lake is depicted in **Figure 9**.

During the period of investigation, copper at the surface was observed within the range of 0.001–0.009 μg/liter. The minimum value of copper was recorded at many stations, and the maximum value of copper was recorded at many stations as well.

3.4.2 Bottom water

Variation in copper in bottom water during various months of 2020–2021 at different stations of Lower Lake is depicted in **Figure 10**.

During the period of investigation, copper at bottom water was observed within the range of 0.002–0.009 μg/liter. The minimum value of copper was recorded in many places, and the maximum value of copper was recorded at many stations. During this period, the maximum value of copper was observed in June 2020.

3.5 Nickel

3.5.1 Surface water

Variation in nickel in surface water during various months of 2020–2021 at different stations of Lower Lake is depicted in **Figure 11**.

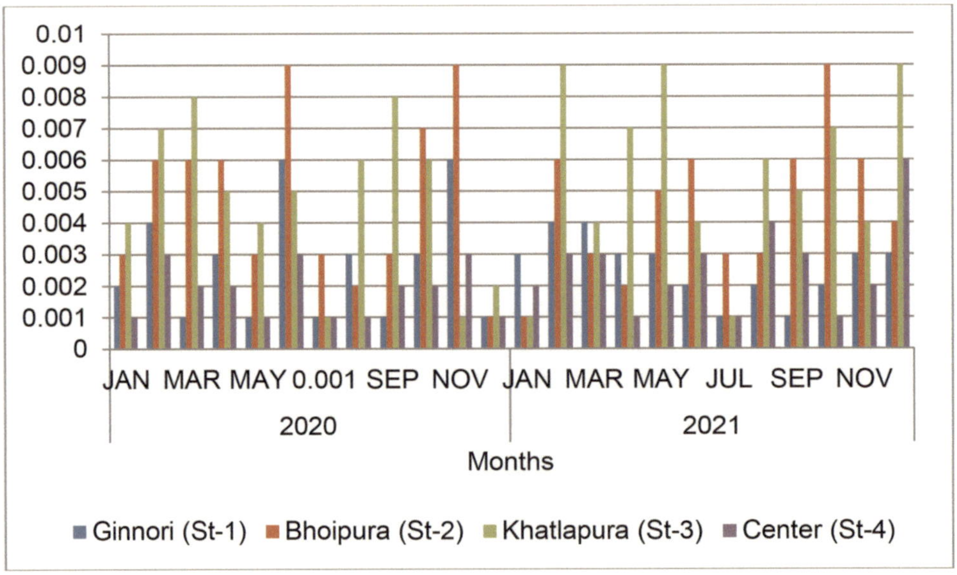

Figure 9.
Heavy metals in surface water of lower Lake: copper (μg/liter) during 2020–2021.

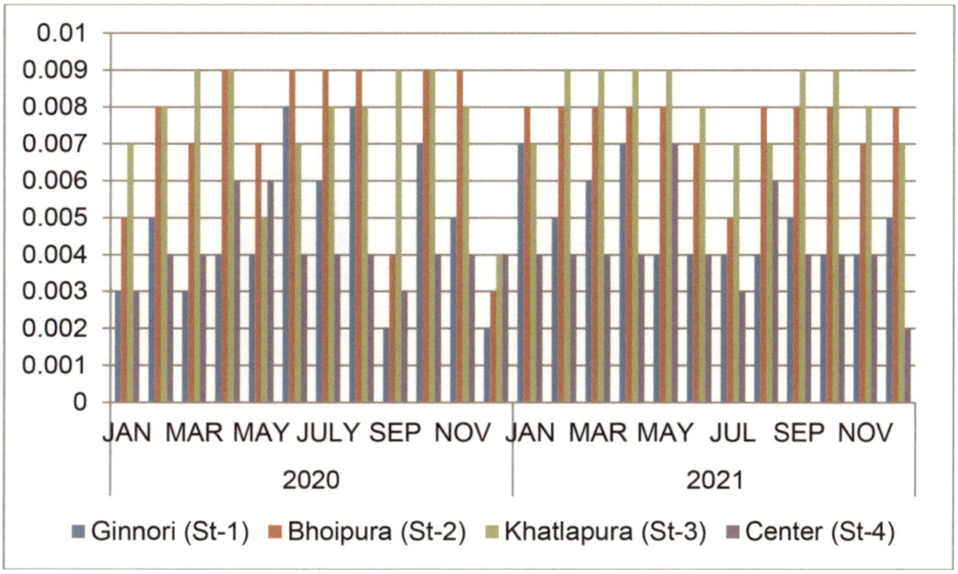

Figure 10.
Heavy metals in bottom water of lower Lake: copper (μg/liter) during 2020–2021.

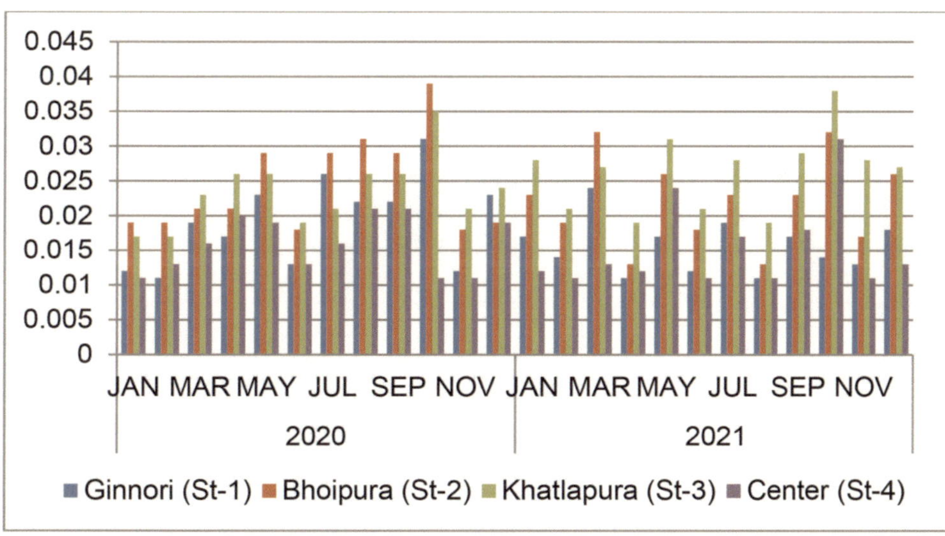

Figure 11.
Heavy metals in surface water of lower Lake: nickel (μg/liter) during 2020–2021.

During the period of investigation, nickel in surface water was observed within the range of 0.011–0.039 μg/liter. The minimum value of nickel was recorded in many places, and the maximum value of nickel was recorded at station 2. During this period, the maximum value of nickel was observed in October 2020, followed by October 2021. Higher values of nickel were recorded at station 2 and station 3 compared to other stations.

3.5.2 Bottom water

Variation in nickel in bottom water during various months of 2020–2021 at different stations of Lower Lake is depicted in **Figure 12**.

During the period of investigation, nickel at bottom water was observed within the range of 0.009–0.064 μg/liter. The minimum value of nickel was recorded at station 1, and the maximum value of nickel was recorded at station 3. During this period, the maximum value of nickel was observed in November 2020. Higher values of nickel were recorded at stations 3 and 2 compared to other stations.

3.6 Iron

3.6.1 Surface water

Variation in iron in surface water during various months of 2020–2021 at different stations of Lower Lake is depicted in **Table 3**.

3.6.2 Bottom water

Variation in iron in bottom water during various months of 2020–2021 at different stations of Lower Lake is depicted in **Table 4**.

3.7 Zinc

3.7.1 Surface water

Variation in zinc in surface water during various months of 2020–2021 at different stations of Lower Lake is depicted in **Table 5**.

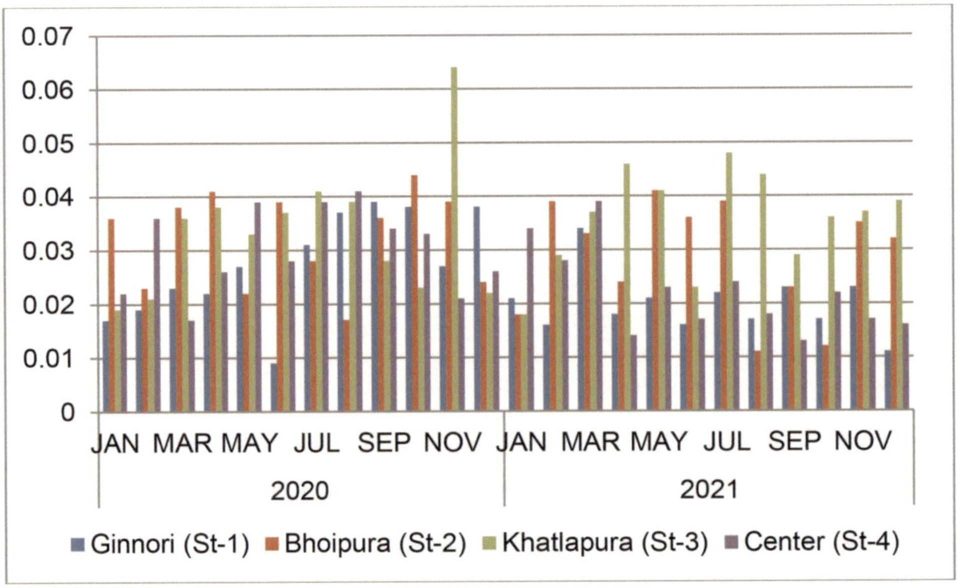

Figure 12.
Heavy metals in bottom water of lower Lake: nickel (μg/liter) during 2020–2021.

	2020												2021											
	JAN	FEB	MAR	APR	MAY	JUN	JUL	AUG	SEP	OCT	NOV	DEC	JAN	FEB	MAR	APR	MAY	JUN	JUL	AUG	SEP	OCT	NOV	DEC
Ginnori (St-1)	0.493	0.128	1.014	1.011	0.984	0.246	0.856	0.644	0.781	1.013	0.9	0.961	0.344	0.473	0.236	0.144	0.852	0.657	0.519	0.746	0.358	0.419	0.387	0.562
Bhoipura (St-2)	0.473	1.236	1.265	1.244	1.237	2.036	0.947	0.634	1.108	1.037	0.985	0.486	1.032	0.987	1.261	0.982	0.457	0.63	0.147	1.089	1.052	1.044	1.026	1.027
Khatlapura (St-3)	0.659	0.478	1.2	1.11	0.948	0.965	0.874	0.865	0.743	0.759	0.968	0.875	0.981	0.694	0.698	1.264	1.042	1.263	1.118	1.096	0.927	0.694	0.466	0.781
Center (St-4)	0.479	0.685	0.986	0.368	0.485	0.463	0.274	0.163	0.318	0.269	0.846	0.873	0.639	0.711	0.279	0.468	0.468	0.561	0.693	0.278	0.237	0.269	0.746	0.478
Min	0.473	0.128	0.986	0.368	0.485	0.246	0.274	0.163	0.318	0.269	0.846	0.486	0.344	0.473	0.236	0.144	0.457	0.561	0.147	0.278	0.237	0.269	0.387	0.478
Max	0.659	1.236	1.265	1.244	1.237	2.036	0.947	0.865	1.108	1.037	0.985	0.961	1.032	0.987	1.261	1.264	1.042	1.263	1.118	1.096	1.052	1.044	1.026	1.027
Mean	0.5393	0.6485	1.1193	0.8908	0.896	0.9987	0.6953	0.5557	0.7293	0.7307	0.92167	0.7737	0.7287	0.7208	0.6618	0.711	0.7197	0.8225	0.6237	0.7638	0.6438	0.6232	0.673	0.7255

Table 3.
Heavy metals in surface water of lower Lake: iron (μg/liter).

	2020												2021											
	JAN	FEB	MAR	APR	MAY	JUN	JUL	AUG	SEP	OCT	NOV	DEC	JAN	FEB	MAR	APR	MAY	JUN	JUL	AUG	SEP	OCT	NOV	DEC
Ginnori (St-1)	1.497	1.129	1.017	1.014	1.584	1.236	1.356	1.444	1.481	1.013	1.194	1.991	1.441	1.479	1.436	1.244	1.752	1.677	1.549	1.746	1.358	1.415	1.387	1.542
Bhoipura (St-2)	1.474	1.239	1.275	1.264	1.281	1.039	1.977	1.637	1.128	1.437	0.989	0.586	1.632	0.988	1.267	1.983	1.458	1.634	1.145	1.189	1.252	1.143	1.127	1.125
Khatlapura (St-3)	1.669	1.678	1.221	1.24	1.448	1.661	0.976	1.867	1.773	1.779	1.988	1.879	1.984	1.894	1.691	1.264	1.242	1.273	1.418	1.092	1.947	1.694	1.476	1.581
Center (St-4)	1.471	1.681	1.982	1.364	1.483	1.461	1.271	1.164	1.317	1.261	1.845	1.871	1.633	1.712	1.277	1.467	1.461	1.563	1.694	1.271	1.233	1.266	1.741	1.472
Min	1.471	1.129	1.017	1.014	1.281	1.039	0.976	1.164	1.128	1.013	0.989	0.586	1.441	0.988	1.267	1.244	1.242	1.273	1.145	1.092	1.23	1.143	1.127	1.125
Max	1.669	1.681	1.982	1.364	1.584	1.661	1.977	1.867	1.773	1.779	1.988	1.991	1.984	1.894	1.691	1.983	1.752	1.677	1.694	1.746	1.95	1.694	1.741	1.581
Mean	1.5418	1.4228	1.4157	1.21	1.4435	1.3495	1.4222	1.5238	1.4333	1.3803	1.49883	1.484	1.6858	1.4925	1.4382	1.5308	1.4845	1.5162	1.4408	1.356	1.5	1.3925	1.4332	1.4043

Table 4.
Heavy metals in bottom water of lower Lake: iron (μg/liter).

DOI: http://dx.doi.org/10.5772/intechopen.1004579

	2020												2021											
	JAN	FEB	MAR	APR	MAY	JUN	JUL	AUG	SEP	OCT	NOV	DEC	JAN	FEB	MAR	APR	MAY	JUN	JUL	AUG	SEP	OCT	NOV	DEC
Ginnori (St-1)	0.501	0.12	0	0.22	0.033	0.02001	0.017	0	0	0.25	0	0.003	1.02	0.008	0.302	0.242	0	1.236	0.045	0.006	0.254	0.324	0.546	0.002
Bhoipura (St-2)	0.54	0.54	0	0.204	0.006	0	0.02	0.001	0	0.31	0.1	0	0.525	0.504	0.245	0.01	0.23	0	0.035	0.002	0.04	0.354	0.12	0
Khatlapura (St-3)	0.06	0.06	0.126	0.23	0	0	0.003	0.12	0.031	0.002	0.202	0.301	0.086	BDL	0.094	1.2	0.65	0	BDL	0.145	0.132	0.008	0.303	0.225
Center (St-4)	0.056	0.009	0.425	0.364	0	0.02	0.001	0	0.002	0	0.025	0.01	0.525	0.504	0.245	0.01	0.23	BDL0	0.035	0.002	0.04	0.354	0.12	0
Min	0.056	0.009	0	0.204	0	0	0.001	0	0	0	0	0	0.086	0.008	0.094	0.01	0	0	0.035	0.002	0.04	0.008	0.12	0
Max	0.54	0.54	0.425	0.364	0.033	0.02	0.02	0.12	0.031	0.31	0.202	0.301	1.02	0.504	0.302	1.2	0.65	1.236	0.045	0.145	0.254	0.354	0.546	0.225
Mean	0.2426	0.1476	0.1102	0.2444	0.0078	0.008	0.0084	0.0242	0.0066	0.1124	0.0654	0.0628	0.4484	0.256	0.196	0.2944	0.222	0.309	0.0375	0.0314	0.1012	0.2096	0.2418	0.0454

Table 5.
Heavy metals in surface water of lower Lake: zinc (μg/liter).

	2020												2021											
	JAN	FEB	MAR	APR	MAY	JUN	JUL	AUG	SEP	OCT	NOV	DEC	JAN	FEB	MAR	APR	MAY	JUN	JUL	AUG	SEP	OCT	NOV	DEC
Ginnori (St-1)	0.521	0.112	0.147	0.325	0.231	0.221	0.178	0.214	0.209	0.252	0.218	0.303	0.402	0.308	0.307	0.342	0.247	0.246	0.445	0.406	0.414	0.521	0.571	0.202
Bhoipura (St-2)	0.545	0.547	0.322	0.219	0.286	0.317	0.024	0.021	0.217	0.329	0.417	0.216	0.425	0.514	0.243	0.401	0.323	0.319	0.135	0.202	0.604	0.454	0.412	0.317
Khatlapura (St-3)	0.506	0.406	0.326	0.423	0.447	0.361	0.203	0.212	0.231	0.302	0.252	0.309	0.586	0.461	0.194	0.249	0.265	0.471	0.367	0.245	0.432	0.508	0.307	0.225
Center (St-4)	0.156	0.309	0.221	0.234	0.327	0.302	0.101	0.119	0.302	0.279	0.325	0.101	0.325	0.204	0.145	0.201	0.123	0.342	0.235	0.202	0.041	0.324	0.312	0.178
Min	0.156	0.112	0.147	0.219	0.231	0.221	0.024	0.021	0.209	0.252	0.218	0.101	0.325	0.204	0.145	0.201	0.123	0.246	0.135	0.202	0.04	0.324	0.307	0.178
Max	0.545	0.547	0.326	0.423	0.447	0.361	0.203	0.214	0.302	0.329	0.417	0.309	0.586	0.514	0.307	0.401	0.323	0.471	0.445	0.406	0.604	0.521	0.571	0.317
Mean	0.40483	0.33883	0.248167	0.30717	0.32817	0.29717	0.12217	0.1335	0.245	0.2905	0.307833	0.22317	0.4415	0.3675	0.2235	0.29917	0.234	0.34917	0.29367	0.27717	0.356	0.442	0.41333	0.23617

Table 6.
Heavy metals in bottom water of lower Lake: zinc (μg/liter).

	2020												2021											
	JAN	FEB	MAR	APR	MAY	JUN	JUL	AUG	SEP	OCT	NOV	DEC	JAN	FEB	MAR	APR	MAY	JUN	JUL	AUG	SEP	OCT	NOV	DEC
Ginnori (St-1)	0.022	0.025	0.08	0.052	0.035	0.002	0.001	0.015	0.012	0.02	0.005	0.1	0.024	0.02	0.052	0.021	0.065	0.024	0.21	0.013	0.002	0.008	0	0.32
Bhoipura (St-2)	0.11	0.112	0.025	0.21	0.305	0.01	0.02	0.024	0.024	0.002	0.004	0.02	0.14	0.154	0.04	0.243	0.335	0.04	0.062	0.058	0.045	0.01	0.009	0.05
Khatlapura (St-3)	0.321	0.322	0.4	0.054	0.12	0.107	0.001	0.002	0.001	0.001	BDL	BDL	0.21	0.421	0.326	0.035	0.23	BDL	0.012	BDL	0.023	BDL	0.5	0.31
Center (St-4)	0.08	0.035	0.23	0.002	0.021	0.003	0.1	0.112	0.04	BDL	0.002	0.002	0.14	0.154	0.04	0.243	0.335	0.04	0.062	0.058	0.045	0.01	0.009	0.05
Min	0.022	0.025	0.025	0.002	0.021	0.002	0.001	0.002	0.001	0.001	0.002	0.002	0.024	0.02	0.04	0.021	0.065	0.024	0.012	0.013	0.002	0.008	0	0.05
Max	0.321	0.322	0.4	0.21	0.305	0.107	0.1	0.112	0.04	0.02	0.005	0.1	0.21	0.421	0.326	0.243	0.335	0.04	0.21	0.058	0.045	0.01	0.5	0.32
Mean	0.111	0.1038	0.152	0.064	0.1004	0.0248	0.0246	0.031	0.0156	0.006	0.00325	0.031	0.1076	0.1538	0.0996	0.1126	0.206	0.032	0.0716	0.0355	0.0234	0.009	0.1036	0.156

Table 7.
Heavy metals in surface water of lower Lake: chromium (μg/liter).

	2020												2021											
	JAN	FEB	MAR	APR	MAY	JUN	JUL	AUG	SEP	OCT	NOV	DEC	JAN	FEB	MAR	APR	MAY	JUN	JUL	AUG	SEP	OCT	NOV	DEC
Ginnori (St-1)	0.329	0.218	0.127	0.211	0.145	0.215	0.041	0.091	0.207	0.013	0.017	0.042	0.247	0.154	0.113	0.127	0.144	0.124	0.327	0.125	0.109	0.115	0.132	0.141
Bhoipura (St-2)	0.117	0.119	0.125	0.214	0.325	0.201	0.202	0.224	0.224	0.202	0.204	0.202	0.114	0.157	0.104	0.249	0.345	0.104	0.162	0.1058	0.145	0.101	0.109	0.105
Khatlapura (St-3)	0.327	0.329	0.412	0.154	0.129	0.137	0.211	0.142	0.201	0.211	0.231	0.171	0.217	0.423	0.329	0.135	0.234	0.319	0.112	0.278	0.123	0.231	0.217	0.316
Center (St-4)	0.083	0.039	0.237	0.142	0.121	0.103	0.116	0.118	0.104	0.178	0.102	0.102	0.145	0.159	0.304	0.246	0.375	0.404	0.262	0.158	0.145	0.116	0.109	0.154
Min	0.083	0.039	0.125	0.142	0.121	0.103	0.041	0.091	0.104	0.013	0.017	0.042	0.114	0.154	0.104	0.127	0.144	0.104	0.112	0.1058	0.11	0.101	0.109	0.105
Max	0.329	0.329	0.412	0.214	0.325	0.215	0.211	0.224	0.224	0.211	0.231	0.202	0.247	0.423	0.329	0.249	0.375	0.404	0.327	0.278	0.15	0.231	0.217	0.316
Mean	0.1878	0.1488	0.2052	0.1726	0.1682	0.1518	0.1222	0.1332	0.168	0.1234	0.1142	0.1118	0.1674	0.2094	0.1908	0.1768	0.2484	0.211	0.195	0.1545	0.13	0.1328	0.1352	0.1642

Table 8.
Heavy metals in bottom water of lower Lake: chromium (μg/liter).

	2020												2021											
	JAN	FEB	MAR	APR	MAY	JUN	JUL	AUG	SEP	OCT	NOV	DEC	JAN	FEB	MAR	APR	MAY	JUN	JUL	AUG	SEP	OCT	NOV	DEC
Ginnori (St-1)	0.002	0.004	0.001	0.003	0.001	0.006	0.001	0.003	0.001	0.003	0.006	0.001	0.003	0.004	0.004	0.003	0.003	0.002	0.001	0.002	0.001	0.002	0.003	0.003
Bhoipura (St-2)	0.003	0.006	0.006	0.006	0.003	0.009	0.003	0.002	0.003	0.007	0.009	0.001	0.001	0.006	0.003	0.002	0.005	0.006	0.003	0.003	0.006	0.009	0.006	0.004
Khatlapura (St-3)	0.004	0.007	0.008	0.005	0.004	0.005	0.001	0.006	0.008	0.006	0.001	0.002	0.001	0.009	0.004	0.007	0.009	0.004	0.001	0.006	0.005	0.007	0.004	0.009
Center (St-4)	0.001	0.003	0.002	0.002	0.001	0.003	0.001	0.001	0.002	0.002	0.003	0.001	0.002	0.003	0.003	0.001	0.002	0.003	0.001	0.004	0.003	0.001	0.002	0.006
Min	0.001	0.003	0.001	0.002	0.001	0.003	0.001	0.001	0.001	0.002	0.001	0.001	0.001	0.003	0.003	0.001	0.002	0.002	0.001	0.002	0.001	0.001	0.002	0.003
Max	0.004	0.007	0.008	0.006	0.004	0.009	0.003	0.006	0.008	0.007	0.009	0.002	0.003	0.009	0.004	0.007	0.009	0.006	0.003	0.006	0.006	0.009	0.006	0.009
Mean	0.0025	0.005	0.0043	0.004	0.0023	0.0058	0.0017	0.0032	0.0038	0.0045	0.00483	0.0013	0.0018	0.0057	0.0035	0.0035	0.005	0.0038	0.0017	0.0038	0.0037	0.0048	0.0038	0.0057

Table 9.
Heavy metals in surface water of lower Lake: copper (μg/liter).

	2020												2021											
	JAN	FEB	MAR	APR	MAY	JUN	JULY	AUG	SEP	OCT	NOV	DEC	JAN	FEB	MAR	APR	MAY	JUN	JUL	AUG	SEP	OCT	NOV	DEC
Ginnori (St-1)	0.003	0.005	0.003	0.004	0.004	0.008	0.006	0.008	0.002	0.007	0.005	0.002	0.007	0.005	0.006	0.007	0.004	0.004	0.004	0.004	0.005	0.004	0.004	0.005
Bhoipura (St-2)	0.005	0.008	0.007	0.009	0.007	0.009	0.009	0.009	0.004	0.009	0.009	0.003	0.008	0.008	0.008	0.008	0.008	0.007	0.005	0.008	0.008	0.008	0.007	0.008
Khatlapura (St-3)	0.007	0.008	0.009	0.009	0.005	0.007	0.008	0.008	0.009	0.009	0.008	0.004	0.007	0.009	0.009	0.009	0.009	0.008	0.007	0.007	0.009	0.009	0.008	0.007
Center (St-4)	0.003	0.004	0.004	0.006	0.006	0.004	0.004	0.004	0.003	0.004	0.004	0.004	0.004	0.004	0.004	0.004	0.007	0.004	0.003	0.006	0.004	0.004	0.004	0.002
Min	0.003	0.004	0.003	0.004	0.004	0.004	0.004	0.004	0.002	0.004	0.004	0.002	0.004	0.004	0.004	0.004	0.004	0.004	0.003	0.004	0.004	0.004	0.004	0.002
Max	0.007	0.008	0.009	0.009	0.007	0.009	0.009	0.009	0.009	0.009	0.009	0.004	0.008	0.009	0.009	0.009	0.009	0.008	0.007	0.008	0.009	0.009	0.008	0.008
Mean	0.0047	0.0062	0.0058	0.0068	0.0055	0.0068	0.0067	0.007	0.0048	0.007	0.0065	0.0032	0.0063	0.0065	0.0067	0.0068	0.0068	0.0058	0.0048	0.0062	0.0065	0.0063	0.0058	0.005

Table 10.
Heavy metals in bottom water of lower Lake: copper (μg/liter).

DOI: http://dx.doi.org/10.5772/intechopen.1004579

	2020												2021											
	JAN	FEB	MAR	APR	MAY	JUN	JUL	AUG	SEP	OCT	NOV	DEC	JAN	FEB	MAR	APR	MAY	JUN	JUL	AUG	SEP	OCT	NOV	DEC
Ginnori (St-1)	0.012	0.011	0.019	0.017	0.023	0.013	0.026	0.022	0.022	0.031	0.012	0.023	0.017	0.014	0.024	0.011	0.017	0.012	0.019	0.011	0.017	0.014	0.013	0.018
Bhoipura (St-2)	0.019	0.019	0.021	0.021	0.029	0.018	0.029	0.031	0.029	0.039	0.018	0.019	0.023	0.019	0.032	0.013	0.026	0.018	0.023	0.013	0.023	0.032	0.017	0.026
Khatlapura (St-3)	0.017	0.017	0.023	0.026	0.026	0.019	0.021	0.026	0.026	0.035	0.021	0.024	0.028	0.021	0.027	0.019	0.031	0.021	0.028	0.019	0.029	0.038	0.028	0.027
Center (St-4)	0.011	0.013	0.016	0.02	0.019	0.013	0.016	0.021	0.021	0.011	0.011	0.019	0.012	0.011	0.013	0.012	0.024	0.011	0.017	0.011	0.018	0.031	0.011	0.013
Min	0.011	0.011	0.016	0.017	0.019	0.013	0.016	0.021	0.021	0.011	0.011	0.019	0.012	0.011	0.013	0.011	0.017	0.011	0.017	0.011	0.017	0.014	0.011	0.013
Max	0.019	0.019	0.023	0.026	0.029	0.019	0.029	0.031	0.029	0.039	0.021	0.024	0.028	0.021	0.032	0.019	0.031	0.021	0.028	0.019	0.029	0.038	0.028	0.027
Mean	0.0148	0.015	0.0197	0.0212	0.0242	0.0158	0.0228	0.0253	0.0247	0.0277	0.01567	0.0213	0.02	0.0162	0.0235	0.0142	0.0243	0.0157	0.022	0.014	0.0222	0.0278	0.018	0.02

Table 11.
Heavy metals in surface water of lower Lake: nickel (μg/liter).

3.7.2 Bottom water

Variation in zinc in bottom water during various months of 2020–2021 at different stations of Lower Lake is depicted in **Table 6**.

3.8 Chromium

3.8.1 Surface water

Variation in chromium in surface water during various months of 2020–2021 at different stations of Lower Lake is depicted in **Table** 7.

3.8.2 Bottom water

Variation in chromium in bottom water during various months of 2020–2021 at different stations of Lower Lake is depicted in **Table 8**.

3.9 Copper

3.9.1 Surface water

Variation in copper in surface water during various months of 2020–2021 at different stations of Lower Lake is depicted in **Table 9**.

3.9.2 Bottom water

Variation in copper in bottom water during various months of 2020–2021 at different stations of Lower Lake is depicted in **Table 10**.

3.10 Nickel

3.10.1 Surface water

Variation in nickel in surface water during various months of 2020–2021 at different stations of Lower Lake is depicted in **Table 11**.

3.10.2 Bottom water

Variation in nickel in bottom water during various months of 2020–2021 at different stations of Lower Lake is depicted in **Table 12**.

4. Results and discussion

In India, the most anthropogenic sources of metals are industrial, petroleum contamination, and sewage disposal Vishwakarma et al. [30]. During the present

	2020												2021											
	JAN	FEB	MAR	APR	MAY	JUN	JUL	AUG	SEP	OCT	NOV	DEC	JAN	FEB	MAR	APR	MAY	JUN	JUL	AUG	SEP	OCT	NOV	DEC
Ginnori (St-1)	0.017	0.019	0.023	0.022	0.027	0.009	0.031	0.037	0.039	0.038	0.027	0.038	0.021	0.016	0.034	0.018	0.021	0.016	0.022	0.017	0.023	0.017	0.023	0.011
Bhoipura (St-2)	0.036	0.023	0.038	0.041	0.022	0.039	0.028	0.017	0.036	0.044	0.039	0.024	0.018	0.039	0.033	0.024	0.041	0.036	0.039	0.011	0.023	0.012	0.035	0.032
Khatlapura (St-3)	0.019	0.021	0.036	0.038	0.033	0.037	0.041	0.039	0.028	0.023	0.064	0.022	0.018	0.029	0.037	0.046	0.041	0.023	0.048	0.044	0.029	0.036	0.037	0.039
Center (St-4)	0.022	0.036	0.017	0.026	0.039	0.028	0.039	0.041	0.034	0.033	0.021	0.026	0.034	0.028	0.039	0.014	0.023	0.017	0.024	0.018	0.013	0.022	0.017	0.016
Min	0.017	0.019	0.017	0.022	0.022	0.009	0.028	0.017	0.028	0.023	0.021	0.022	0.018	0.016	0.033	0.014	0.021	0.016	0.022	0.011	0.013	0.012	0.017	0.011
Max	0.036	0.036	0.038	0.041	0.039	0.039	0.041	0.041	0.039	0.044	0.064	0.038	0.034	0.039	0.039	0.046	0.041	0.036	0.048	0.044	0.029	0.036	0.037	0.039
Mean	0.0245	0.0257	0.0282	0.0317	0.0303	0.0268	0.0347	0.032	0.034	0.0342	0.03933	0.0283	0.0238	0.0278	0.0358	0.027	0.0313	0.024	0.0338	0.0242	0.0217	0.0225	0.0277	0.0247

Table 12.
Heavy metals in bottom water of lower Lake: nickel (μg/liter).

investigation, comparatively in monsoon and post-monsoon months, metal ions can be incorporated into food chains and concentrated in aquatic organisms to a level that affects their physiological state. Of the effective pollutants are the heavy metals, which have a drastic environmental impact on all organisms. Trace metals such as Pb, Fe, Zn, Hg, Cr, Cd, Co, and Ni can play a biochemical role in the life processes of all aquatic plants and animals; therefore, they are essential in the aquatic environment in trace amounts of Mason [31].

During the present investigation, it was observed that the metals in the Lake water attained their maximum values during monsoon and post-monsoon seasons. Station 2, followed by stations 3 and 1, ranked first in the accumulation of metals, while station 4 ranked last. This may be attributed to the discharge of sewage effluents from the adjacent habitation clusters where a well-developed and operational Sewage and Effluent Treatment Plant has yet to be fully established.

The maximum mean values of the measured metals were recorded at stations 2 and 3. This may be attributed to the huge amounts of raw sewage and industrial wastewater (automobile repairing and service center) discharged into the Lake from the adjacent catchment area. Higher values of heavy metals due to the discharge of raw sewage and industrial effluents were also reported by Abdel-Moat and El-Samar [32]. The high levels of Zn and Fe in the lake water can be attributed to industrial and agricultural discharge, which was also reported by Mason [31].

The health hazards associated with exposure to heavy metals depend on their oxidation state, ranging from the low toxicity of the metal form to the high toxicity. While comparing the values of the above physicochemical and heavy metal parameters, it can be concluded that the deterioration in the water quality of Lake is mainly due to the inflow of sewage and urban wastes from it is densely habitation.

5. Conclusions

In general, most of the parameters during the period of investigation, the range values of different heavy metals in surface and bottom waters, were compared with the EPA standards [29] referred by the Central Pollution Control Board (CPCB), India (**Table 2**). At the same time, comparing the values, it was found that most of the heavy metals were below the permissible limits, except Fe, which was alarming. During the monsoon season, there was a noticeable increase in heavy metal concentrations in both surface and bottom waters. This can be attributed to the runoff from adjacent areas, which carry pollutants and sediments into the Lake. The post-monsoon period showed a relatively lower level of heavy metal contamination, indicating a decrease in external inputs during this time. However, the pre-monsoon season witnessed a resurgence in heavy metal levels, potentially due to increased anthropogenic activities and sediment disturbances. Although presently there is no alarming level of heavy metal pollution in this water body, and fish that are being consumed is safe as per the present analysis however, in the future, if the inflow of untreated domestic sewage and dumping of solid wastes continues, the quality of Lake water may further deteriorate to alarming level. The state of water bodies that are utilized for the main reasons must be preserved, while those utilized for secondary uses must be enhanced. For the rehabilitation, protection, and maintenance of this water body at the governmental and private levels, appropriate preservation and administration plans and tactics must be developed and implemented because this Lake is the backbone of the city of Bhopal.

DOI: http://dx.doi.org/10.5772/intechopen.1004579

Acknowledgements

The authors are thankful to the Department of Zoology, Government Motilal Vigyan Mahavidyalaya, Bhopal, and EPCO, Bhopal, India, for providing the necessary facilities to carry out the sample analysis.

Conflict of interest

The authors declare that there is no competing or conflict of interest.

Nomenclature

Cu	copper
Cr	chromium
Fe	iron
Ni	nickel
Cd	cadmium
Hg	mercury
Zn	zinc
Pb	lead
Ppm	parts per million
St-1	Station 1
St-2	Station 2
St-3	Station 3
St-4	Station 4
S	surface
B	bottom
AAS	Atomic absorption spectroscopy
BOD	Biochemical oxygen demand
COD	Chemical oxygen demand
mg/l	milligrams per liter
µg	microgram
µg/liter	microgram per liter
CPCB	Central Pollution Control Board

A. Appendices

See **Figure A1**.

Figure A1.
Photographs of the sites selected for sampling.

Author details

Aarefa Jan[1*], Suchitra Banerjee[2], Rajendra Chouhan[3], Subrata Pani[4] and Saima Syed[5]

1 Department of Zoology and Applied Aquaculture, Barkatullah Vishwavidyalaya, Bhopal, Madhya Pradesh, India

2 Department of Biotechnology, Institute for Excellence in Higher Education, Bhopal, Madhya Pradesh, India

3 Department of Zoology, Government Motilal Vigyan Mahavidyalaya, Bhopal, Madhya Pradesh, India

4 Environmental Research Laboratory, Environmental Planning and Co-ordination Organization, Bhopal, Madhya Pradesh, India

5 Department of Zoology, Government Degree College for Women, Pulwama, Jammu and Kashmir, India

*Address all correspondence to: aarifaganilone97@gmail.com

References

[1] Adoni AD, Joshi G, Ghosh K, Chourasia SK, Vaishya AK, Yadav M, et al. Work Book on Limnology. Sagar, India: Pratibha Publishers; 1985. 127 pp

[2] Alam MZ, Ahmad S, Malik A. Genotoxic and mutagenic potential of agricultural soil irrigated with tannery effluents at Jajmau (Kanpur), India. Archives of Environmental Contamination and Toxicology. 2009; **57**(3):463-476

[3] Alam MZ, Ahmad S, Malik A, Ahmad M. Mutagenicity and genotoxicity of tannery effluents used for irrigation at Kanpur, India. Ecotoxicology and Environmental Safety. 2011;**73**(7):1620-1628

[4] Lingaswamy, Saxena. Assessment of the water quality of Hussain Sagar, Fox Sagar and Kattamysamma Lakes of Hyderabad, Telangana State, India: Using Water Quality Index (WQI). Current World Environment. 2016;**11** (2):537-543

[5] Yadav KK, Gupta N, Kumar V, Sharma S, Arya S. Water quality assessment of Pahuj River using water quality index at Unnao Balaji, M.P., India. International Journal of Basic and Applied Sciences. 2015;**19**(1):241-250

[6] Padma Priya KT, Seeta Y, Manikya Reddy P. Assessment of water quality in Saroornagar Lake, Hyderabad. International Journal of Recent Scientific Research. 2015;**6**(11):7536-7540

[7] Srinivas L, Seeta Y, Manikya Reddy P. Assessment of Water Quality Index in Lower Manair Dam, Karimnagar district, Telangana. International Journal of Recent Research Aspects. 2017;**4**(4): 6-10

[8] Brack W, Dulio V, Ågerstrand M, et al. Towards the review of the European Union Water Framework Directive: Recommendations for more efficient assessment and management of chemical contamination in European surface water resources. Science of the Total Environment. 2017;**576**:720-737

[9] Chabukdhara M, Nema AK. Assessment of heavy metal contamination in Hindon River sediments: A chemometric and geochemical approach. Chemosphere. 2012;**87**(8):945-953

[10] Liu WX, Li XD, Shen ZG, et al. Multivariate statistical study of heavy metal enrichment in sediments of the Pearl River Estuary. Environmental Pollution. 2003;**121**(3):377-388

[11] Uluturhan E, Kucuksezgin F. Heavy metal contaminants in Red Pandora (Pagellus erythrinus) tissues from the Eastern Aegean Sea, Turkey. Water Research. 2007;**41**(6):1185-1192

[12] Yi Y, Yang Z, Zhang S. Ecological risk assessment of heavy metals in sediment and human health risk assessment of heavy metals in fishes in the middle and lower reaches of the Yangtze River basin. Environmental Pollution. 2011;**159**(10):2575-2585

[13] Laws EA. Aquatic Pollution-An Introductory Text. New York: Wiley; 2000

[14] Fatoki OS, Lujiza N, Ogunfowokan AO. Levels of Cd, Hg and Zn in some surface waters from the Eastern Cape Province, South Africa. Water SA. 2003; **29**(4). ISSN 0378-4738

[15] Waqar AS, Hasan J. Effect of rain water harvesting on the amounts of

some heavy metals in the ground water of East Dwarka subcity of Delhi. Current World Environment: An International Research Journal of Environmental Sciences. 2006;**2**:145-150

[16] Storelli MM, Storelli A, D'addabbo R, et al. Trace elements in loggerhead turtles (Carettacaretta) from the eastern Mediterranean Sea: Overview and evaluation. Environmental Pollution. 2005;**135**(1):163-170

[17] Dural M, Bickici E. Distribution of trace elements in the tissues of Upeneuspori and Upeneusmolucensis from the eastern cost of Mediterranean, Iskenderun Bay, Turkey. Journal of Animal and Veterinary Advances. 2010; **9**:1380-1383

[18] Hasan MR, Khan MZ, Khan M, et al. Heavy metals distribution and contamination in surface water of the Bay of Bengal coast. Cogent Environmental Science. 2016;**2**(1): 1140001. DOI: 10.1080/23311843.2016.1140001

[19] Khound NJ, Bhattacharyya KG. Multivariate statistical evaluation of heavy metals in the surface water sources of Jia Bharali river basin, North Brahmaputra plain, India. Applied Water Science. 2017;**7**(5):2577-2586

[20] Keesstra S, Nunes J, Novara A, et al. The superior effect of nature based solutions in land management for enhancing ecosystem services. Science of the Total Environment. 2018;**610**:997-1009

[21] Paller MH, Littrell JW. Long-term changes in mercury concentrations in fish from the middle Savannah River. Science of the Total Environment. 2007; **382**(2–3):375-382

[22] Saha N, Rahman MS, Ahmed MB, et al. Industrial metal pollution in water and probabilistic assessment of human health risk. Journal of Environmental Management. 2017;**185**:70-78

[23] Bhattacharya PT, Misra SR, Hussain M. Nutritional aspects of essential trace elements in oral health and disease: An extensive review. Scientifica. 2016;**2016**: 1-12. DOI: 10.1155/2016/5464373

[24] Nair IV, Kailash S, Arumugam M, Gangadhar K, Clarson D. Trace metal quality of Meenachil river at Kottayam, Kerala (India) by principal component analysis. World Applied Sciences Journal. 2010;**9**(10):1100-1107

[25] Dinakaran J, Abbas NS, Shvetambriarora SB, Kaula BC. Assessment of heavy metals in ground water of different locations of National Capital Region, Delhi, India. Current World Environment. 2021;**16**(1):143-150. DOI: 10.12944/CWE.16.1.14

[26] Bangotra P, Jakhu R, Mukesh Prasad RS, Aswal AA, Mushtaq Z, Mehra R. Investigation of heavy metal contamination and associated health risks in groundwater sources of southwestern Punjab, India. Environmental Monitoring and Assessment. 2023;**195**:Article number: 367

[27] Mawari G, Kumar N, Sarkar S, Frank AL, Daga MK, Singh MM, et al. Human Health Risk Assessment due to Heavy Metals in Ground and Surface Water and Association of Diseases With Drinking Water Sources: A Study From, Maharashtra, India. Environmental Health Insights. 2022;**2022**(16): 11786302221146020. DOI: 10.1177/11786302221146020

[28] APHA. Standard Methods for the Examination of Water and Waste Water.

20th ed. American Water Works Association Water Environmental Federation: American Public Health Association; 2010

[29] Environment (Protection) Rules, 1989 Ministry Of Environment And Forests, S.O. 8(E) dated the 3rd January, 1989. Clauses (aa), (ee) and (ff) inserted by Notification NO. G.S.R. 931(E) dated 27.10.89 published in the Gazette No. 564 dated 27.10.89

[30] Vishwakarma S, Varma A, Saxena G. Assessment of water quality of lower lake river, Madhya Pradesh, India. International Journal of Water Resources and Environmental Engineering. 2013;**5** (4):217-222. DOI: 10.5897/IJWREE2012.0376

[31] Mason CF. Biology of Freshwater Pollution. 4th ed. England: Essex Univ; 2002. 387 p

[32] Abdel-Moati MA, El-Sammak AA. Man-made impact on the geochemistry of the Nile Delta Lakes. A study of metals concentrations in sediments. Water, Air and Soil Pollution. 1997;**97**:413-429

Chapter 5

Determination of the Physicochemical and Bacteriological Parameters of the Waters of Lake Sonfonia Commune of Ratoma (Republic of Guinea) 2021

Mohamed Lamine Komara and Kandé Bangoura

Abstract

Lake Sonfonia located in the commune of Ratoma in the Republic of Guinea is under the influence of various anthropogenic activities. Field visits and interviews with the local population made it possible to identify the sources of pollution and the different stations for taking water samples. The results of the analysis of physicochemical and bacteriological parameters showed that the chemical oxygen demand (COD) is >20 mg/l WHO limit value for surface waters indicating the presence of a high material load organic and a drop in the dissolved oxygen level < 10 mg/l, the WHO limit value. The concentration of fecal coliforms and fecal streptococci exceed the WHO maximum admissible concentration, respectively (200–1000 CFU/100 ml) and (62 CFU/100 ml), and the presence of *Escherichia coli* in water indicates contamination by feces. Given its advanced level of pollution, it is no longer used by the Guinea Water Company (GWC) as a source of drinking water. Stata 15 software was applied for statistical analysis of the data. The correlation test between the physicochemical parameters gave highly significant correlations. It was found that the spatiotemporal variations of bacteriological parameters actually depend on the incubation conditions.

Keywords: Lake Sonfonia, physicochemical parameters, bacteriological parameters, anthropogenic activities, fecal bacteria

1. Introduction

Guinea, like other developing countries, is not immune to the problems of non-sanitation of wastewater. Indeed, most buildings are not connected to sanitation networks and directly discharge domestic wastewater into septic tanks or into courtyards, rivers, and lakes [1].

Currently and for many years, the situation of Lake Sonfonia is alarming, it has become an illegal dump, and all liquid waste and household wastewater are constantly dumped there [2].

Given its advanced level of pollution, it is no longer used as a source of drinking water by the Société des Eaux de Guinée (SEG). As a contribution to monitoring and protection efforts against pollution, this study was chosen: *Determination of the physicochemical and bacteriological parameters of the waters of Lake Sonfonia Commune of Ratoma (Republic of Guinea) 2021* [2].

To carry out this study, we carried out field visits, interviews with resource people, and taking photos. This approach made it possible to identify the different sampling points while taking into account the representativeness of the pollution sources and operational feasibility [2].

This interview above all made it possible to know the potential sources of pollution and the various uses made of them by the local population of the waters of the lake, namely: watering market gardening crops, fishing, washing rolling machines, swimming, etc., exposing them to a health risk.

The main objective: Contribute to the evaluation of the water quality of Lake Sonfonia by determining the physicochemical and bacteriological characteristics for sustainable use.

Specific objectives:

1. Appreciate the presence of invasive aquatic plants covering the surface at certain times of the year to determine the natural aging process of Lake Sonfonia or Eutrophication.

2. Determine the concentration of physicochemical and bacteriological parameters of water samples from Lake Sonfonia in the dry season and in the rainy season.

3. Determine the level of significance of the variation in concentration of the parameters studied in dry weather and wet weather.

4. Determine the degree of correlation between the physicochemical parameters of water samples from the Lake Sonfonia.

5. Identify the sources of pollution and the different uses made of them by the populations living near the waters of Lake Sonfonia.

Study methodology: Longitudinal study data are collected in the dry season and in the rainy season to better observe spatiotemporal changes.

Carrying out this study will include the following steps:

- Survey: visits to the study site, interviews with resource people and populations, local locations will be organized as well as photo taking;

- Collection of water samples: water samples will be collected at different stations in Lake Sonfonia to cover a variety of areas and depths; physicochemical analyses of water samples to determine the parameters such as pH, temperature, conductivity, TSD concentration, turbidity, MES, COD, dissolved oxygen, nitrates, nitrites, phosphates;

- Bacteriological analyses of water samples to determine fecal pollution indicator bacteria such as fecal coliforms, fecal streptococci, and *E. coli* and pathogenic bacteria such as *salmonella*;

- Statistical processing of data: significance test and correlation test; and

- Interpretation of the results/compared to international standards, to determine the state of the water quality of Lake Sonfonia by proposing appropriate measures to limit this pollution.

What is the research question?

What is the water quality of Lake Sonfonia in Guinea and what are the factors that influence this quality?

What are the research hypotheses?

"Does the low level of local sanitation systems influence surface water pollution"

Verification of the research hypothesis.

2. Materials and methods

2.1 Presentation of the study area

Sonfonia Lake is located between the Sonfonia and Foulah-Madina districts, in the commune of Ratoma in Conakry. The population of the municipality of Ratoma is estimated at 697,744 people. Lake Sonfonia has an area of 62 km^2, a length of approximately 2.5 km, and a width varying between 100 to 200 m. The depth varies between 6 and 7 m in the rainy season and between 3 and 4 m in the dry season. The lake is fed by natural groundwater sources and is home to the SEG drinking water treatment point [2].

See **Figure 1**.

2.2 Sampling

Sampling covered six stations taking into account the representativeness of pollution sources and operational feasibility. Each station was geolocated as shown in **Table 1**.

At each station, two sampling campaigns were carried out during the low-flow period corresponding to the dry season (March 2021) and two others during the high-water period corresponding to the rainy season (August 2021) [2].

The collection, transport, and storage of the samples complied with the protocol are defined by AFNOR and Rodier 2009 [3].

Figure 1.
The map of Lake Sonfonia on Google maps [1].

Stations	North latitude	West longitude
SEG station	NO: 09040′ 28.6″ E	WO: 13034′ 55.5" N
YAT station	NO: 09040′ 26.4″ E	WO: 13034′ 58.1" N
FLM station	NO: 09040′ 26.5″ E	WO: 13034′ 54.0"N
SOC station	NO: 09040′ 43.1″ E	WO: 13034′ 25.6"N
BLZ station	NO: 09040′ 40.9″ E	WO: 13034′ 47.9"N
USF station	NO: 09040′ 46.1″ E	WO: 13034′ 46.2"N

Table 1.
Stations and geographic coordinates.

3. Determination of physicochemical parameters

3.1 Material

The device shown in **Figures 2–7** was applied to determine the physicochemical parameters [4].

3.2 Analysis methods

The analyses were carried out *in situ* and in the laboratory using the techniques of Rodier 2009 [3]. The values obtained were compared to WHO standards.

Measurements carried out *in situ*: They concerned the following parameters: temperature, pH, electrical conductivity, total dissolved solids, and dissolved oxygen.

Laboratory analyses: A Lovibond Model TB210 IR Turbidimeter was used to measure turbidity by the optical method. A turbidity sensor shines light into the water

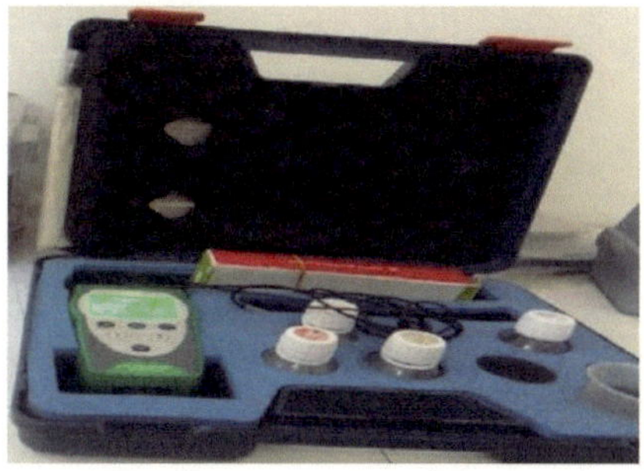

Figure 2.
pH 70 + DHS for measuring temperature and pH.

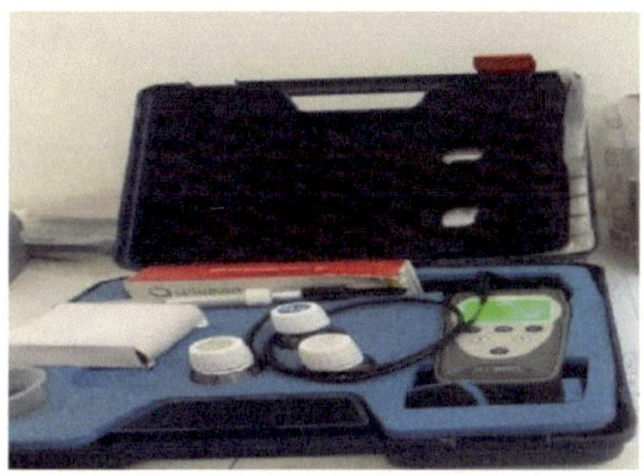

Figure 3.
COND 70+ for measuring electrical conductivity and total dissolved solids.

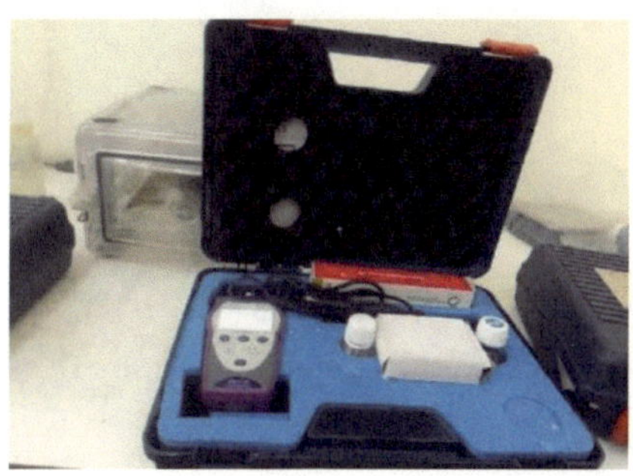

Figure 4.
OXY 70+ for measuring dissolved oxygen.

and measures the amount of light that passes through the water versus the amount of light that is reflected by particles in the water. Results are expressed in nephelometric turbidity units (NTU).

The DR/850 HACH Spectrophotometer was used to determine the concentration of suspended matter by measuring absorbance (Beer-Lambert law).

The colorimetric assay to determine COD, nitrates, nitrites, total iron, sulfates, and phosphates was carried out using the DR/850 HACH brand Spectrophotometer.

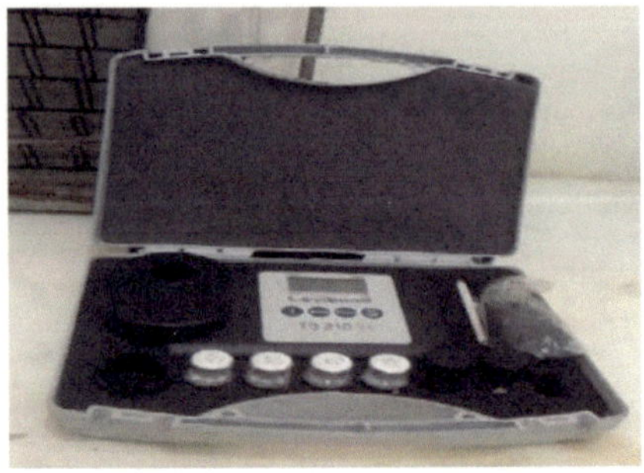

Figure 5.
TURD LOVIBOND for measuring turbidity.

Figure 6.
HANNA thermoreactor for heating water samples.

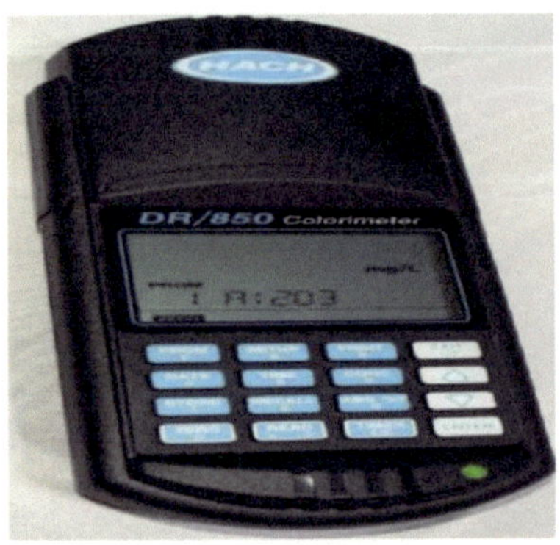

Figure 7.
DR/850 colorimeter HACH for colorimetric assay.

The principle of colorimetry consists of mineralization in a digester and the measurement of the intensity of the coloring which is proportional to the concentration of the element to be measured in the colorimeter. Mineralization takes place in an acidic environment at a temperature of 150°C for 2 hours [5].

The analyses of the water samples were carried out in the laboratory of the environmental study and research center (CERE), of the Gamal Abdel Nasser University of Conakry.

4. Determination of bacteriological parameters

4.1 Material

See **Figure 8**.

4.2 Analysis methods

The number of germs sought was carried out using the membrane filter technique and the culture of the germs was carried out on culture media on cardboard (NKS) (see **Figure 9**). Culture media following were used for incubation of germs: azide on green background for the isolation of fecal streptococci (SF) (**Figures 10** and **11**); chromocult on a white background for the isolation of total coliforms (CT), fecal coliforms(CF), and *Escherichia coli* (*E. coli*) (**Figures 12** and **13**); and finally bismuth sulfite for the isolation of *salmonella*.

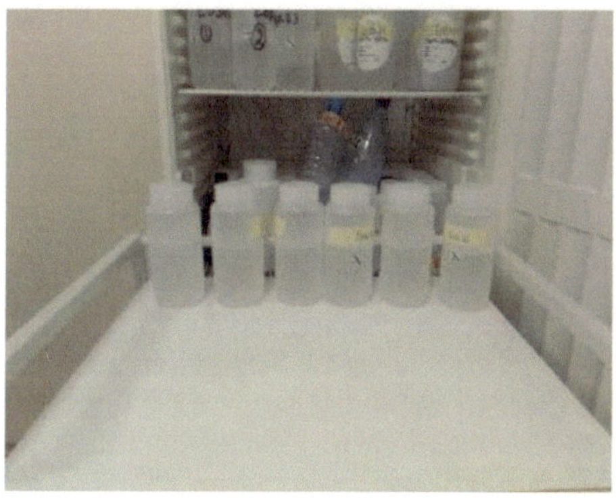

Figure 8.
Bottles containing the water samples to be analyzed.

Figure 9.
Equipment and consumables for the determination of bacteriological parameters. Top from left to right: The Pasteur pipette, bottle of physiological water, filtration device with an electric vacuum cleaner, a pair of scissors, forceps, and a Bunsen nozzle. Bottom from left to right: the filter membrane, the Petri dish containing culture media in cardboard [2].

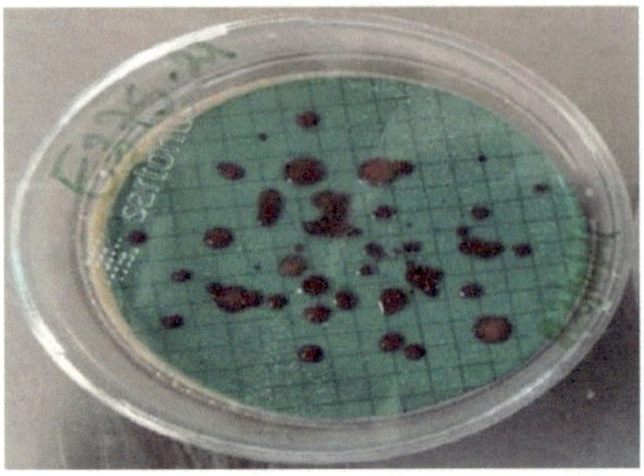

Figure 10.
A Petri dish containing azide culture medium on a green background for the isolation of fecal streptococci [2].

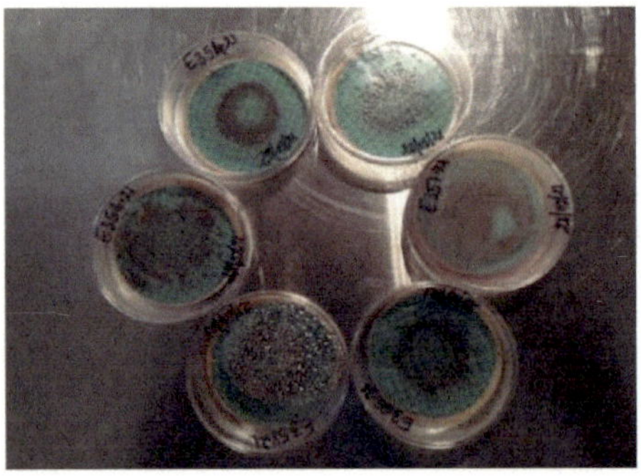

Figure 11.
Six Petri dishes containing azide culture medium on a green background for the isolation of fecal streptococci [2].

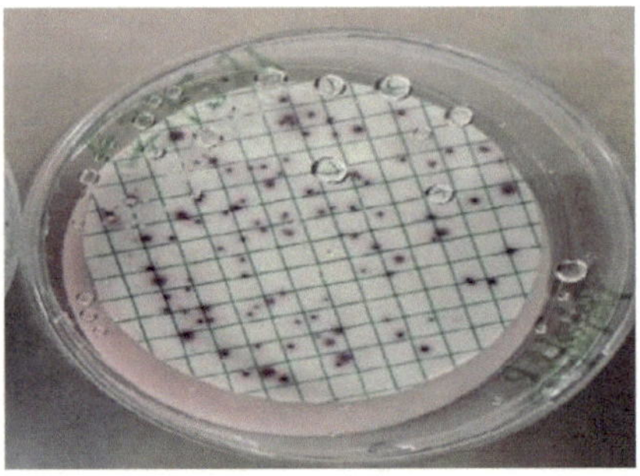

Figure 12.
A Petri dish containing the chromocult culture medium on a white background for the isolation of fecal coliforms and Escherichia coli [2].

Before use, the NKS are moistened with 3.5 ml of distilled water using a Pasteur pipette.

The water sample taken comes from an environment that appears to be contaminated, which requires successive dilutions.

The 250 ml water sample to be analyzed was diluted successively to 1/10th, 1/100th, and 1/1000th. During this technique, 100 ml of the water sample was filtered.

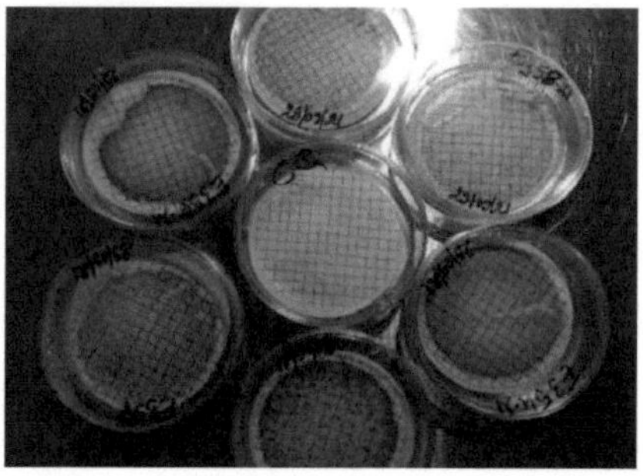

Figure 13.
Six Petri dishes containing chromo cult culture medium on a white background corresponding to the water samples taken in the different stations for the isolation of fecal coliforms and Escherichia coli *and a Petri dish in the center [2].*

The number of bacteria is obtained by the following relationship = Number of bacteria counted x Dilution factor (10) /Inoculum volume (0.1) for a 1/10 dilution.

All bacteria present in the sample are retained on the surface of the membrane, which is then placed on the sterile SCN using forceps. The SCNs incubated in the Petri dishes with the lid are placed in the incubator. The reading is taken in 24 or 48 hours at a temperature varying between 22 and 44°C [2].

Fecal streptococci were enumerated by counting typical red to brown-red smooth-edged colonies, and fecal coliforms form pink-red colonies. *Escherichia coli* grew as dark blue to purple colonies. Confirmation of *Escherichia coli* was made by adding a drop of Erlich Kovacs reagent. The principle of this reaction is based on the ability of *Escherichia coli* to cleave tryptophan into an organic compound, indole, using the enzyme tryptophanase. This compound reacts with the dimethylaminobenzaldehyde reagent to form a dark red color, indicating the presence of indole, and thus confirms the Indole/Tryptophanase-positive character of the *Escherichia coli* colony [2].

Salmonella will form lightly colored colonies with a brown to black center surrounded by a black halo (fish eye).

All these operations take place under the hood in the presence of bunsen nozzle flames.

The number of colonies counted is expressed in colony-forming units per 100 ml of water (UCF/100 ml). The analyses of the water samples were carried out at the Guinéo-German Medical Laboratory BP: 4529 Conakry, E-mail: Info@lga-guinee.com [2].

5. Results and discussion

5.1 Results of physicochemical parameters

See Ref. [4].

6. Temperature

Tables 2 and **3** show that the water temperature varies depending on the stations and the sampling period. The temperatures observed are above 25°C, the WHO limit value for surface waters.

Settings Stations	**T°C**	**pH**	**THIS (µs /cm)**	**TDS (mg/l)**	**MY (mg/l)**	**Turbidity (NTU)**
SEG station	32.90	7.01	1350	623	7	10
YAT station	31.90	7.02	1680	994	9	32
FLM Station	31.80	7.05	1464	702	15	20
SOC station	32.80	7.02	1570	784	31	18
BLZ station	33.00	7.01	1650	887	6	14
USF station	33.90	7.00	1486	802	28	31
Average	32.71	7.01	1533.33	798.66	16.00	20.83
WHO limit value for surface water	25	6.5 < pH < 8.5	200–1000	<600	<10	< 5

Table 2.
Physical parameters measured in March 2021 at the different stations of the lake.

Settings Stations	**T°C**	**pH**	**THIS (µs /cm)**	**TDS (mg/l)**	**MY (mg/l)**	**Turbidity (NTU)**
SEG station	27	7.50	1306	653	11	18.5
YAT	28.5	7.30	1607	803	39	38.3
FLM	27.4	7.20	1550	775	35	40
SOC	28	7.40	1465	732	38	46
BLZ	27	7.20	1468	734	20	34
USF	27.6	7.50	1529	764	32	47.5
Average	27.58	7.35	1487.50	743.50	29.16	37.38
WHO limit value for surface waters	25	6.5 < pH < 8.5	200–1000	<600	<10	< 5

Table 3.
Physical parameters measured in August 2021 at the different stations of the lake.

The analysis of water samples also shows that during the dry season temperatures vary between 33.80°C (FLM) and 33.90°C (USF) with a thermal difference of 2.1°C, while during the rainy season, temperatures vary between 27°C (SEG and BLZ) and 28.50°C (YAT) with a thermal difference of 1.5°C [4].

This rise in the water temperature of Lake Sonfonia is a limiting factor; in fact, the higher the water temperature rises, the more the quantity of dissolved oxygen decreases, leading to the disappearance of species [6].

6.1 pH

It appears in **Tables 2** and **3** that the minimum value of pH 7.00 was observed at the station (USF) during the dry season and the maximum value of pH 7.50 at the station (USF) during the rainy season. The average pH values obtain with the WHO recommendation for surface waters (6.5<pH<8.5) where life will develop optimally [4].

The pH of water depends on its origin. A drop in pH can lead to the dissolution of trace metals, which are often harmful and sometimes fatal at low doses for aquatic organisms. High pH increases the concentration of ammonia toxic to fish [7].

6.2 Electrical conductivity (EC) and total dissolved solids (TDS)

The conductivity of water is a measure of its ability to conduct electric current. It allows you to quickly assess the mineralization of water. Electrical conductivity is proportional to the overall content of chemical species with decreasing pH. The conductivity increases gradually from upstream to downstream of the lakes. Fresh water has a low electrical conductivity of less than 100 (μs/cm) [7].

TDS is a measure of the total amount of dissolved solids in water, including salts, mineral ions, and organic compounds. The salts in lakes and rivers come from rocks that degrade over time and other compounds resulting from human activity.

Electrical conductivity values vary between 1350 μs/cm (on the SEG site) and 1680 μs/cm (on the YAT site) during the dry season and between 1306 μs/cm (on the SEG site) and 1607 μs/cm (on the YAT website) during the rainy season. The average values found during both seasons are higher than the WHO standard (200 to 1000 μs/cm). This strong mineralization of the waters of Lake Sonfonia could be explained by a probable influx of saline water and discharges of untreated wastewater discharged directly into the lake.

Total dissolved solids (TDS) provide information on the salt content of water. We see that the highest values are obtained on the sites (YAT) 994 mg/L and (BLZ) 887 mg/L. The high level of salts in the waters of Lake Sonfonia indicates the presence of contaminants such as heavy metals, chemicals, or other substances potentially harmful to human health or the environment.

This state of affairs is positively correlated with the high conductivities measured in the different stations [4].

6.3 The concentration of suspended solids (MES) and turbidity

Suspended matter: refers to insoluble solid matter, visible to the naked eye, present in suspension in water.

They include clays, sands, silts, planktons, or other water microorganisms. The quantity of suspended matter varies according to the season and the water flow regime; resulting from erosion and leaching of soils in rainy weather, particularly when withdrawals are intense and in the event of flooding.

Biodegradable suspended matter contributes significantly to the oxygen demand and causes a decrease in the concentration of dissolved oxygen in the aquatic environment. These materials affect the transparency of the water and reduce the penetration of light, therefore photosynthesis, they can also hinder the breathing of fish [8].

Turbidity is the reduction in the transparency of a liquid due to the presence of colloidal matter and suspended matter. It is due to the presence of finely divided suspended matter: clay, silica grains, silt, organic matter, etc. [9]

Concentration of suspended solids fluctuate between 7 mg/L (SEG) and 31 mg/L (SOC) during the dry season (March) and between 11 mg/L (SEG) and 39 mg/L (SOC) during the rainy season (August). The average values found are greater than 10 mg/l WHO limit value.

Turbidity varies between 10 NTU (SEG) and 32 NTU (YAT) during the dry season and between 18.5 NTU (SEG) and 47.5 NTU (USF) during the rainy season. The average values obtained are greater than 5 NTU limit values accepted by WHO.

The peak values observed during the rainy season at the stations (USF, YAT, and SOC) are probably the consequence of the presence of discharges around the lake. In these, landfills are formed by the natural fermentation of leachate [4]. This turbid water would explain the low level of dissolved oxygen observed (**Tables 2** and **3**).

6.4 Dissolved oxygen

The analysis in **Tables 4** and **5** also shows that, during the hot period in March, a maximum of 6.5 mg/l was recorded at the YAT station and a minimum of 5.6 mg/l at the USF station, with an average of 6 mg/l. The cold period in August saw a maximum of 7.5 mg/l at the YAT station and a minimum of 6.9 mg/l at the stations (FLM, SOC, and USF), with an average of 7.08 mg/l [4]. However, in humid periods, the concentrations of dissolved oxygen observed are relatively higher than in hot periods, mainly due to the cooling of the water during the cold season, which maintains oxygen in the water mass [4]. Too high or too low dissolved oxygen levels can harm aquatic life and water quality.

These average values obtained are less than 10 mg/L, the limit value accepted by the WHO. This low level of dissolved oxygen affects biological diversity, in particular the death of fish by asphyxiation. It should be noted that temperature and dissolved oxygen are negatively correlated, which means that as the temperature increases, there is less oxygen available.

Settings Stations	**Phosphates mg/l**	**Nitrates mg/l**	**Nitrites mg/l**	**Sulfates mg/l**	**Total iron mg/l**	**O_2 dissolved mg/l**	**COD mg/l**
SEG station	24	47.2	0.03	16	0.8	6	31
YAT station	23.4	48.3	0.05	20	0.7	6.5	36
FLM station	27.2	52	0.07	13	0.5	5.8	54
SOC station	35	51.6	0.05	6	0.6	5.7	95
BLZ station	54	49	0.03	11	0.5	6.4	35
USF station	15	22.5	0.08	18	0.7	5.6	62
Average	29.76	45.1	0.05	14	0.63	6	52.16
WHO limit value for surface water	≤ 0.5	≤ 50	≤ 0.1	—	—	> 10	≤ 20

Table 4.
Chemical parameters measured in March 2021 at the different stations of the lake.

Settings Stations	Phosphates mg/l	Nitrates mg/l	Nitrites mg/l	Sulfates mg/l	Total iron mg/l	O_2 dissolved mg/l	COD mg/l
SEG	15.5	20.3	0.02	7	0.6	7	23
YAT	24	33	0.06	11	0.6	7.5	28
FLM	24	21.8	0.06	7	0.6	6.9	30
SOC	26	18.6	0.05	7	0.4	6.9	75
BLZ	25	17.5	0.02	7	0.5	7.3	25
USF	16.6	29.2	0.06	9	0.6	6.9	43
Average	21.85	23.4	0.04	8	0.54	7.08	37.33
WHO limit value for surface water	≤ 0.5	≤ 50	≤ 0.1	—	—	> 10	≤ 20

Table 5.
Chemical parameters measured in August 2021 at the different stations of the lake.

6.5 Chemical oxygen demand (COD)

COD is the consumption of oxygen by strong chemical oxidants to oxidize organic and mineral substances in water. It is one of the most used methods to assess the overall load of organic pollutants in water [10].

High COD values are recorded during the hot period with extreme values: the SEG station (31 mg/L) and the SOC station (95 mg/l). During humid periods, extreme values are observed: the SEG station (23 mg/L) and the SOC station (75 mg/L). The average values obtained are >20 mg/l WHO standard for surface water, which shows a shortage of dissolved oxygen in the lake, compromising the future of the lake. The high COD value observed indicates the presence of a significant load of organic matter in the lake, particularly in the SOC and USF stations, and is linked to deposits of branches and wastewater from surrounding localities [4].

6.6 Nitrates, nitrites, phosphorus, and eutrophication

Eutrophication causes the proliferation of microscopic algae and aquatic plants on the water surface, which gives a green color to the water. These algae release oxygen to the water surface, but have a short lifespan. Once dead and on the bottom, these algae still need a lot of oxygen for their decomposition. This is massively withdrawn from the water [11].

Nitrates (NO3-): Nitrates are anions representing the most oxygenated and stable form of nitrogen. Their presence in water is due to leaching from agricultural land following the spreading of fertilizers, livestock effluents (slurry), domestic, and industrial discharges. Nitrates come from the mineralization of organic nitrogen and the oxidation of the ammonium ion, present in the soil which is oxidized to nitrite and then to nitrate by nitrification bacteria. In water, nitrates react with amines to form potentially carcinogenic nitrosamines [12].

Nitrites (NO2-): Compounds result from the reduction of nitrates in water or soil by microorganisms.

In the blood, nitrites cause the formation of methemoglobin, and in large quantities reduces the oxygenation of cells, by inability to transport oxygen from the lungs to the tissues [13].

Phosphorus (P): Phosphorus in the natural environment is found in the form of phosphates (calcium, iron, and aluminum) in volcanic and sedimentary rocks. Its passage into the water occurs through the erosion of soil and rocks. Plants take up the phosphates and thus solubilized for photosynthesis. It is transferred along the food chain through consumption of plants by animals. Phosphorus is solubilized again thanks to the decomposition of dead matter by microorganisms [14].

The analysis in **Tables 4** and **5** show that the values of total phosphorus, nitrates, and nitrites vary according to the season and the station. For total phosphorus, the average values are higher than the WHO standard which is 0.5 mg/l. During the dry season, the average values are 29.76 and 21.85 in the rainy season; the average nitrate values are all lower than the WHO standard which is 50 mg/l for surface water. During the dry season, the average value is 45.1 and 23.4 mg/L in the rainy season. For nitrites also, the averages are slightly lower than the limit accepted by the WHO which is 0.1 mg/l. In the dry season, the average value is 0.05 and 0.04 mg/L in the rainy season [4].

A phosphate concentration greater than 0.5 mg/L of water is sufficient, in the presence of nitrate and ammonium, to trigger excessive growth of vegetation, that is, eutrophication [15].

6.7 Sulfates and total iron

Iron is generally found in surface waters as iron(III)-containing salts when the pH is >7. Sulfates can be found in almost all natural waters. They come from the oxidation of sulfite ores, the presence of shales, or industrial waste **Tables 4** and **5** show that the average values observed in the dry season are 0.63 and 0.54 mg/l in the rainy season. Sulfates have an average value of 14 mg/l in the dry season and 8 mg/l in the rainy season [4].

For total iron and sulfates in surface waters, including lakes, WHO does not provide a specific limit value.

However, high levels of iron in water can cause water discoloration, a metallic taste, and iron precipitation problems affecting water quality. The presence of sulfates is an indicator of the presence of potential contaminants in water, such as heavy metals or organic compounds.

It is important to monitor total iron and sulfate levels in surface waters to ensure they remain at levels acceptable for human and environmental health.

7. Statistical studies

Stata 15 software was applied for statistical analysis of the data. For this, the Shapiro-Wilk test for data normality showed that the values observed after the *in situ* and laboratory analyses are normally distributed. Thus, for the comparison of the different means of the parameters studied between the dry season and the rainy season, the Student t test was applied.

For the following parameters, such as temperature (P-value = 0.000), pH (P-value = 0.002), turbidity (P-value = 0.0149), dissolved oxygen

(P-value = 0.0002), the P-values are less than 5%, and we admit that the variations observed between the two seasons are statistically significant at 5%. On the other hand, for the EC (P-value = 0.5035), TDS (P-value = 0.3613), MES (P-value = 0.0669), and COD (P-value = 0.2722), the P-values are greater than 5%. We only admit that the variations in the values observed between the dry season and the rainy season are not statistically significant at 5% [4].

The correlation test between physicochemical parameters.

The level of correlation between the physicochemical parameters is given by the correlation matrixes following (**Figures 14** and **15**) [4].

Correlation matrixes between physicochemical parameters.

The correlation test between the parameters studied showed, among other things, highly significant correlations between electrical conductivity and total dissolved solids (r = 0.9454 and P-value = 0.0044).

The correlation test between the parameters studied gave, among other things, highly significant positive correlations between electrical conductivity and total dissolved solids (r = 0.9454 and P-value = 0.0044) in March 2021; chemical oxygen demand and suspended solids concentration (r = 0.9293 and P-value = 0.0073) March 2021; turbidity and concentration of suspended solids (r = 0.8524 and P-value = 0.0311) August 2021; but also strong negative correlations between pH and temperature (r = – 0.8331 and P-value = 0.0395) March 2021; temperature and dissolved oxygen (r = – 0.4190 and P-value = 0.4083) March 2021; and between chemical oxygen demand and dissolved oxygen (r = – 0.7077 and P-value = 0.1156) [4].

	Temper~e	PH	Conduc~c	TDS_mgL	MES_mgL	Turbid~U	Ox_dis~L
Temperature	1.0000						
PH	-0.8331	1.0000					
	0.0395						
Conductivi~c	-0.2271	-0.0436	1.0000				
	0.6653	0.9346					
TDS_mgL	-0.1474	-0.1849	0.9454	1.0000			
	0.7806	0.7258	0.0044				
MES_mgL	0.4129	-0.0848	-0.0723	-0.1046	1.0000		
	0.4159	0.8731	0.8917	0.8437			
Turbidite_~U	-0.0139	-0.0670	0.4135	0.6180	0.3426	1.0000	
	0.9792	0.8996	0.4150	0.1910	0.5062		
Ox_dissous~L	-0.4190	-0.0621	0.6103	0.6319	-0.8100	-0.0119	1.0000
	0.4083	0.9070	0.1982	0.1783	0.0507	0.9821	
DCO_mgLO2	0.1906	0.1157	0.0322	-0.1085	0.9293	0.1319	-0.7077
	0.7175	0.8272	0.9517	0.8379	0.0073	0.8033	0.1156

	DCO_mg~2
DCO_mgLO2	1.0000

Figure 14.
Correlation matrix of physicochemical parameters at the different stations of the lake in March 2021.

	Temper~e	PH	Conduc~c	TDS_mgL	MES_mgL	Turbid~U	Ox_dis~L
Temperature	1.0000						
PH	0.0123	1.0000					
	0.9815						
Conductivi~c	0.6781	-0.4836	1.0000				
	0.1388	0.3312					
TDS_mgL	0.6754	-0.4869	1.0000	1.0000			
	0.1410	0.3274	0.0000				
MES_mgL	0.8364	-0.2390	0.8475	0.8460	1.0000		
	0.0380	0.6484	0.0331	0.0338			
Turbidite_~U	0.5537	-0.1125	0.7328	0.7311	0.8524	1.0000	
	0.2543	0.8319	0.0975	0.0987	0.0311		
Ox_dissous~L	0.3429	-0.4247	0.3444	0.3449	0.0290	-0.1999	1.0000
	0.5058	0.4013	0.5038	0.5031	0.9566	0.7042	
DCO_mgLO2	0.4158	0.2939	0.0641	0.0609	0.5299	0.6424	-0.4730
	0.4123	0.5718	0.9040	0.9087	0.2795	0.1690	0.3434

	DCO_mg~2
DCO_mgLO2	1.0000

Figure 15.
Correlation matrix of physicochemical parameters at the different stations of the lake in August 2021.

8. Results of bacteriological parameters

The analysis of **Tables 6** and **7** shows that the levels of bacteria indicative of fecal contamination differ from one station to another and depending on the sampling period [2].

During the dry season with low water flow, high loads of fecal bacteria were observed especially at the YAT station: 3000 CFU/100 ml in SF; 4000 CFU/100 ml/l in CF, 50,000 CFU/100 ml in CT, and 1200 CFU/100 ml in *E. coli* at the SOC station [2].

On the other hand, in the rainy season, the maximum values observed are respectively, 3200 CFU/100 ml in SFat the station (BLZ); 3000 CFU/100 ml in CF (YAT); 9600 CFU/100 ml in CT at the SEG station; and 802 CFU/100 ml in *E. coli* at the (SOC) station [2].

According to other authors, there is not necessarily a clear trend for bacterial load between the dry season and the rainy season, as it depends on many factors such as precipitation, presence of point sources of pollution such as wastewater discharges, the level of human and animal activity in the area, agricultural and industrial practices [2].

The presence *of E. coli* in the different stations effectively confirms contamination by fecal matter but also the possible presence of other pathogenic germs [2]. YAT and SOC stations close to residential areas receiving sewage and domestic wastewater discharges recorded peaks in bacterial load values [2].

Comparison of the levels of fecal contamination bacteria in this study with a previous one carried out in 2018 on the waters of the same lake indicates an increase

Settings	Fecal streptococci		Total coliforms		Fecal coliforms	*Escherichia coli*		*Salmonella*
Incubation conditions	**22 °C 24H**	**37°C 48H**	**22°C 37°C 24H**	**48H**	**44°C 24H**	**22°C 24H**	**37°C 48H**	**37°C 48H**
SEG station	80	1200	2,001,000	>	1500	15	120	0
YAT station	100	3000	5,005,000		4000	34	180	0
FLM station	50	2000	4,509,500	800	2000	150	200	0
SOC station	300	1000	9000 150	3000	3000	322	1200	0
BLZ station	40	3000			900	53	80	0
USF station	30	750	120	2000	1000	18	45	0

Table 6.
Concentrations of fecal streptococci, total coliforms, fecal coliforms, Escherichia coli, *and* salmonella *found in CFU/100 ml at the different stations of the lake in March 2021.*

Settings	Fecal streptococci		Total coliforms		Fecal coliforms	*Escherichia coli*		*Salmonella*
Incubation	22°C	37°C	22°C	37°C	44°C	22°C	37°C	37°C
conditions	24H	48H	24H	48H	24H	24H	48H	48H
SEG station	8	800	100	9600	1800		42	0
YAT station	26	2600	71	6600	3000	9	86	0
FLM station	12	1200	92	8700	1600	22 54	90	0
SOC station	18	1800	69	6400	2000	416	802	0
BLZ station	32	3200	31	2600	700	10	22	0
USF station	6	600	83	7800	78	4	32	0

Table 7.
Concentrations of fecal streptococci, total conforms, fecal conforms, Escherichia coli, *and* salmonella *found in CFU/100 ml at the different stations of the lake in August 2021.*

in the content of fecal coliforms and fecal streptococci which exceed the maximum concentration accepted by the WHO and the MDDEFP (62 CFU/100 ml in fecal streptococci and 200 to 1000 CFU/100 ml in fecal coliforms) for surface water. This could be linked to the increase in anthropogenic and demographic activities over the last 4 years [2].

The R = CF/SF ratio is a first-order informative element to determine whether fecal pollution was of animal or human origin.

In the dry period, in March, the R ratios from the different stations give 67%, R > 1, 16.5% R < 1, and 16.5%R = 1.

During the wet period, in August, the R ratios of the different stations give 67%, R > 1 and 33% R < 1 [2].

Overall, this study gives the ratio R = CF/SF > 1, allowing us to assume that the fecal pollution of the waters of Lake Sonfonia in 2021 is 67% human in nature, due to domestic wastewater (toilet water, kitchen wastewater, washing machine wastewater, etc.); but also, the transformation of certain parts of the lake bank into a place of waste and garbage deposits and defecation for the local population. Also, with 33% R < 1, the values indicate contamination of animal origin, due to the presence of animal breeding such as cattle near the lake or even other animals such as margouillats, migratory birds, cat, dog, which are frequently encountered there. Their excreta will contaminate the body of water through runoff, rain, and leaching from agricultural land laden with manure [2].

The search for pathogenic bacteria such as *salmonella* in water usually requires a concentration step. As *salmonella* may be present in low numbers and may have been altered in the aqueous environment, their detection in water requires pre-enrichment [2].

Analysis of the results indicates a notable absence of *salmonella* from the waters of Lake Sonfonia. The lack of reagent for pre-enrichment could partly explain the absence of *salmonella* in the different stations [2].

8.1 Statistical studies

Stata 15 software was used for statistical analysis of the bacteriological data. As the sample size was small, the Shapiro-Wilk normality test indicated that the variables studied were not normally distributed. It was preferable to use a nonparametric test. For this, we used the Mann-Whitney U test to compare the value of the means of each of the variables observed during the two seasons as shown in **Table 8** [2].

The test results reveal that there are, in their respective incubation conditions, statistically significant differences between the values of the two seasons (March and August) for the following parameters: SF (22 °C in 24 hours), CT (22°C in 24 hours), and CF (44°C in 24 hours). On the other hand and under the indicated incubation conditions, it appears from our study that there are no statistically significant differences between the values of these two seasons (March and August) for the SF parameters (37°C in 48 hours), CT (37°C in 48 h), and *E. coli* (22°C in 24 h and 37°C in 48 h) [2].

This study shows that variations in the concentration of fecal bacteria between the dry season and the rainy season actually depend on the incubation conditions. SF and CT taken under different incubation conditions in dry weather and humid weather give statistically significant differences at 5% (22°C in 24 hours) and statically insignificant differences at 5% (37°C in 48 hours) [2].

Settings	Incubation conditions	P-value	Decision making
SF	T = 22 ° C in 24 hours	0.0065	*P-value* < 0.05; therefore, the differences observed during the two seasons are statistically significant at 5%
CT	T = 22° C in 24 hours	0.0039	
CF	T = 44° C in 24 hours	0.0133	
SF	T = 37° C in 48 hours	0.8095	*P-value* > 0.05; therefore, the observed differences are considered statistically insignificant at 5%.
CT	T = 37° C in 48 hours	0.4233	
E-coli	T = 22 ° C in 24 hours	0.3367	
	T = 37° C in 48 hours	0.1495	

Table 8.
Significance test results between the bacteriological parameters of the waters of Lake Sonfonia in March and August 2021.

The statistical analysis of the data was carried out at the National Institute of Statistics of Guinea.

9. Verification of the research hypothesis

The following approach was adopted:

1. Analysis of contaminants:

- The presence of *E. coli* indicates contamination of the water by feces; the phosphate content exceeding the WHO limit value, the excess of this nutrient justifies the eutrophication observed, due to the poor management of agricultural inputs, in the bank and in the lakeside neighborhoods.

2. Analysis of some indicators:

- High COD;
- High turbidity;
- Reduction of dissolved oxygen levels.

The level of these parameters indicates the pollution of water by organic matter.

- The absence of laboratory equipment did not allow the research of toxic chemicals, such as POPs.

3. Surveys on sanitation practices:

Interviewing local residents to understand existing sanitation practices showed the non-existence of specialized structures not only for household solid waste management, but also this interview revealed that most local residents were completely unaware of the impact of poor management of liquid and solid waste on surface waters. The banks of the lake have become a garbage dump, and the images of the lake above illustrate this sad situation.

4. Identification of potential sources of pollution:

Like most developing countries, Guinea does not have a sanitation or sewer network allowing the treatment and disposal of household wastewater. Water from toilets and latrines is sometimes discharged directly into nature. In the rainy season, the flow of rainwater causes runoff of domestic wastewater, industrial, agricultural, hospital waste, and manure, coming from livestock trucks and slaughterhouses.

The bank of the lake has become a place of defecation for some local residents and the feces are carried into the lake in the rainy season.

This assertion becomes plausible because the low level of the sanitation system can lead to increased discharge of pollutants into the water through untreated wastewater.

In light of the above, the study actually shows that poor urban waste management practice negatively influences the quality of surface water.

10. The answer to the research question on the water quality of Lake de Sonfonia

By comparing the values of the main pollution indicators in this study to WHO standards. It was observed that the content of *E. coli* is high;

- Electrical conductivity is high;
- The COD is high;
- Turbidity is high;
- The dissolved oxygen concentration is low;
- The presence in places of layers of greenish algae covering Lake Sonfonia, an old or eutrophic lake, confirms the eutrophication process.

These indicators were measured using biochemical tests.

Consequently, the waters of Lake Sonfonia, according to our studies, are of poor quality and unsuitable as a source of drinking water production. They are also not suitable as irrigation water, cleaning water, or for other human activities.

Efforts must be made to improve waste management or regulate polluting activities.

11. General conclusion

The results of research carried out in this study aimed at evaluating the water quality of Lake Sonfonia 2021 showed that this water is of doubtful quality for the life of the species present as well as for the health of users. The presence of *E. coli*, indicates contamination by feces. The levels of fecal coliforms and fecal streptococci exceed the maximum concentration accepted by the WHO for surface waters. It was noted that the average temperature exceeds the limit value admitted by the World Health Organization which is 25°C.

High COD and turbidity indicate pollution by organic and chemical materials. This situation promotes a drop in dissolved oxygen levels which are detrimental to aquatic life. The pH and nutrients in solution are generally within acceptable limits.

Stata 15 software was used for statistical analyses of the data. The results obtained showed that variations in the concentration of fecal bacteria between the dry season and the rainy season actually depend on the incubation conditions. The difference is statistically significant at 5% between the two seasons for CF and SF under the temperature of 22°C in 24 hours. On the other hand, the SF and CT under a temperature of 37°C in 48 hours gives a statistically insignificant difference at 5%.

The correlation test between the physicochemical parameters gave highly correlated correlations significant between electrical conductivity and total dissolved solids ($r = 0.9454$ and P-value = 0.0044); there chemical oxygen demand and suspended solid concentration ($r = 0.9293$ and P-value = 0.0073) in the dry season and between the chemical demand for oxygen and dissolved oxygen ($r = -0.7077$ and P-value = 0.1156). The same is true between the temperature and the concentration of solid matter in suspension ($r = -0.8364$ and P-value = 0.0380) in rainy season among others. The main source of pollution is due to anthropogenic activities, that is, the discharge of household waste and wastewater domestic animals in the lake without any prior treatment. This creates long-term environmental and health problems. This pollution of surface water has significant impacts on human and environmental health. Chemicals (heavy metals) and waste can reduce biodiversity and disrupt the food chain. Polluted water may contain pathogens such as bacteria, viruses, etc. which are responsible for waterborne diseases such as cholera, dysentery and typhoid. Agricultural, industrial, livestock and fishing activities are the most incriminated. Strategies to limit lake pollution must be put in place by decision-makers at all levels in order to avoid eco-health problems. Domestic, agricultural, industrial, and fishing activities artisanal industries are the most incriminated ones. Strategies to limit lake pollution must be put in place by decision makers at all levels in order to avoid eco-health problems.

Some images of Lake Sonfonia are shown (**Figures 16–20**) [1, 3].

Figure 16.
Overview of Lake Sonfonia.

Figure 17.
Gutters discharging wastewater and toilet water into the lake.

Figure 18.
The author on a field visit: Illegal dumping near Lake Sonfonia.

Figure 19.
Green algae on the surface of the lake water: Eutrophication.

Figure 20.
Line fishing.

Author details

Mohamed Lamine Komara[1*] and Kandé Bangoura[2]

1 Department of Biology, Gamal Abdel Nasser University of Conakry, Republic of Guinea

2 Conakry Scientific Research Center (CERESCOR), Republic of Guinea

*Address all correspondence to: komaramohamedlamine@gmail.com

References

[1] Malouono M. Problem of rainwater in the peripheral districts of cities in subSaharan Africa (bibliographical search, 2014). Environmental and Water Sciences, Public Health and Territorial Intelligence Journal. 2018;**2**(1):77-85

[2] Komara ML, Camara CAT, Bangoura NNK, Bangoura K, Traore S. Evaluation of the bacteriological pollution of the waters of the Lake of Sonfonia commune of Ratoma (Republic of Guinea) 2021. Journal of Agricultural Chemistry and Environment. 2023;**12**(2):142-151

[3] Gnagne YA, Yapo BO, Meite L, Kouame VK, Gadji AA, Mambo V, et al. Physico-chemical and bacteriological characterization of raw wastewater from the sewer network of the city of Abidjan. International Journal of Biological and Chemical Sciences. 2015;**9**(2):1082-1093

[4] Komara ML, Sacko AM, Diallo A, Camara CAT, Bah MK. Monitoring of the PhysicoChemical parameters of the waters of Lake Sonfonia, municipality of Ratoma (Republic of Guinea) 2021. Journal of Environment Sciences and Engineering B. 2023;**12**:105-114

[5] Brémaud I. Colorimetry of wood: Diversity of the material and complexity in quantifying visual perception. In: Ninth Study Day of the Wood Group of the SFIIC (French Section of the International Conservation Institute) – "Wood Colors, Paint Support, Stained Wood", October 11, 2019. Aubervilliers: INP (National Heritage Institute); 2019

[6] Kpidi YH, Yapo OB, Ballet TG, Ohou-Yao MJ. Daily variability of the physicochemical quality of Lake M'koa in Jacqueville (Ivory Coast). International Journal of Biological and Chemical Sciences. 2017;**11**(2):901-910

[7] Tfeil H, Mahfoudh M, Moulay Mhamed BA, Aliyen A, Yarba L, Vall Hmeyada AM. Physico-chemical characterization of surface waters and study of the ichthyological diversity of some continental wetlands in Mauritania. European Scientific Journal. 2018;**14**(6):83-101

[8] Buhungu S, Montchowui E, Barankanira E, Sibomana C, Ntakimazi G, Bonou CA. Spatio-temporal characterization of the water quality of the Kinyankonge River, a tributary of Lake Tanganyika, Burundi. International Journal of Biological and Chemical Sciences. 2018;**12**(1):576595

[9] N'Diaye AD, Ousmane T, Namr KI. Turbidity and suspended matter in water: Application to the evaluation of metals contained in water from the right bank of the Senegal River. LARHYSS Review. P-ISSN 1112-3680/E-ISSN 2521-9782, (14)

[10] Chadli K, Boufala M'h. Assessment of water quality using Moroccan WQI and multivariate statistics in the Sebou watershed (Morocco). Arabian Journal of Geosciences. 2021

[11] Ferd K. Contribution to the study of the relationship between excessive use of agricultural fertilizers and the eutrophication phenomenon [doctoral dissertation]; 2022

[12] Announ S, Bounouar K. Dynamics of mineral nitrogen in a classic wastewater treatment plant, case of the eastern station of the city of Tizi-Ouzou [doctoral dissertation]. Mouloud Mammeri University; 2017

[13] Chaima G, Nour el Houda B, Chaima N. Effect of high nitrite levels on the rabbit brain [doctoral dissertation]. University of Larbi Tébessi; 2022

[14] Morel C, Le Clech B, Linères M, Pellerin S. Beware of a drop in the bioavailability of phosphorus. Alter Agri. 2006;**79**(3):21-23

[15] Dimon F, Dovonou F, Adjahossou N, Chouti W, Mama D, Alassane A, Boukari M. Physico-chemical characterization of Lake Ahémé (South Benin) and demonstration of sediment pollution by lead, zinc and arsenic. J.Soc. West Africa. Chim, 2014;**37**:36-42

Chapter 6

Hydropower Reservoirs as Arbiters of Climate Change

Thomas Shahady

Abstract

Hydropower is an important source of renewable energy worldwide. In 2022, hydropower was estimated to produce 15% of the world's electricity with pump-storage an integral part of this production. Generating hydropower mitigates the use of fossil fuels thus reducing Green House Gas emissions from some of the most polluting industries such as Coal Fired Powerplants. However, reservoirs used for this type of energy production may be highly polluting themselves. Production of methane and CO_2 may be extensive from storage reservoirs. Current changes in precipitation patterns will bring in more organic material and nutrients to these reservoirs causing increases in GHG production as this material is broken down. And in the case of pump-storage reservoirs, artificial generation hydrology may be exacerbating this problem. In this chapter, I analyze current literature on the impact of reservoirs on GHG emissions. Further, I analyze my research on reservoir water quality looking at how this problem is worsening through time and how this may not be a sustainable energy when considering CO_2 and methane production from these reservoirs. Ideas related to the unique operation of hydropower reservoirs, changes in water quality, precipitation norms and weather patterns are discussed.

Keywords: hydropower, water quality, climate change, methane, reservoirs

1. Introduction

The importance of energy driving world economic productivity and growth is universally accepted [1, 2]. Thus, how future energy needs are met coupled with the importance of sustainability is developing into a critical question. Continued reliance on fossil fuels, while efficient, generates levels of pollution the global climate may not further assimilate without change. Hence, with increasing climate change concerns and a growing need to meet energy demand sustainably, solutions using renewables are needed moving forward.

Currently, renewable energy sources account for about 20% of global energy production with hydropower accounting for 80% of that total or 15% of all renewables [3]. Hydropower is the generation of electricity by flowing water over a turbine. This may occur when a powerplant is constructed near a river diverting flow directly through that power plant (**Figure 1**). The drawback with this type of energy production is the one-way flow of water and reliance on river dynamics and precipitation

Figure 1.
Hydropower plant located along Rio Aranjuez in Costa Rica. This power plant diverts water from the river above to drive turbines for electricity production. Power generation is completely dependent on adequate flow from the river.

patterns. During periods of drought and dry season the production of electricity may be limited.

By constructing a reservoir above the power plant, water is summarily stored for a more consistent production of energy. To provide even greater storage and efficiency of use, pump storage reservoir projects are built providing an upper reservoir to generate electricity and a lower reservoir for storage (**Figure 2**). In this system, the upper reservoir water is released over turbines to generate electricity during peak demand with water pumped back from the lower reservoir during low demand or off peak. This type of production is very attractive because of uneven precipitation distribution [1]. Reservoir storage has shown to be a very resilient power source even during drought [4].

Currently over 1 million dams are operating globally [5] with future hydropower development concentrated in developing and emerging economies [3]. In 2022, hydro power plant (HPP) production of electricity accounted for 28.7% of US renewables with pumped storage hydropower (PSH) representing 96% of all storage capacity [6]. Worldwide, hydroelectricity accounted for 15.6% of the world demand in 2019, ranking third behind coal and natural gas [7]. A 17% growth in hydropower by 2030 is expected [2]. As world energy demand increases and the need for renewables heightens to meet that demand it is believed hydropower will become the predominate renewable source of energy.

The construction of new hydropower facilities has been relatively dormant since the 1990s with most of the projects built in the 1970s [8]. It is expected however that hydropower project construction will rapidly increase into the future to meet demand

Figure 2.
Smith Mountain Lake reservoir dam shown from the pump storage reservoir Leesville Lake located in Central Virginia. This pump storage operation pumps water from Leesville Lake during off peak to generate power during peak from Smith Mountain Lake.

with an expectation of lessening fossil fuel reliance. Large projects while incurring considerable capital costs still have a good breakeven point due to low operation and maintenance costs [9]. And pump storage reservoirs offer some of the best options for energy production and storage. Small hydropower is becoming an ideal energy source due to a cheaper price and ability to site in multiple locations [10].

So, while hydropower offers tremendous future growth potential and generates electricity without burning fossil fuels, the water source for generation must become a point of focus. The building of reservoirs can displace communities, consume a considerable amount of land, disrupt river flow and impact biodiversity and managed improperly can be significant emitters of Green House Gasses (GHGs). Thus, this energy source presents a dichotomy using a sustainable energy source of water to produce electricity while presenting environmental problems with how where the water is stored and managed. How storage water is used and managed makes these projects arbiters of climate change in the communities where they operate.

2. Reservoir ecology

Reservoirs have unique characteristics that impact their functionality. Water stored in these reservoirs have physical, chemical and biological characteristics impacting overall water chemistry and quality. By understanding and managing the ecology of reservoirs, greater sustainability can be achieved in the overall process of energy production.

2.1 Stratification and spatial structure

Stratification occurs as the reservoir surface water warms creating a less dense mass of water that floats to the surface. Cooler and denser water sinks lower creating layers. These stratified layers have differing water qualities and characteristics. The epilimnion is a term used to describe the upper layer and hypolimnion the lower layer. The term metalimnion is used to describe the layer of changing conditions between the other two layers. Temperature is the most common measure used to define these layers but oxygen concentration is a very good delineator as the metalimnion becomes hypoxic and the hypolimnion becomes anoxic when eutrophic reservoirs stratify (**Figure 3**).

Reservoirs are also structured spatially from dam to headwaters. This gradient begins where river inputs generate patterns similar to a river. In this section, high concentrations of nutrients, bacteria and sediment enter from river transport. As water travels further into the reservoir, these riverine conditions begin to lessen and more lake qualities (defined as lacustrine) influence water quality. This middle portion of the reservoir is considered a transition zone as the riverine and lacustrine portions of the reservoir mix. This area may have the highest overall productivity in the reservoir as sediments associated with river flow settle from the water column yet

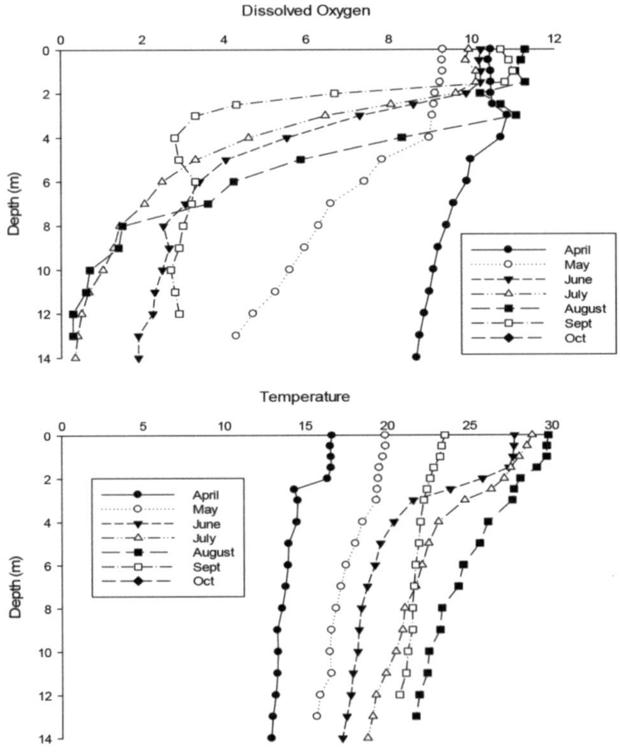

Figure 3.
Stratification due to temperature (degrees C) and resultant oxygen (mg/L) concentrations in Leesville Lake - a pump storage reservoir in Central Virginia USA. Note the rapid loss of oxygen through the metalimnion (2–6 meters) developing into anoxic conditions at depths greater than 6 meters in the hypolimnion during the 3 summer months (July-august). This pattern typically occurs from may – October each year with turnover soon after causing low oxygen conditions throughout the reservoir.

nutrient concentrations remain plentiful. The final sections of a reservoir are considered lacustrine and resemble lake qualities. This area often is lower in productivity due to settling of particulates and lower in nutrient concentrations. If stratification is continuous, upper layers become very isolated from lower portions of the reservoir further isolating nutrients and other pollutants. The best water quality for the reservoir is located near the dam.

In any reservoir, water quality is best evaluated along both gradients. Headwater often mix due to storm inputs eliminating stratification with resultant poor water quality due to heavy river impacts. The remaining portions of the reservoir develop extensive areas of oxygen loss in the hypolimnion from the transition zone through the dam. The epilimnion becomes productive with algae growth (phytoplankton that are floating forms of algae) with increases in pH and supersaturation of oxygen due to photosynthesis. These reservoirs become two water masses between the productive and highly oxygenated epilimnion and the isolated hypoxic or even anoxic hypolimnion.

Pump storage reservoirs can be an exception to this pattern. These reservoirs receive water input from the upper reservoir that mixes often in an unpredictable way with river inputs from various sources (**Figure 4**). Upper dam release is often good water quality but may have very low oxygen content if release is from the upper reservoir hypolimnion. Secondary sources from river inputs can bring in high concentrations of nutrients, sediment and bacteria during stormwater flow. The mix of these water sources create both a unique hydrology and water quality in these reservoirs.

Figure 4.
Unique inflow characteristics in the headwater region of a pump-storage reservoir. Here, turbid river water inflow mixes with dam release from the upper reservoir. Water becomes fully mixed several kilometers below this area adopting water quality resembling these river inputs.

One additional pattern occurs in pump storage reservoirs. Water is periodically pumped back into the upper reservoir. This generates two distinct patterns. Primarily it provides water for the next round of energy production in the upper reservoir. Unfortunately, when considering ecology of the upper reservoir it will degrade water quality. Such decisions to pump water from lower to upper are almost exclusively energy demand driven. Secondarily, it creates reverse flow that can move water throughout the lower reservoir headwater region in the opposite direction. This causes turbid river water to predominate lower reservoir water quality. It also disturbs the river bottom causing even greater turbidity to flow. This water may also move back and forth as a slug – a patch of water along a spatial gradient moving back and forth due to pump-storage operations. These scenarios create unique hydrology and water quality in both reservoirs.

2.2 Eutrophication

Eutrophication is the excessive loading of nutrients into a body of water. When this occurs, algae growth is stimulated and a multitude of problems occur that are well documented in the literature occur [11]. Reservoirs are prone to eutrophication. There are multiple reasons for this. High catchment to surface area ratios generates high loading rates from rivers of both organic material and nutrients [12, 13]. Reservoirs unlike lakes may behave similar to rivers during certain times of the year with water input moving rapidly through. This reality renders the watershed extremely important toward resultant water quality. Activities such as farming, urban development, deforestation or other land disturbing activities directly influence the river and hence the reservoir water quality [14, 15].

Secondly, storm events play a critical role. Storms are responsible for the bulk (up to 90%) of nutrient loading to river systems and into these reservoirs [16–18]. And more importantly, the watershed attributes (land use) and not necessarily the strength of the storm cause the greatest impacts [19]. Additionally, reservoir orientation into prevailing winds creates fetch that impacts mixing and how nutrients are ultimately incorporated into lake productivity. Reservoirs tend to have a greater fetch (length exposed to an air mass) because they are elongate being built into stream channels. As fetch increases, mixing becomes more pronounced and storm impact intensifies.

Ultimately, the excessive inputs of nutrients stimulate algal growth to the point of generating harmful algal blooms (HABs) and loss of oxygen in the hypolimnion [20]. As these blooms grow, they are very resistant to grazing or other losses generating very problematic conditions [21]. As this biomass builds up, it generates organic matter sinking into the lower levels of the reservoir. The amount of organic material is proportional to oxygen loss in the lower layers during stratification.

2.3 Loss of oxygen

A consequence of both stratification and eutrophication, hypoxia (low oxygen) and anoxia (no oxygen) develop throughout the lower stratified layers [22]. Oxygen cannot be replenished due to isolation from the upper oxygenated layers and lack of light allowing photosynthesis. This phenomenon occurs during the summer and fall periods of stratification. And during fall turnover as air temperature cools, the hypolimnion and epilimnion mix due to equilibration of temperature with resultant low oxygen concentrations throughout the entire reservoir.

There are multiple ramifications from oxygen loss such as habitat reduction and environmentally adverse chemical reactions. Most importantly, a reducing environment is created. This decouples complexes formed with sediment such as iron and phosphorus releasing available phosphorus. Phosphorus is understood to be a limiting nutrient for algae growth and implicated as the greatest contributor to eutrophication of reservoirs [22]. Release of available phosphorus can stimulate algal growth further complicating the problem of oxygen loss. Other compounds also develop in this environment such as sulfide and manganese that can be toxic to aquatic life.

Predicting oxygen loss as it is related to other variables is more difficult. Nürnberg [23, 24] suggest it is best predicted by phosphorus concentration and shape of the lake or reservoir. For hydropower reservoirs, this is more difficult due to the nature of reservoir operation considering both the release though the dam and inputs from rivers. Ultimately, oxygen loss is the greatest problem limiting their sustainable operation.

2.4 Methane and carbon dioxide production

Both carbon dioxide (CO_2) and methane (CH_4) are produced by reservoirs. Carbon dioxide is a water-soluble gas that quickly dissolves in water. It is a byproduct from the breakdown of organic material in the hypolimnion or from gas exchange with the atmosphere. It is mediated through photosynthesis where CO_2 is consumed through the growth of algae. It can also build-up in the hypolimnion as it is produced through the breakdown of organics [22]. Methane is generally insoluble in water and is formed in the hypolimnion as a byproduct of organic material breakdown. Some of the methane is oxidized to CO_2 but the majority of production occurs as bubbles in the sediment water interface.

Whether reservoirs act as sinks for CO_2 or are net emitters to the atmosphere is variable [23, 24]. Phytoplankton uptake consumes CO_2 but decomposition of algal biomass in the hypolimnion generates it. During summer peaks and stratification, the warm epilimnion acts as a CO_2 sink and all exchange is almost exclusively diffusion. During turnover however, release of CO_2 can be significant [25]. Warmer water and mixing facilitates this CO_2 release. The synergistic effects of warming, nutrient driven eutrophication and development of algal blooms in turn increases CO_2 production. The difference between algal uptake and CO_2 accumulation released during turnover will account for the net emission or consumption of CO_2 in the reservoir.

Methane production operates differently. As generally insoluble, a majority of it is produced along the sediment – water interface as bubbles. This can happen very rapidly as high nutrient loading supports primary production, favorable (hypoxic) conditions develop and the organic substrate necessary is available [26]. Trophic status or increased eutrophication also influences the rates of production as eutrophic reservoirs produce greater amounts of methane [27]. Hence, chlorophyll *a* (the measure of algal biomass) may be the master variable regulating methane production in reservoirs [28, 29]. This methane, produced as CH_4 bubbles rises directly from the sediments and into the atmosphere [30]. Heat waves, algal blooms, fall overturn and strong stratification all impact the effluxes of CH_4 into the atmosphere. In some instances, dissolved methane may accumulate in the hypolimnion [31] but this is a small fraction of all methane production.

3. Greenhouse gas emissions

Conventional hydropower production seeks to optimize energy and producer-economic benefits while operating within ecological and reservoir management constraints [32]. While reservoirs and even more so pump-storage reservoirs provide an abundant supply of energy in the form of stored water, it is the storage and movement of that water that is most concerning. If the reservoirs that store the water emit enough GHGs to negate the benefit of this type of energy, then the industry is ultimately not sustainable and loses its renewable appeal.

The concern with the building of reservoirs and storage of water for energy production is allowing the buildup and ultimate release of GHGs into the atmosphere. The essential question lies in the understanding of how much of these gasses are released into the atmosphere during use or simply as a by-product of the storage and whether it is a significant contribution to overall global carbon budget and climate change. To gain some insight on how these processes impact the atmosphere it is best to understand processes in reservoirs and subsequent release.

3.1 Diffusion

Diffusion occurs when gas is exchanged between the reservoir surface and atmosphere due to concentration differences. Carbon dioxide is very soluble in water and thus diffusion accounts for 99% of CO_2 emissions from reservoirs [33]. Yet, as the epilimnion functions as a CO_2 sink it can offset the overall impact to the atmosphere. As methane is rather insoluble in water, very little diffusion occurs directly to the atmosphere. Some diffusion occurs between sediments where methane is produced and overlying water in the hypolimnion through oxidation. Diffusion of CO_2 and CH_4 accounts for a very small fraction of all GHGs emitted from reservoirs into the atmosphere with the exception of fall turnover [34].

Off-gassing of GHGs during fall overturn presents the greatest diffusive flux to the atmosphere [35]. Methane and CO_2 buildup that has occurred in the hypolimnion during stratification is released via diffusion. Up to 46% of stored methane can be released during overturn which is likely 80% of the diffusive release. Overturn represents a very large GHG diffusive release event in eutrophic lakes.

3.2 Ebullition (bubbling)

As methane gas is generated in the sediment bubbles form. These bubbles may remain in place due to hydrostatic pressure or be oxidized into CO_2. However, as conditions change these methane bubbles may release moving up through the water column and into the atmosphere via ebullition (the physical bubbling of methane from lake and reservoir sediments). Methane flux, and in particular ebullition is recognized as a highly significant source of global GHG [36]. And it is important to recognize that ebullition dominates most of the CH_4 emissions [28, 37]. The generation of methane in the sediment is temperature and trophic state dependent [38] as more methane is produced in eutrophic and warmer reservoirs.

Multiple reservoir conditions can cause methane to release from the sediments and bubble to the surface. Strong patterns in stratification (a well-developed hypolimnion) generates high levels of ebullitive emission [39]. Eutrophic nutrient rich reservoirs tend emit more CH_4 than less productive ones. Shallow reservoirs and those with

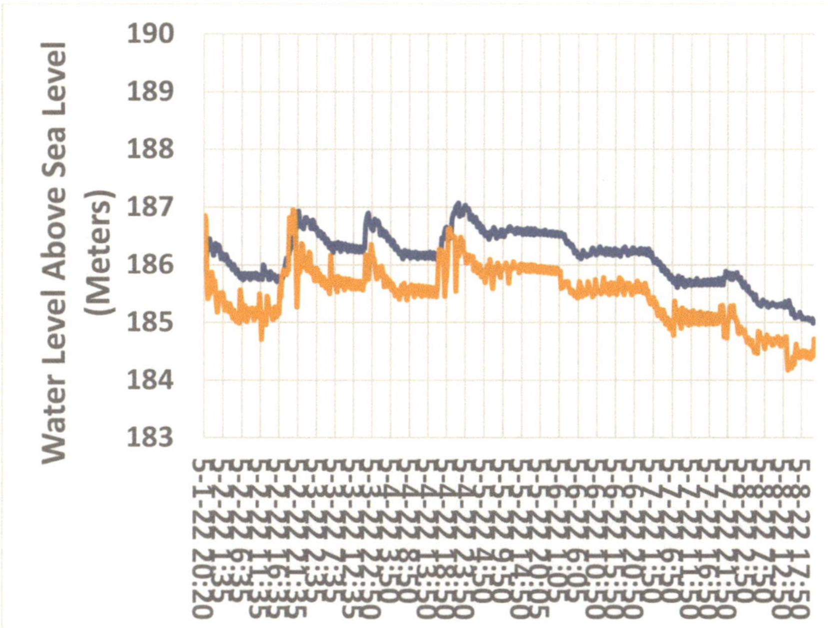

Figure 5.
Changes in water level for storage reservoir Leesville Lake (blue line) (storage reservoir) and upper reservoir Smith Mountain tail water (orange line) over an 8-day period. Data illustrates the fluctuation in water levels (187–185 meters) of the reservoirs in a typical week. The pump storage reservoir is in a constant state of flux as it may fluctuate 1–1.5 meters in a 24-hour period and over 2 meters in a week. The tail water from the upper reservoir also exhibits substantial change.

a long fetch generating wind disturbance and multiple mixing events will emit greater levels of methane [19].

Further, the greatest ebullition rates occur when we drawdown reservoirs [27, 40]. Water movement creates turbulence releasing methane bubbles. Estimates suggest a 1.4–77% increase in ebullitive methane release in just a 24-hour period during draw down [40]. Reservoirs with higher epilimnion chlorophyll *a* experienced larger increases in CH_4 emission in response to drawdown [27].

Pump storage reservoir are very prone to this type of emission. These reservoirs are often shallow making them prone to excessive methane release rather than deep and stable water systems [41]. Lowering of water level decreases hydrostatic pressure releasing bubbles from confinement in the soils. Harrison et al. [27] observed pulses of CH_4 emission via ebullition, via diffusion, and total emission (ebullition plus diffusion) associated with water level drawdown with a 3.6-fold increase in emissions during drawdown. Reservoir drawdown and in particular a shallow system with repeated water movement and drawdown represents a very high potential source of methane release (**Figure 5**).

3.3 During energy production

As power is generated, water runs over the turbine and is released through the tail waters. Most often, hydropower source water is drawn from the hypolimnion of the production reservoir and that water layer is either hypoxic or anoxic. This water may be methane-rich. The hydrostatic pressure change during use and turbulence forces large portions of the CH_4 gas to be released [42, 43]. These emissions can be significant during the summer months when the reservoir is stratified.

Observations suggest methane concentrations at the turbine intake reached their maximum value during greatest stratification and dry weather [39]. This is likely a combination of build-up in the hypolimnion and additional methane that dissolves when bubbling occurs [41]. While ebullition is the most significant source of methane, water traveling over the turbines adds to methane emission by releasing the dissolved fraction [44]. Movement of water causing ebullition and turbulence at the turbines cause an extensive release of methane.

4. Impacts from climate change

As climate change intensifies, it is expected that precipitation patterns will become more variable with more extreme rainfall events and a warming of the planet surface [45]. These two probabilities will have a profound impact on water resources through time. Increasing frequencies of extreme weather will likely bring in more material to lakes and reservoirs increasing eutrophication. Warming temperatures will cause that material to metabolize more intensively. This has the potential to create the conditions where we will see GHG emission from reservoirs intensify.

4.1 Warming temperatures

Warming temperatures and declining water clarity will directly contribute to an increased loss of oxygen in reservoirs [46]. Increasing oxygen loss from rising temperatures will occur as a warmer epilimnion strengthens density differences between layers. Strengthening and lengthening stratification allows the breakdown of more organic material under anoxic conditions and more GHGs particularly methane to be produced [47].

While eutrophication will contribute to this problem, the increasing temperatures will drive greater productivity, potential algal blooms, loss of oxygen and GHG release. Hence, it may be expected that warming temperatures may have significant impact on exacerbating the effects eutrophication and GHG release from reservoirs. Increased GHG emissions can be expected from rising temperatures alone [48]. Thus, the combination of greater production of GHGs and release will certainly have impacts on the global carbon budget.

Further, climate change may force systems from diffusion dominated to ebullition dominated system [37]. Increased methane production generating more methane bubbles will boost CH_4 ebullition. This will be very pronounced in shallow reservoirs and shallow portions of large reservoirs [47]. Further, temperature increases at depth may generate more diffusive methane releases from sediments and enhance methane build-up in the hypolimnion [49]. This has multiple consequences increasing methane release during lake turnover and creating much greater methane release during energy production as water flows over the turbines. Increased warming will be a catalyst to intensify already potent processes for GHG production and release.

4.2 Stormwater and flooding

Inputs from stormwater have profound impacts on reservoirs [50]. It is expected as climate change intensifies so will the intensity and frequency of extreme weather and hence flooding events [51]. Flooding can both increase the oxygen concentration on rainy days and accelerate the consumption of oxygen after the storm [52]. Organic material from storms entrained into the reservoir during rain events will consume oxygen during stable stratification [53]. As flooding increases, so will oxygen loss that will exacerbate production of GHGs.

Coupled with increased oxygen loss, flooding will exacerbate eutrophication with increases in nutrients and organic matter flowing into reservoirs from storms [54, 55]. High winds and precipitation associated with more frequent and intense storms will influence water-column mixing and the behavior, amount, and composition of runoff swept into lakes [19]. Additionally, reservoirs will experience greater flushing disrupting stratification and releasing GHGs.

These small, hydrological changes have the potential to restructure entire phytoplankton communities in both the short- and long-term [56]. Large and even extreme weather events will further disrupt current reservoir dynamics and even generate new patterns [57]. Because reservoirs are so interdependent on watersheds, we may see simple changes in precipitation patterns change a somewhat healthy reservoir into an unhealthy one as climate change persists. Evidence suggest climate driven changes to reservoirs will create ideal conditions for increased production and ultimately emission of GHGs beyond what is occurring now [56].

Looking at patterns in reservoir data to support or refute impacts due to temperature and precipitation, a principal component analysis was conducted on a current data set from a pump-storage reservoir in central Virginia USA (**Table 1**). From this analysis, patterns suggest that reservoir variables at the river and headwater stations are most strongly correlated to precipitation with the remaining reservoir stations correlated to temperature. Most importantly, reservoir oxygen percentages correlated to external temperatures and only in the portions of the reservoir with lake attributes and extensive summer stratification and oxygen loss in the hypolimnion. This suggest

	River	Tail Water	Headwater	Transition	Dam
% Variability	32.70%	26.40%	27.90%	33.60%	29.30%
Precip. (cm)	16.077	8.699	21.139	2.917	3.367
Temp C	0.641	29.305	0.397	25.958	32.535
DO%	0.572	15.545	0.494	24.369	15.720
Turbidity	29.263	1.884	29.464	20.439	15.560
TP	29.069	3.842	4.714	0.473	6.336
Secchi	22.656	28.121	39.033	12.071	23.876
Chl a	1.722	12.603	4.759	13.773	2.607

Table 1.
Results from principal component analysis (PCA) showing similarities among variables using % contribution of each variable in first factor. Principal component analysis is a multivariate technique analyzing data tables to look for patterns of similarity among the observations [58]. The first factor accounts for the greatest similarity among all of the variables tested. Each station represents a station on the pump storage reservoir (Leesville Lake in Central Virginia, USA) with river input at the headwaters (river), release from the upper dam (tail water), portion of reservoir where both inputs converge (headwater), further down the reservoir in the transition area (transition) and then near the dam (dam). Observations with the largest correlations are highlighted. Data analyzed from samples taken on a monthly basis at the end of the month from April-October 2015–2023 for a total of 63 total samples. Temperature is the monthly mean air temperature and precipitation is the total for the month. All other variables are measured in the reservoir.

oxygen loss in these reservoirs may be strongly dependent on temperature in the region and may intensify as we see the effects of climate change.

Conversely, precipitation did correlate with reservoir parameters in the riverine and headwater portions of the reservoir. Importantly, nutrient inputs as total phosphorus (TP) correlated with changes in precipitation. This suggests the importance of river input and eutrophication entering these reservoirs. Water clarity (turbidity and Secchi) was correlated to both temperature and precipitation. This suggests that both temperature and precipitation have influence in these reservoirs and these are the parameters of concern as climate change progresses. Data here have implications with eventual methane flux as it is positively correlated with temperature and lake nutrient status [59].

5. Trending (sustainability of hydropower)

With concerns over the production of GHGs and potential of worsening with climate change, the sustainability of hydropower must be carefully analyzed and improved if this will be the energy source of the future. Jager and Smith [32] define sustainability as the operation of reservoirs to meet societal needs for water and power while protecting long-term health of the river ecosystem. At its most basic sense, hydropower changes the natural flow regime of the river optimizing water storage for release during peak demands for electricity – storage or movement at night and release during the day. And then in a most critical sense, these systems are massive GHG emitting systems that must be controlled. The production of energy through hydropower must be viewed in the context of the entire operation and managed accordingly.

5.1 Managing hydropower holistically

It is critical that when considering the sustainability of hydropower projects all aspects and factors be part of the considerations [60]. Compared to other renewable forms of energy, hydropower provides an almost continuous supply of energy, relatively low cost and with lower emissions compared to conventional forms of energy production [61]. Reservoirs created for these projects provide water sources for recreation, drinking water supply and even fish and wildlife enhancements with multiple societal benefits [62]. And even with concerns in communities during construction, eventually these reservoirs can improve the quality of the landscape and become accepted [63]. Thus, the attractiveness of the power source and environmental enhancements all exist for a favorable evaluation.

Conversely, the building of hydroelectric dams has widespread impacts on populations, ecology and climate. Dams cause fragmentation of many river systems that are now free-flowing impacting the ecology and hydrology of the region [3]. These construction projects are high in capital cost, involve considerable amounts of machinery and disturbance of the areas with loss of arable land, residential neighborhoods and scenic river valleys [32]. The emission of CO_2 and CH_4 are considerable both from the reservoirs and during operations. The three greatest barriers to future progress are (1) the valuation of ecological benefits, (2) understanding the ecological effects of flow releases sufficiently well to quantify them and (3) lack of incentive for power producers to operate sustainably [32].

While our understanding of environmental concerns with hydropower continues to increase, all of this information has not resulted in significant management actions [56]. In fact, a comprehensive and integrative holistic tool for hydropower reservoir management is not available [64]. Thus, two distinctive management needs are apparent as we move forward in a transition to more renewable forms of energy. Proper management of existing hydropower projects and in particular large dam and pump-storage operations to become more sustainable. And potential decentralization of the power projects as breaking these operations into smaller projects lowers the environmental costs.

5.2 Solutions

Small Hydroelectric Power (SHP) may be a solution. There is generally no agreed upon definition of SHP but in general these are small plants that use run of the river flow with minimal impoundment (**Figure 1**). These projects are gaining popularity for a suite of potential benefits. Most of the ecological problems associated with large reservoir operations such as oxygen depletion, increased temperature, decreased flow and GHG emission can be avoided [60, 65]. These projects can be sited in less than traditional hydropower areas and throughout the world to meet expanding demand [66]. They rely on direct river flow but if spaced thought a region (decentralization) some of this problem can be diminished.

Such projects put into sequential order and adequately spaced to maximize ecology ecosystem services (fish, macros and water quality) may be the most preferential hydroelectric projects [67, 68]. Small low head dams have the potential to improve water quality [69] and benefit local governance through adaptation to local economies, local jobs and enhancement of local infrastructure [66]. Localities can assume responsibility for local energy production.

These projects are not without concern. Factors such as mode of operation, the degree of river flow alteration, impacts on habitat connectivity, and the cumulative ecological effects of multiple SHP installations on a single river warrant close consideration and management [70]. Yet even with these concerns, this type of power production is the preferred going into the future as it minimizes large reservoir operation and management. These projects are often too large for local management and operate throughout a region making impacts quite considerable.

Existing large systems need to be reviewed for operational strategy and moved into a more sustainable approach. Several operational strategies such as Run of River (ROR) allowing water to flow continuously or spill flows (non-generating flows not released through turbines) need consideration. Adopting a more natural ROR flow regime meets environmental needs maintaining aquatic ecosystems and human needs embedded in domestic and agricultural usage [71]. It can be adjusted from a consumer peak demand model to meet energy production goals. Cost/benefit analyses are desperately needed to weigh environmental atmospheric costs, health costs and improvement in aquatic resources and fisheries habitats in exchange for energy production.

Sluicing or spill flows ameliorate water quality downstream by re-oxygenating the water avoiding excessive sedimentation behind dams and allowing sediment to flow into downstream areas [72]. This improves watershed natural function. And because of the strong relationship between reservoirs and watersheds, new reservoirs need sited away from land use and anthropomorphic conditions that cause declines in reservoir water quality [28]. Current and new large reservoir projects need careful planning to not only meet energy needs but to minimize environmental impacts.

5.3 Future trends

The damming of rivers has seen less and less popularity and large hydropower projects are not accepted as a clean and renewable energy source [65]. With concerns over meeting growing electricity demand and a desire to grow hydropower to meet this demand, more sustainable approaches need developed and accepted.

Apart from its doubtless advantages, even nominally successful hydropower projects are often associated with negative environmental consequences in the form of biodiversity loss, disruptions to fish migration, potentially large-scale land inundation, the disruption of human resettlement, and many others [73]. Sustainability studies are need to consider hydropower holistically rather than by its individual parts. If it is to be the energy source of the future, the negative aspects must be addressed and understood.

6. Conclusion

Hydropower when used well can provide a very viable renewable energy source to meet the growing demand as we move into the future. Alternatively, if the industry continues to ignore management of reservoirs associated with these projects adhering to schedules of power generation only, viability as a sustainable source of energy will diminish. It is clear that reservoirs are integrators of watersheds and as such must be maintained and managed with this in mind. The assimilation of organic material and processing into GHGs makes management imperative. And as climate change that is expected to exacerbate the GHG emission problem intensifies, management can no longer be ignored as part of these hydropower projects.

As we push for construction of new hydroelectric projects to meet demand, large projects with a myriad of environmental concerns may become less feasible. Small hydropower may be a good alternative. These projects may be sited in areas not amenable for large projects and spread throughout a region to diminish the environmental impact. These projects may fit better under local control helping local economies and providing better management as a small project. These small projects are not without concern but may be a very good alternative.

An integration of small hydropower, large hydropower and sustainable reservoir management would seem to be the best solution. Integration of environmental concerns needs to be at the forefront of these projects instead of an afterthought or retrofit once the project is in operation. The need for more power is increasing and growth using fossil fuels does not provide the path to a sustainable future. We can use hydropower to produce energy and demonstrate environmental stewardship. In this way, hydropower will lead the way toward a sustainable energy future.

Acknowledgements

The author would like to acknowledge the University of Lynchburg for providing the time and resources necessary to complete this project. I would also like to thank the Leesville Lake Association for providing funding for this project and Dr. Tony Capuco and the entire water quality committee for their support and help throughout. To the countless students have assisted me in this work, spent time on the boat and in the laboratory to quantify the water quality of the reservoirs I am very grateful. My family is always at the forefront of support and no project can be completed without them.

Author details

Thomas Shahady[1,2]

1 University of Lynchburg, Lynchburg, VA, USA

2 Biophilia Research Center, Quepos, Costa Rica

*Address all correspondence to: shahady@lynchburg.edu

References

[1] Yüksel I. Hydropower for sustainable water and energy development. Renewable and Sustainable Energy Reviews. 2010;**14**:462-469

[2] Subramanya K, Chelliah T. Capability of synchronous and asynchronous hydropower generating systems: A comprehensive study. Renewable and Sustainable Energy Reviews. 2023;**188**: 113863

[3] Zarfl C, Lumsdon A, Berlekamp J, Tydecks L, Tockner K. A global boom in hydropower dam construction. Aquatic Sciences. 2015;**77**:161-170

[4] Turner S, Voisin N, Nelson K, Tidwell V. Drought Impacts on Hydroelectric Power Generation in the Western United States. U.S. Dept. of Energy. Richland, WA, USA: Pacific Northwest National Laboratory; 2022

[5] Lehner B, Liermann C, Revenga C, Vörösmarty C, Fekete B, Crouzet P, et al. High-resolution mapping of the world's reservoirs and dams for sustainable river-flow management. Frontiers in Ecology and the Environment. 2011;**9**: 494-502

[6] Department of Energy, Office of Energy Efficiency and Renewable Energy. Pumped Storage Hydropower. 2023. Available online: https://www.energy.gov/eere/water/pumped-storage-hydropower [Accessed: December 30, 2023]

[7] Zhang Y, Ma H, Zhao S. Assessment of hydropower sustainability: Review and modeling. Journal of Cleaner Production. 2021;**321**:128898

[8] Yang C, Jackson R. Opportunities and barriers to pumped-hydro energy storage in the United States. Renewable and Sustainable Energy Reviews. 2011; **15**:839-844

[9] Girma T, Chala M. In: Ma'Arof, Sharma R, James RG flay (reviewing editor). Trends in an increased dependence towards hydropower energy utilization—A short review, cogent engineering. 2019;**6**:1

[10] Yah N, Oumer A, Idris M. Small scale hydro-power as a source of renewable energy in Malaysia: A review. Renewable and Sustainable Energy Reviews. 2017;**72**:228-239

[11] Ansari A, Singh S, Lanza R, Rast W, editors. Eutrophication: Causes, Consequences and Control. Berlin, Germany: Springer Science and Business Media; 2010

[12] Walter J, Fleck R, Pace M, Wilkinson G. Scaling relationships between lake surface area and catchment area. Aquatic Sciences. 2020;**82**:47

[13] Hayes N, Corman J, Deemer B, Strock K, Razavi R. Key differences between lakes and reservoirs modify climate signals: A case for a new conceptual model. Limnology and Oceanography Letters. 2017;**2**:47-62

[14] Withers P, Neal C, Jarvie H, Doody D. Agriculture and eutrophication: Where do we go from here? Sustainability. 2014;**6**:5853-5875

[15] Hilton J, O'Hare M, Bowes M, Iwan JJ. How green is my river? A new paradigm of eutrophication in rivers. Sci. Tot. Environ. 2006;**365**:66-83

[16] Royer T, David M, Gentry L. Timing of riverine export of nitrate and phosphorus from agricultural

watersheds in Illinois: Implications for reducing nutrient loading to the Mississippi River. Environmental Science and Technology. 2006;**40**: 4126-4131

[17] Janke B, Finlay J, Hobbie S, Baker L, Sterner R, Nidzgorski D, et al. Contrasting influences of stormflow and baseflow pathways on nitrogen and phosphorus export from an urban watershed. Biogeochemistry. 2014;**121**: 209-228

[18] Carpenter S, Booth E, Kucharik C, Lathrop R. Extreme daily loads: Role in annual phosphorus input into a north temperate lake. Aquatic Sciences. 2015; **77**:71-79

[19] Stockwell J, Doubek J, Adrian R, Anneville O, Carey C, Carvalho L, et al. Storm impacts on phytoplankton community dynamics in lakes. Global Change Biology. 2020;**26**: 2756-2784

[20] Smith V, Tilman G, Nekola J. Eutrophication: Impacts of excess nutrient inputs on freshwater, marine, and terrestrial ecosystems. Environmental Pollution. 1999;**100**: 179-196

[21] Lürling M. Grazing resistance in phytoplankton. Hydrobiologia. 2021; **848**:237-249

[22] Wetzel R. Limnology. London: Academic Press; 2001

[23] Nürnberg G. Quantifying anoxia in lakes. Limnology and Oceanography. 1995;**40**:1100-1111

[24] Beaulieu J, Waldo S, Balz D, Barnett A, Hall W, Platz A, et al. Methane and carbon dioxide emissions from reservoirs: Controls and upscaling. Journal of geophysical research. Biogeosciences. 2020;**125**: e2019JG005474

[25] Bartosiewicz M, Maranger R, Przytulska A, Laurion I. Effects of phytoplankton blooms on fluxes and emissions of greenhouse gases in a eutrophic lake. Water Research. 2021; **196**:116985

[26] DelSontro T, Beaulieu J, Downing J. Greenhouse gas emissions from lakes and impoundments: Upscaling in the face of global change. Limnology and Oceanography Letters. 2018;**3**:64-75

[27] Harrison J, Deemer B, Birchfield M, O'Malley M. Reservoir water-level drawdowns accelerate and amplify methane emission. Environmental Science and Technology. 2017;**51**: 1267-1277

[28] Deemer B, Harrison R, Li J, Beaulieu S, DelSontro J, Barros T, et al. Greenhouse gas emissions from reservoir water surfaces: A new global synthesis. Bioscience. 2016;**66**:949-964

[29] Deemer B, Holgerson M. Drivers of methane flux differ between lakes and reservoirs, complicating global upscaling efforts. Journal of geophysical research. Biogeosciences. 2021;**126**:e2019JG0 05600

[30] Bastviken D, Cole J, Pace M, Tranvik L. Methane emissions from lakes: Dependence of lake characteristics, two regional assessments, and a global estimate. Global Biogeochemical Cycles. 2004;**18**: GB4009

[31] McClure R, Hamre K, Niederlehner B, Munger Z, Chen S, Lofton M, et al. Metalimnetic oxygen minima alter the vertical profiles of carbon dioxide and methane in a

managed freshwater reservoir. Science of The Total Environment. 2018;**636**: 610-620

[32] Jager H, Smith B. Sustainable reservoir operation: Can we generate hydropower and preserve ecosystem values? River Research and Applications. 2008;**24**:340-352

[33] Rosa L, dos Santos M, Matvienko B, dos Santos E, Sikar E. Green house gas emissions from hydroelectric reservoirs in tropical regions. Climatic Change. 2004;**66**:9-21

[34] Bastviken D, Cole J, Pace M, Van de Bogert M. Fates of methane from different lake habitats: Connecting whole-lake budgets and CH4 emissions. Journal of Geophysical Research: Biogeosciences. 2008;**113**:1-13

[35] Fernández E, Peeters J, Hofmann H. Importance of the autumn overturn and anoxic conditions in the hypolimnion for the annual methane emissions from a temperate lake. Environmental Science and Technology. 2014;**48**:7297-7304

[36] Bastviken D, Tranvik L, Downing J, Crill P, Enrich-Prast A. Freshwater methane emissions offset the continental carbon sink. Science. 2011;**331**:50

[37] Aben R, Barros R, Van Donk E, Frenken T, Hilt S, Kazanjian G, et al. Cross continental increase in methane ebullition under climate change. Nature Communications. 2017;**8**:1682

[38] Delsontro T, Boutet L, St-Pierre A, del Giorgio P, Prairie Y. Methane ebullition and diffusion from northern ponds and lakes regulated by the interaction between temperature and system productivity. Limnology and Oceanography. 2016;**61**:62-77

[39] Kemenes A, Forsberg B, Melack J. Methane release below a tropical hydroelectric dam. Geophysical Research Letters. 2007;**34**:L12809

[40] Beaulieu J, Balz D, Birchfield M, Harrison J, Nietch C, Platz M, et al. Effects of an experimental water-level drawdown on methane emissions from a eutrophic reservoir. Ecosystems. 2018; **21**:657-674

[41] McGinnis D, Greinert J, Artemov Y, Beaubien S, Wuest A. Fate of rising methane bubbles in stratified waters: How much methane reaches the atmosphere? Journal of Geophysical Research. 2006;**111**:1-15

[42] Abril G, Guérin F, Richard R, Delmas C, Galy-Lacaux P, Gosse A, et al. Carbon dioxide and methane emissions and the carbon budget of a 10-year old tropical reservoir (petit Saut, French Guiana). Global Biogeochemical Cycles. 2005;**19**:GB4007

[43] Giles J. Methane quashes green credentials of hydropower. Nature. 2006;**444**:524-525

[44] Diem T, Koch S, Schwarzenbach S, et al. Greenhouse gas emissions (CO2, CH4, and N2O) from several perialpine and alpine hydropower reservoirs by diffusion and loss in turbines. Aquatic Sciences. 2012;**74**:619-635

[45] Yasarer L, Sturm B. Potential impacts of climate change on reservoir services and management approaches. Lake and Reservoir Management. 2016; **32**:13-26

[46] Jane S, Hansen G, Kraemer B, et al. Widespread deoxygenation of temperate lakes. Nature. 2021;**594**:66-70

[47] Edlund M, Almendinger J, Fang X, Ramstack Hobbs J, VanderMeulen D,

Key R, et al. Effects of climate change on Lake thermal structure and biotic response in northern wilderness lakes. Water. 2017;**9**:678

[48] Scharfenberger U, Jeppesen E, Beklioğlu M, Søndergaard M, Angeler D, Aİ C, et al. Effects of trophic status, water level, and temperature on shallow lake metabolism and metabolic balance: A standardized pan-European mesocosm experiment. Limnology and Oceanography. 2019;**64**:616-631

[49] Peeters F, Encinas Fernandez J, Hofmann H. Sediment fluxes rather than oxic methanogenesis explain diffusive CH4 emissions from lakes and reservoirs. Scientific Reports. 2019; **9**:243

[50] Vogel R, Lane M, Ravindiran R, Kirshen P. Storage reservoir behavior in the United States. Journal of Water Resources Planning and Management. 1999;**125**:245-254

[51] Wang X, Huang G, Liu J. Projected increases in intensity and frequency of rainfall extremes through a regional climate modeling approach. Journal of Geophysical Research Atmospheres. 2014;**119**:213-271

[52] Ma W, Huang T, Li X, Zhang H, Ju T. Impact of short-term climate variation and hydrology change on thermal structure and water quality of a canyon-shaped, stratified reservoir. Environmental Science and Pollution Research. 2015;**22**:18372-18380

[53] Liu M, Zhang Y, Shi K, Zhang Y, Zhou Y, Zhu M, et al. Effects of rainfall on thermal stratification and dissolved oxygen in a deep drinking water reservoir. Hydrological Processes. 2020; **34**:3387-3399

[54] Nazari-Sharabian M, Ahmad S, Karakouzian M. Climate change and eutrophication: A short review. Engineering, Technology and Applied Science Research. 2018;**8**:3668

[55] Dhillon G, Inamdar S. Extreme storms and changes in particulate and dissolved organic carbon in runoff: Entering uncharted waters? Geophysical Research Letters. 2013;**40**: 1322-1327

[56] Meerhoff M, Audet J, Davidson T, De Meester L, Hilt S, Kosten S, et al. Feedback between climate change and eutrophication: Revisiting the allied attack concept and how to strike back. Inland Waters. 2022;**12**:187-204

[57] Paerl H, Hall N, Hounshell A, Luettich R, Rossignol K, Osburn L, et al. Recent increase in catastrophic tropical cyclone flooding in coastal North Carolina, USA: Long-term observations suggest a regime shift. Scientific Reports. 2019;**9**:1-9

[58] Abdi H, Williams L. Principal component analysis. In: In R for Political Data Science. Boca Raton, FL, USA: Chapman and Hall/CRC; 2020. pp. 375-393

[59] Rasilo T, Prairie Y, del Giorgio P. Large-scale patterns in summer diffusive CH4 fluxes across boreal lakes, and contribution to diffusive C emissions. Global Change Biology. 2015;**21**:1124-1139

[60] Nautiyal H, Goel V. Sustainability assessment of hydropower projects. Journal of Cleaner Production. 2020;**265**: 121661

[61] Varun BI, Prakash I. LCA of renewable energy for electricity generation systems—A review. Renewable and Sustainable Energy Reviews. 2009;**13**:1067-1073

[62] Winter P, Selin S, Cerveny L, Bricker K. Outdoor recreation, nature-based tourism, and sustainability. Sustainability. 2020;**12**:81

[63] Zdankus N, Vaikasas S, Sabas G. Impact of a hydropower plant on the downstream reach of a river. Journal of Environmental Engineering and Landscape Management. 2008;**16**: 128-134

[64] Daus M, Koberger K, Koca K, Beckers F, Encinas Fernández J, Weisbrod B, et al. Interdisciplinary reservoir management—A tool for sustainable water resources management. Sustainability. 2021;**13**: 4498

[65] Pang M, Zhang L, Ulgiati S, Wang C. Ecological impacts of small hydropower in China: Insights from an emergy analysis of a case plant. Energy Policy. 2015;**76**:112-122

[66] Warren G. Small hydropower, big potential: Considerations for responsible global development. Idaho Law Review. 2017;**53**:149

[67] Santucci V, Gephart S, Pescitelli S. Effects of multiple low-head dams on fish, macroinvertebrates, habitat, and water quality in the Fox River, Illinois. Journal of North American Fisheries and Management. 2005;**25**:975-992

[68] Jager H, Efroymson R, Opperman J, Kelly M. Spatial design principles for sustainable hydropower development in river basins. Renewable and Sustainable Energy Reviews. 2015;**45**:808-816

[69] Shahady T, Cleary W. Influence of a low-head dam on water quality of an urban river system. Journal of Environmental Management. 2021;**297**: 113334

[70] Couto T, Olden J. Global proliferation of small hydropower plants – Science and policy. Frontiers in Ecology and the Environment. 2018;**6**: 91-110

[71] Suen J, Eheart J. Reservoir management to balance ecosystem and human needs: Incorporating the paradigm of the ecological flow regime. Water Resources Research. 2006;**42**: W03417

[72] Kondolf G, Gao Y, Annandale G, Morris G, Jiang E, Zhang J, et al. Sustainable sediment management in reservoirs and regulated rivers: Experiences from five continents. Earth Future. 2014;**2**:256-280

[73] Tahseen S, Karney B. Reviewing and critiquing published approaches to the sustainability assessment of hydropower. Renewable and Sustainable Energy Reviews. 2017;**67**:225-234

Chapter 7

The European REFORM Project for Hydromorphological Quality in River Basin Management

Erik Mosselman, Massimo Rinaldi and Diego García de Jalón

Abstract

The Water Framework Directive commits European Union member states to achieve good ecological and chemical status of all water bodies. As hydromorphology is a key factor for ecological status, a consortium of 26 partners from 15 countries studied the role of hydromorphological pressures and measures in the REFORM project. Its main objective was to answer the question: How to make river restoration successful? The project developed guidance for this by structuring the information along the different stages of restoration projects and river basin management plans, posing a logical sequence of questions: How does my river work? What's wrong? How to improve? Things can be wrong for ecological status as a result of morphological alterations. These alterations form pressures that can be countered or mitigated by measures that improve sedimentological and morphological features. We present two specific results of REFORM that focus on river morphology. First, we provide an overview of methods to assess morphological quality and diagnose alteration. Second, we present systematic cause–effect relationships for restoration measures.

Keywords: hydromorphology, river morphology, ecological status, river restoration, water framework directive, river basin management

1. Introduction

The Water Framework Directive commits European Union member states to achieve good ecological and chemical status of all water bodies. Its adoption and publication in 2000, however, caused widespread concern among river engineers and fluvial geomorphologists about the way the directive addresses hydromorphology in rivers. The directive defines hydromorphological quality in terms of visibility of static anthropogenic features, without considering hydromorphological functioning relevant for ecology, and without considering the physical processes of water flow, sediment transport, erosion, and sedimentation. Neither did the directive address temporal variability [1, 2], nor spatial variability in the light of habitat diversity and connectivity [3, 4]. The European Union was receptive to the criticisms and accordingly called for a project to resolve this within its 7th Framework Programme. It granted the project to a consortium of 26 partners from 15 countries

IntechOpen

under the name "REFORM", an acronym standing for "Restoring rivers FOR effective catchment Management". The partners executed the project from November 2011 to October 2015.

REFORM's main objective was to answer the question: How to make river restoration successful? This requires that river restoration practitioners understand the complex systems of hydromorphology and ecology. Processes operate at different scales, different disciplines play a role, and different species depend on hydromorphology in different ways. It is not easy to find a way in this complexity when developing an integrated design. The project developed guidance by structuring the information along the different stages of restoration projects and river basin management plans, following the cycle of the iterative PDCA management method: Plan – Do – Check – Act. This fits in a logical sequence of questions guiding river basin management planning: How does my river work? What's wrong? How to improve? (**Figure 1**). The guidance was made accessible through an online wiki: wiki.reformrivers.eu.

Things can be wrong for ecological status as a result of morphological alterations. Examples are alteration of instream habitat, alteration of riparian vegetation, channelization, cross-section alteration, embankments (levees and dikes), impoundment (dams), loss of vertical connectivity, mining of sand and gravel, sedimentation, and sediment input. In Water Framework Directive terminology these alterations are called "pressures". They can be countered or mitigated by measures that improve sedimentological and morphological features such as sediment flow quantity, longitudinal connectivity, lateral connectivity, riverbed depth variation, width variation, in-channel bed structure, substrate, riparian zone, and floodplains. REFORM compiled information on both pressures and measures. Furthermore, it sought to enhance awareness of the importance of sediment and morphology by summarizing insights and results into one-liners with down-to-earth messages, so-called tiles of wisdom (**Figure 2**). The online wiki presents and explains all these pressures, measures, and tiles of wisdom.

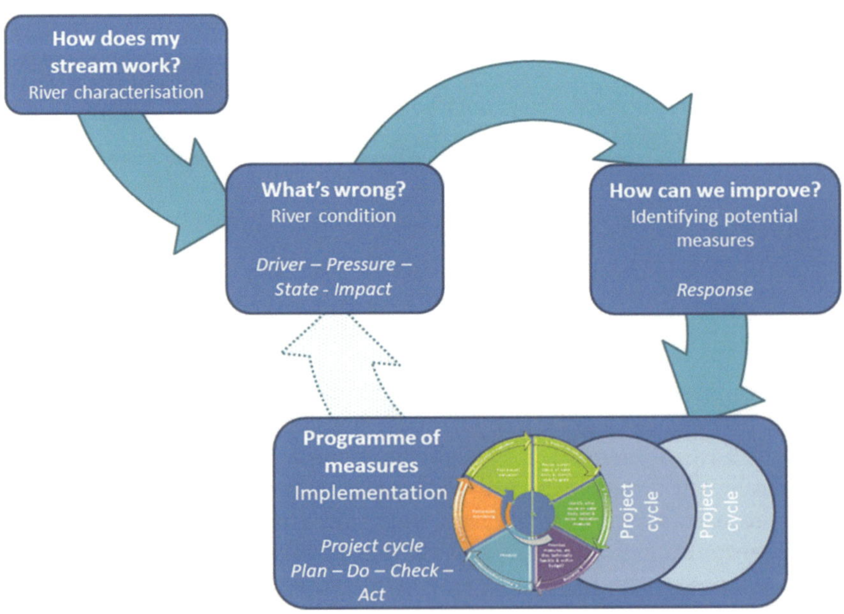

Figure 1.
Cycles of overall river basin management plans and individual river restoration projects.

Figure 2.
Tiles of wisdom.

In this chapter, we present two specific results of REFORM that focus on river morphology. First, we present methods to assess hydromorphological quality and diagnose alteration. Second, we present systematic cause–effect relationships for diagnosis and for restoration measures.

2. Hydromorphological quality and alteration

Evaluation of hydromorphological quality and alteration requires a careful assessment that considers physical processes and resulting fluvial forms and physical habitats at appropriate spatial and temporal scales. This type of stream assessment has significantly expanded, with numerous methods in different EU member states that vary widely in terms of their aims, scales, and approaches [5, 6]. REFORM grouped these methods into four broad categories, based on the focus and objectives of each method: (1) physical habitat assessment; (2) riparian habitat assessment; (3) morphological assessment; and (4) assessment of hydrological regime alteration [6]. Their suitability for diagnosing alteration, however, was found to be limited because the methods insufficiently considered the physical processes of natural fluvial systems that maintain or recreate fluvial forms and thereby physical habitats. This contrasted with recent scientific developments that base the interpretation of current conditions on attempts to understand river functioning and evolution, see for instance [7–11].

REFORM therefore developed methods for a process-based hydromorphological assessment that considers how the character and dynamics of river reaches are affected by small-scale and large-scale natural and human-induced changes, in the past and in the present. Gurnell et al. [12] developed a multi-scale framework like previous hierarchical frameworks [7, 10, 13] but tuned to the European context. The framework is open-ended, allowing European member states to incorporate their own datasets, methods, and modeling tools. It distinguishes spatial units at region, catchment, landscape unit, segment, reach, geomorphic unit, hydraulic unit, and river element scales. Rinaldi et al. [14] refined the framework with four stages of assessment (**Table 1**), in accordance with the structure of existing frameworks [7, 15].

Each stage contains a series of procedural steps for consistent assessment of river conditions. The key spatial scale is the reach, defined as river sections along which present valley setting, channel slope, imposed flow, and sediment load are sufficiently uniform [7]. Channel morphology is a fundamental feature for delineating reaches. A first simple level of classification regards the number of river channel threads and

Stage	Definition	Description	Main outputs
I	Catchment-wide delineation and spatial characterization of the fluvial system	delineates, characterizes, and analyzes the catchment and the river system in their current conditions	(i) spatial units; (ii) character of spatial units, including hydrology, sediment sources and delivery, and assemblages of geomorphic units; (iii) main physical pressures and impacts at catchment scale; (iv) spatial patterns of morphological parameters and their control on channel morphology
II	Assessment of temporal changes and current conditions	reconstructs the history and evolutionary trajectories of morphological changes that have resulted in the current river conditions	(i) natural and human factors in historical times; (ii) evolutionary trajectories of channel changes; (iv) catchment-scale maps of pressures and critical reaches; (v) hydrological, morphological, and riparian vegetation state; (vi) geomorphic units; (vii) identified problems and most critical reaches; (viii) reports on monitored parameters or indicators along with their temporal changes
III	Assessment of scenario-based future trends	identifies possible future scenarios of hydromorphological modification	(i) catchment-scale maps of sensitivity and morphological potential; (ii) past channel evolution, current conditions, and possible future trends
IV	Management	identifies possible hydromorphological restoration or management actions	(i) one or more scenarios of management actions or restoration interventions; (ii) potential effects of proposed interventions on physical processes and hydromorphological conditions

Table 1.
Stages and main outputs of the REFORM overall framework, abbreviated from [14].

the channel planform pattern in the context of the valley setting (confinement). This basic river typology (BRT: [16]) defines seven river types using readily available information, mainly remotely sensed imagery. The initial delineation of the river reaches is followed by collecting additional information on reach properties and indicators. Then the river type may be defined during a field survey, according to an extended river typology (ERT) that comprises 22 river types [16].

The temporal context (Stage II) is linked to the concept of evolutionary trajectory [17, 18], expressing that river systems are dynamic and follow a complex trajectory of changes in response to driving variables at various spatial and temporal scales. The specific characteristics of a river result from its historical evolution, including climatic variations, human interventions, and unique sequences of large flood events. Assessment of current conditions and possible future scenarios and adjustments thus requires proper interpretation of temporal adjustments in morphology.

The framework allows selecting representative reaches or sites for monitoring river conditions and for upscaling or downscaling of information. It helps in classifying and understanding current conditions, in assessing the potential for morphological changes, and in supporting prioritization of actions and selection of sustainable management strategies.

The hydromorphological framework contains a set of more specific assessment procedures [14]. One of them is the extended European version of the Morphological

Quality Index (MQI), originally developed in Italy [10] and revised and tested within REFORM [19]. The MQI can be called 'process-based' because:

1. It takes processes into account that go beyond mere channel forms, as it includes indicators linked to the functioning of basic processes such as sediment continuity, wood flux continuity, bank erosion, and lateral channel mobility;

2. It explicitly accounts for the temporal component through indicators for adjustments of channel form through time;

3. Reference conditions are defined in terms of dynamic processes and functions that are expected for each physical context. This differs significantly from most current hydromorphological methods which define reference conditions in terms of channel configuration or channel characteristics.

The spatial scale of MQI application, in accordance with the multi-scale hierarchical framework, is the reach (i.e., a sufficiently uniform section of river, commonly a few kilometers in length). This is generally seen as the most appropriate and meaningful scale for assessing hydromorphology [7, 12]. The MQI includes twenty-eight indicators [10, 15, 19], falling within the following three classes:

1. Geomorphological functionality: indicators to evaluate whether artificial elements or channel adjustments prevent or alter the processes and related forms that are responsible for the correct functioning of the river;

2. Artificiality: indicators to assess the presence and frequency of occurrence of artificial elements, pressures, interventions, and management activities, irrespective of their effects on channel forms and processes;

3. Channel adjustment: indicators to assess morphological changes over about the last 100 years that can indicate systematic instability related to human factors.

Operators with sufficient background and training in fluvial geomorphology collect data by integrating remote sensing, GIS analysis, and field survey. The evaluation is based on a scoring system. The total score is equal to the sum of the scores for all components and aspects. The Morphological Quality Index is then defined as $MQI = 1 - S_{tot} / S_{\max}$, where S_{tot} is the calculated total score, and $S_{\max}$ is the maximum score that could be reached. The index thus increases with quality of the reach and decreases with the level of alteration, varying from 0 (minimum quality) to 1 (maximum quality), allowing investigation of the full range of morphological conditions.

The MQI assessment can be integrated with specific indices of hydrological alteration, such as IARI [20] and IAHRIS [21]. These indices align with the indicators of hydrological alteration (IHA) proposed by [22]. Furthermore, a Morphological Quality Index for monitoring (MQIm) was specifically designed to account for small changes and short time scales. This index is therefore suitable for monitoring and environmental impact assessment of interventions [19].

Rinaldi et al. [10] tested the original version and then applied it to many river reaches in Italy. Within REFORM, Belletti et al., [23] extended the method and tested it on several European streams of types that were underrepresented or entirely unrepresented in the Italian context. The indicators and scores were the same as in the original MQI to ensure data comparability, but with some modification or integration of aspects not covered fully previously. Further extensive application of the MQI at catchment scale was recently carried out on the Guadalquivir River in Southern Spain [24].

3. Cause-effect diagrams for diagnosis and restoration measures

Proper selection and design of restoration measures to improve fluvial ecosystem services require identifying which pressures affect the river and which are the limiting factors causing degradation. The intensity of the limiting hydromorphological processes can be assessed by quantitative measurement of the variables affected by these processes. REFORM proposes conceptual diagrams that relate different types of hydromorphological pressures to fluvial system functioning, accounting for hydromorphological processes and variables that result from both degradation and restoration. The aim is to identify the main hydromorphological effects of different pressure types across spatial and temporal scales, especially those that have a significant impact on aquatic biology.

Often, human pressures affecting rivers do not come alone. Multiple pressures affect rivers simultaneously, stressing many components of the hydrological cycle which have different time-scale responses within fluvial ecosystems. However, for practical reasons the effects of hydromorphological pressures and their most direct impacts on ecosystems were analyzed separately. The results were synthetized in diagrams that show the direct effects on processes and state variables, but also which process changes these effects induce with respect to hydromorphological variables. Corresponding quantitative variables are to be measured in order to monitor river changes and evaluate pressure effects.

The pressures have been grouped into hydrological regime alterations, river fragmentation, morphological alterations, and other elements and processes affected. The hydrological regime may be altered by water abstractions or by flow regulation through temporary storage in reservoirs. Rivers are fragmented by discontinuity in the river's longitudinal, lateral, or vertical dimensions. Such spatial discontinuities disrupt hydrological connectivity [25] and interrupt transfers of water, mineral sediment, organic matter, and organisms, thus affecting the biotic and physical components of the river [26]. Morphological alterations include impoundment, reservoirs with large dams, channelization, alteration of riparian vegetation and instream habitats, embankments, bank reinforcement, extraction of sand and gravel, and floodplain soil sealing and compaction. Other elements and processes include physicochemical pressures such as thermal changes, eutrophication, and overloads of organic material.

Hydromorphological pressures alter structure and composition of fluvial systems through changes in the natural hydromorphological processes, which can be characterized by changes in hydromorphological variables. Hydromorphological processes transform physical components of the fluvial system. These transformations change the morphology and the structure of the river, but they also create different environments that promote changes in the biological communities. Understanding the relationships between physical components and subsequent biological responses is therefore essential for process-based analysis of impacts. Hydrodynamic processes are characterized by variables and parameters. Their effects can be evaluated through selected state variables (**Table 2**). Usually, a modified variable triggers processes

Processes	Variables
• Water flow dynamics • Sediment dynamics a. entrainment b. transport c. deposition d. armoring • Riverbank dynamics a. erosion and failure b. stabilization c. accretion • Vegetation dynamics a. encroachment b. uprooting c. recruitment • Large-wood dynamics a. entrainment b. transport c. deposition • Aquifer dynamics a. recharge b. discharge • Other processes: primary production, heat exchanges, REDOX	1. Hydrological regime variables: • Flow regime (magnitude, variability, floods, and droughts) • Sediment regime • Connection to groundwater 2. Longitudinal river continuity variables 3. Morphological condition variables: • Channel dimensions • Planform • Thalweg • Riverbed structure and grain sizes • Riverbank • Water • Structure of the riparian zone • Structure of the floodplain 4. Physicochemical variables: • Nutrient concentration • Water temperature • Dissolved oxygen

Table 2.
Hydromorphological processes considered and their associated variables for evaluating their effects.

which in turn transform the values of that or other variables. REFORM grouped the hydromorphological variables into flow, flood, flow variability, drought, sediment flow, hydraulic, groundwater connection, longitudinal connectivity, channel dimensions, thalweg, planform, bed substrate, bank, hydraulic energy, riparian, floodplain, and physicochemical variables. Physicochemical variables were included because some impacts of hydromorphological pressures (e.g., large dams) cannot be understood without them.

Anthropogenic impacts of hydromorphological pressures reduce biodiversity by interfering with fluvial succession trajectories, habitat diversification, migratory pathways, and other processes [27]. For each pressure REFORM developed a theoretical diagram of the effects on the system of fluvial hydromorphological interactions [28]. This system is described by the processes involved, the altered variables, and the possible impacts on the biological elements responsible for changes of ecological status.

As an example from the complete set of diagrams [28], **Figure 3** shows the diagram for water abstraction. The red arrows indicate direct effects of pressures on hydromorphological variables, without processes. Water can be abstracted from a river channel by direct surface abstraction or indirect groundwater abstraction.

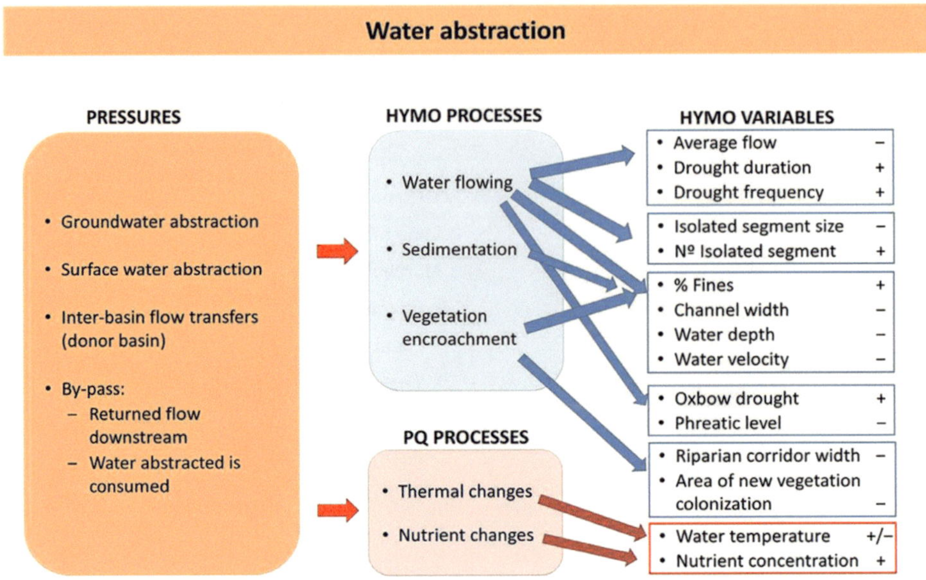

Figure 3.
Conceptual framework representing water abstraction effects on hydromorphological processes and variables that are responsible for their ecological impacts. HYMO stands for hydromorphological and PQ for physicochemical.

Over-abstraction of groundwater can lower the groundwater levels within aquifers or severely reduce the flow in rivers. Surface seepage from aquifers supports groundwater-fed ecosystems such as wetlands and springs. If phreatic levels decline, riparian vegetation rapidly shows signs of water stress and in extreme cases widespread death. Water abstraction alters water flow processes by reducing average flows, reducing flow velocities, increasing duration and frequency of droughts, and lowering phreatic levels. It enhances sedimentation, leading to more fines on the substrate and less water depth and channel width. It reduces the riparian corridor because it forces vegetation to retract at its outer edge, whereas it inhibits invasion of gravel bars due to drought conditions. Finally, water abstraction also alters physicochemical processes by making water temperature more dependent on the air temperature and by increasing eutrophication due to higher concentrations of nutrients in the water.

Changes in the normal functioning of natural and free-flowing rivers occur by natural disturbances, such as floods, droughts, or geological events (**Figure 4**). These disturbances alter the hydromorphological processes that produce changes in habitats and consequently in the biota. However, the resilience capacity of the ecosystem will produce a reversal tendency. Thus, the system follows an oscillatory trajectory that represents the natural variability of the ecosystem and forms an important aspect of its natural biodiversity. The ecosystem services provided by this natural river functioning may be used as a reference.

The non-natural disturbances due to anthropic pressures, however, degrade the status of the fluvial ecosystem and affect its ecosystem services. Some anthropic pressures are hydromorphological as they alter the hydromorphological processes that regulate river functioning. Hydromorphological pressures can change the habitats of biological communities into environments they are not adapted to.

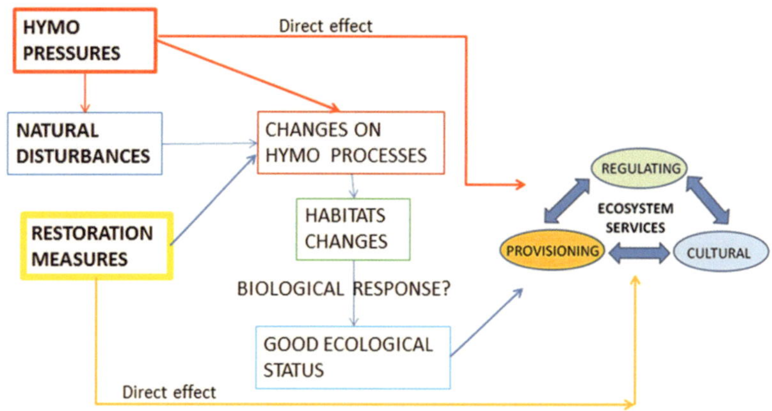

Figure 4.
Mechanisms by which natural disturbances, hydromorphological pressures, and restoration measures may affect fluvial ecosystem services. Direct effects are simple to predict, whereas much more science still needs to be developed for precise predictions of overall interactions affecting hydromorphological processes, habitat changes, and biological response.

This reduces biodiversity or promotes invasive and alien species. Some ecosystem services may also be directly affected by hydromorphological pressures (**Figure 4**). Restoration or mitigation measures are designed to improve habitats, through structural measures or through the recovery of lost hydromorphological processes. Sometimes restoration measures target the recovery of certain ecosystem services. River management thus faces the challenge of understanding how a naturally varying river works, simultaneously subjected to different pressures and programmes of measures.

Main types of restoration measures can be classified according to the hydromorphological elements of the Water Framework Directive [29]:

Water flow quantity improvement:

- Reduce surface water abstraction, with or without return. Improve water retention in upstream catchment. Reduce groundwater extraction. Improve or create water storage. Increase minimum flows. Divert or transfer water. Recycle used water. Reduce water consumption.

Sediment flow quantity improvement:

- Add sediment. Reduce sediment input. Prevent sediment accumulation in reservoirs. Reduce erosion. Improve sediment transport continuity. Manage dams for sediment flow. Trap sediments.

Flow dynamics (water and sediment) improvement:

- Ensure minimum flows. Establish environmental flows. Reduce hydropeaking. Increase frequency and duration of flooding in riparian zones or floodplains. Reduce anthropogenic flow peaks (urban runoff). Favor morphogenic flows. Shorten the length of impounded reaches. Connect flood reduction with ecological restoration. Manage aquatic vegetation.

Longitudinal connectivity or continuity improvement:

- Remove barrier (e.g., weir, dam). Install fish pass or bypass for upstream migration. Facilitate downstream migration. Modify culverts, siphons, and piped streams (e.g., daylighting). Manage sluice and weir operation for fish migration. Apply fish-friendly turbines and pumping stations.

Riverbed depth and width variation improvement:

- Re-meander or widen water courses. Make water courses less deep. Allow or increase lateral channel migration. Make water courses narrower. Create low-flow channels in oversized channels.

In-channel structure and substrate improvement:

- Initiate natural channel dynamics to promote natural regeneration. Remove sediments (e.g., eutrophic, polluted, fine). Modify aquatic vegetation maintenance. Introduce large wood. Add sediments (gravel, sand). Remove bank protection. Re-create gravel bars and riffles. Remove or modify in-channel hydraulic structures. Reduce the impact of dredging.

Riparian zone improvement:

- Adjust land use (e.g., buffer strips) to develop riparian vegetation or to reduce nutrient input, sediment input, or bank erosion. Revegetate riparian zones. Remove non-native substratum. Develop riparian forest.

Floodplains/off-channel/lateral connectivity habitats improvement:

- Make riverbanks or floodplains lower to enlarge inundation. Set back embankments, levees, or dikes. Improve or reconnect backwaters (oxbow lakes) and wetlands. Remove hard engineering structures that impede lateral connectivity. Restore or create wetlands. Retain floodwater (e.g., through local sluice management).

Figure 5 presents the main fluvial ecosystem services that are affected by hydromorphological pressures. Ecosystem services are the benefits that human populations obtain from ecosystems. They can be altered when pressures and water management affect fluvial systems [30]. Three main types have been considered: provisioning, regulating, and cultural ecosystem services. Provisioning services refer to products obtained from ecosystems. Regulating services refer to the benefits obtained from regulating ecosystem processes. Cultural services refer to the nonmaterial benefits that people obtain from ecosystems through spiritual enrichment, cognitive development, reflection, recreation, and esthetic experiences.

Figure 6 shows the effects of water abstraction and corresponding restoration measures on ecosystem services. It shows that water abstraction reduces provisioning services. Reduction of sediment transport affects the provision of mineral raw materials, declined fish habitats affect the provision of aquatic food, and reduced flows affect cooling systems and energy renewal. On the other hand, water abstraction enhances the provisioning service of terrestrial food production as it is mainly done for irrigation and as the remaining low flow rates in the river allow cultivation

Provisioning Service	
Biological raw materials	forestry products poplar plantations genetic resources natural medicines reed and willows used for thatching
Mineral raw materials	drinking water irrigation water construction gravel, construction sand clay for construction, bricks and pottery
terrestrial Food	agricultural dairy and fruit trees crops on terraces
aquatic food	commercial fisheries, Fish yield
Fluvial transport	Fluvial transport
Cooling system	Cooling system
renewal energy	hydropower,

Regulating Service	
Climate Regulation	Local Climatic Regulation Carbon sequestration in riparian woodland
Water regulation	Peak flows reduction flood energy dissipation Soil moisture and aquifer recharge
self-purification	Reduction of organic and inorganic pollutant load Riparian nutrient trap
soil formation	flood retention in floodplain (water, sediment, nutrients) flooding sedimentation
Channel maintenance	Reshaping and adjustments after disturbances
Biological recovery	Dispersion and recolonization mechanims by drift
Biological Control	invasive species control pest/disease control

Cultural Service	
Recreation & Ecoturism	trout and salmon* fly fishing, angling rafting, kayaking, yachting, sailing, sunbathing, swimming, hiking, waterfowl hunting, hunting,
Landscape & Aesthetic values	Scenic beauty of the landscape Nature art
Environmental education	wildlife and biodiversity
Scientific knoledge	
Spirtual & Religious values	

Figure 5.
List of main fluvial ecosystem services that are affected by hydromorphological pressures, classified according to provisioning, regulating, and cultural services.

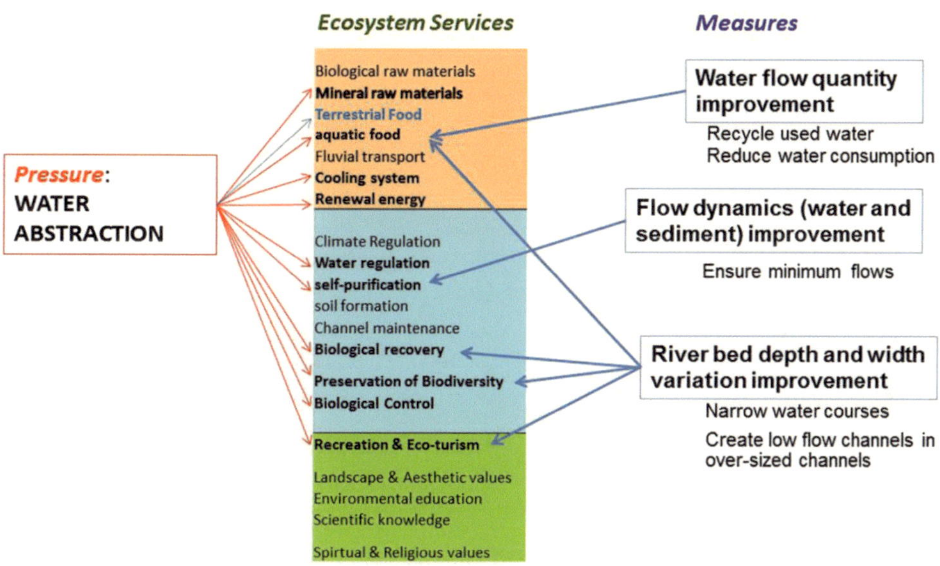

Figure 6.
Scheme of interactions among water abstraction and restoration measures and their effects on ecosystem services. Red arrows indicate inhibit services and blue ones improve them.

of the riverbanks. The figure shows that regulating services are reduced too: water regulation, self-purification, biological recovery, preservation of biodiversity, and biological control. Finally, among the cultural services, flow reduction affects mainly recreation and ecotourism.

Figure 6 shows three possible restoration measures for mitigating the effects of water abstraction:

1. Water flow quantity improvement, by recycling used water and reducing water consumption;

2. Flow dynamics improvement, by ensuring environmental flows;

3. Channel depth and width improvement (although the structural measure of concentrating reduced water flows in narrow water courses and low-flow channels is not sustainable from a geomorphological point of view).

None of these measures mitigates the degraded services of mineral raw materials, cooling, energy renewal, and water regulation. Restoration of these services would have to be the focus of further research and innovation.

4. Conclusion

This chapter has presented methods to assess hydromorphological quality and diagnose alteration in rivers, as well as systematic cause–effect relationships for diagnosis and for river restoration measures. They constitute guidance and tools for addressing hydromorphology in the implementation of the Water Framework Directive.

Acknowledgements

The work of this chapter was funded through the European Union's FP7 programme under Grant Agreement No. 282656 (REFORM).

Author details

Erik Mosselman[1]*, Massimo Rinaldi[2] and Diego García de Jalón[3]

1 Deltares and Delft University of Technology, Delft, The Netherlands

2 University of Florence, Florence, Italy

3 Technical University of Madrid, Madrid, Spain

*Address all correspondence to: erik.mosselman@deltares.nl

References

[1] Junk WJ, Bayley PB, Sparks RE. The flood pulse concept in river-floodplain systems. Canadian Special Publication of Fisheries and Aquatic Sciences. 1989;**106**:110-127

[2] Tockner K, Malard F, Ward JV. An extension of the flood pulse concept. Hydrological Processes. 2000;**14**(16-17):2861-2883

[3] Ward JV, Stanford JA. The serial discontinuity concept of lotic ecosystems. Dynamics of Lotic Ecosystems. 1983;**10**:29-42

[4] Ward JV, Stanford JA. The serial discontinuity concept: Extending the model to floodplain rivers. Regulated Rivers: Research & Management. 1995;**10**:159-168

[5] Fernández D, Barquin J, Raven PJ. A review of river habitat characterisation methods: Indices vs. characterisation protocols. Limnetica. 2011;**30**(2):217-234

[6] Belletti B, Rinaldi M, Gurnell AM, Buijse AD, Mosselman E. A review of assessment methods for river hydromorphology. Environmental Earth Sciences. 2015;**73**:2079-2100

[7] Brierley GJ, Fryirs KA. Geomorphology and River Management: Applications of the River Style Framework. UK, 398 p: Blackwell, Oxford; 2005

[8] Ollero OA, Ballarín FD, Díaz BE, Mora MD, Sánchez FM, Acín NV, et al. Un indice hydrogeomorfologico (IHG) para la evaluacion del estado ecologico de sistemas fluviales. Geographical. 2007;**52**:113-141

[9] Ollero A, Ibisate A, Gonzalo LE, Acín V, Ballarín D, Díaz E, et al. The IHG index for hydromorphological quality assessment of rivers and streams: Updated version. Limnetica. 2011;**30**:255-262

[10] Rinaldi M, Surian N, Comiti F, Bussettini M. A method for the assessment and analysis of the hydromorphological condition of Italian streams: The morphological quality index (MQI). Geomorphology. 2013;**180-181**:96-108

[11] Fryirs KA. Developing and using geomorphic condition assessments for river rehabilitation planning, implementation and monitoring. WIREs Water. 2015;**2**:649-667

[12] Gurnell AM, Belletti B, Bizzi S, Blamauer B, Braca G, Buijse AD, et al. A multi-scale hierarchical framework for developing understanding of river behaviour. Aquatic Sciences. 2016;**78**:1-16

[13] Frissell CA, Liss WJ, Warren CE, Hurley MD. A hierarchical framework for stream habitat classification: Viewing streams in a watershed context. Environmental Management. 1986;**10**:199-214

[14] Methods, models, tools to assess the hydromorphology of rivers - Part 1 Main report | REFORM [Internet]. Reformrivers.eu. 2015. Available from: https://www.reformrivers.eu/methods-models-tools-assess-hydromorphology-rivers-part-1-main-report [cited 2024 May 15]

[15] Rinaldi M, Surian N, Comiti F, Bussettini M. A methodological framework for hydromorphological assessment, analysis and monitoring (IDRAIM) aimed at promoting integrated river management. Geomorphology. 2015;**251**:122-136

[16] Rinaldi M, Gurnell AM, González Del Tánago M, Bussettini M, Hendriks D. Classification of river morphology and hydrology to support management and restoration. Aquatic Sciences. 2015;**78**:17-33

[17] Brierley GJ, Fryirs KA, Boulton A, Cullum C. Working with change: The importance of evolutionary perspectives in framing the trajectory of river adjustment. In: Brierley G, Fryirs KA, editors. River Futures: An Integrative Scientific Approach to River Repair. Washington, DC, USA: Society for Ecological Restoration International, Island Press; 2008. pp. 65-84

[18] Dufour S, Piégay H. From the myth of a lost paradise to targeted river restoration: Forget natural references and focus on human benefits. River Research and Applications. 2009;**25**:568-581

[19] Rinaldi M, Belletti B, Bussettini M, Comiti F, Golfieri B, Lastoria B, et al. New tools for the hydromorphological assessment and monitoring of European streams. Journal of Environmental Management. 2017;**202**:363-378

[20] ISPRA. Implementazione della Direttiva 2000/60/CE - Analisi e valutazione degli aspetti idromorfologici. Versione 1.1. Roma: Istituto Superiore per la Protezione e la Ricerca Ambientale; 2009. 85 p

[21] Indicators of Hydrologic Alteration in Rivers User's Manual -Version 1.0 [Internet]. 2008. Available from: https://moodle.upm.es/en-abierto/pluginfile.php/993/mod_label/intro/iahris-users-manual-1.0.pdf [cited 2024 May 15]

[22] Richter BD, Baumgartner JV, Powell J, Braun DP. A method for assessing hydrologic alteration within ecosystems. Conservation Biology. 1996;**10**:1163-1174

[23] Belletti B, Nardi L, Rinaldi M, Poppe M, Brabec K, Bussettini M, et al. Assessing restoration effects on river hydromorphology using the process-based morphological quality index in eight European river reaches. Environmental Management. 2018;**61**:69-84

[24] Rinaldi M, Baena RE, Nardi L, Guerrero-Amador IC, García-Martínez B. An assessment of the hydromorphological conditions of the middle and lower Guadalquivir River (southern Spain). Physical Geography. 2019;**41**:254-271. DOI: 10.1080/02723646.2019.1653668

[25] Pringle C. What is hydrologic connectivity and why is it ecologically important? Hydrological Processes. 2003;**17**:2685-2689

[26] Bunn SE, Arthington AH. Basic principles and ecological consequences of altered flow regimes for aquatic biodiversity. Environmental Management. 2002;**30**:492-507

[27] Ward J. Riverine landscapes: Biodiversity patterns, disturbance regimes, and aquatic conservation. Biological Conservation. 1998;**83**:269-278

[28] Review on effects of pressures on hydromorphological variables and ecologically relevant processes | REFORM [Internet]. Reformrivers.eu. 2013. Available from: https://www.reformrivers.eu/deliverables/d1-2 [cited 2024 May 15]

[29] Category:Measures - REFORM wiki [Internet]. wiki.reformrivers.eu. . Available from: https://wiki.reformrivers.eu/index.php/Category:Measures [cited 2024 April 15]

[30] MEA. Ecosystems and Human Well-Being: Synthesis. Millennium Ecosystem Assessment. Washington: Island Press; 2005. 155 p

Section 3

Ecology and Limnology

Chapter 8

Experimental Study of a Fish Behavioral Barrier Based on Bubble Curtains for a River Water Intake

Gabriela Circiumaru, Rares-Andrei Chihaia,
Lucia-Andreea El-Leathey and Andreea Voina

Abstract

The protection of fish habitat near water diversion or hydrotechnical facilities is of particular interest for researchers working in various fields. The chapter is focused on an experimental study of a fish behavioral barrier consisting of a bubble curtain operating along a river water intake scale model. It approaches elements of fish hydrodynamics, river water intakes, as well as physical and non-physical barriers for fish diversion from hydrotechnical facilities. The design, development, and testing of the fish behavioral barrier as well as the results of the experimental analysis are discussed. The proposed experimental setup was based on a barrier placed in the proximity of a river water intake scale model fitted and tested in a closed-circuit hydraulic stand. The intake chamber is provided with perforated orifices which communicate through an inlet with a lower tank for discharging the water into another tank located beneath. A certain water velocity within the hydraulic stand was set to assess the tandem operation of the bubble curtain and water intake. A sharp velocity gradient was found in most cases, indicating local velocity changes and creating the premises for impacting the fish behavior. Conclusions and future research development are also envisaged in the chapter.

Keywords: behavioral barrier, fish guidance systems, river water intake, bubble curtain, induced velocity

1. Introduction

As human activity development has a significant impact on biodiversity, there is an increasing need of environmental protection solutions and new legislative measures. According to World Wide Fund for Nature [1], due to the accelerated decline of fish populations and aquatic fauna, new and important measures have to be undertaken in order to reduce the anthropogenic effects caused by hydrological changes [2] as well as the environmental degradation. In this regard, national and international legislation in the field of aquatic habitat protection has been continuously updated, taking into account more and more of its requirements and providing rules for the design and elaboration of sustainable integrated watercourse management regulations.

These rules [3] provide the inclusion of alternatives for the fish protection in the passage, diversion, and catchment constructions. The fish protection involves not only their transit or removal from the construction but also the prevention of their injury or death as a result of the facility components' operation. The hydrotechnical constructions interrupt the rivers longitudinal and lateral connectivity, making it almost impossible for fish to migrate or to have access to primary resources that ensure their welfare.

Therefore, in order to improve or restore the rivers connectivity, proper fishways must be designed and deployed, as components of complex hydraulic infrastructures, as reported by Bunt et al. [4]. In addition, to ensure proper living conditions for fish, the water of the streams should have certain dissolved oxygen (DO) concentrations. The oxygen is present in surface water bodies due to the free surface aeration, achieved as a result of the turbulence. The DO is permanently influenced by a series of processes like diffusion, photosynthesis, and decomposition. The aquatic fauna as well as the decaying organic matter consume oxygen. Also, its concentration varies with water temperature, salinity, pressure, and season. Consequently, all these factors lead to a DO variation from 1 to 20 mg/L. Thus, due to its importance to the fish welfare, it should be monitored in freshwaters such as lakes and rivers.

According to European and national legislation [3, 5], the design of an ecological water intake must take into account the following fish habitat characteristics: the migratory species' presence on medium and long distances on the river sector of interest, the fish migration duration, the migration period, the species' swimming abilities, and their behavior to various external stimuli. Over time, both physical (structural/constructive) and nonphysical (behavioral guidance systems based on the natural fish attraction and repelling behavior) barriers have been developed and used. Some traditional physical systems (such as exclusion screens and grates, with or without screens) may be ineffective for certain smaller species, for juvenile specimens, or for species with special swimming performance. Hence, the need to design and develop appropriate systems for fish guidance in the proximity of river water intakes by also considering the local species characteristics is outlined. The fish guidance systems have the role either to discourage the migration of some species or to direct the fish away from certain hydropower facilities or river water intakes. Therefore, the design criteria of the protection systems have to consider their size, biomechanics, and age, as well as the understanding of the needs, psychology, and behavior of each species [6].

In order to improve the efficiency of fishways, fish ladders, or fish guidance systems, several research has been undertaken. Among the literature studies that address this topic is the study conducted by Bunt et al. [4], which analyzes and compares the results of 19 studies on different types of fish passages. The analysis included 26 species of anadromous and potamodromous fishes, and the results showed that the passages efficiency depends mainly on the fish biological characteristics and less on the type of fish ladder, slope, and so forth. Moreover, in order to increase the fishway efficiency, the need for rigorous studies related to each fish species and for different types of fish passages was emphasized.

Also, to improve the means of fish attraction or diversion to river infrastructure, Kerr and Kemp [7] performed laboratory studies on whether the hydraulic noise generated by water turbulence can mask certain stimuli such as a velocity gradient detecting. The fish response analysis showed mixed results; still, in some cases, it was concluded that the detection is lower when the turbulence is high. Some studies analyzed the importance and influence of water body hydrodynamics on fish behavior and their swimming efficiency. By numerical modeling and on-site data collection (bathymetry, acoustic tag fish location, measured data from an acoustic Doppler current profiler, ADCP), Silva et al. [8] studied the fish movement and the hydrodynamics effect in the proximity of a hydropower plant turbines. Hockley et al. [9] analyzed the modification of fish behavior with the water flow conditions (water velocity, turbulence), in correlation with their size and sex. For the analyzed species (guppies - *Poecilia reticulata*), swimming behavior and different preferences were identified in a heterogeneous flow from a free surface channel, by using boulders to change the flow conditions. Thus, it was detected that larger size fish prefer areas with high velocity and low turbulence, and those of smaller size spend more time in areas characterized by low velocity and high turbulence degree (behind the boulders). According to Mogdans [10], fish can detect velocity and pressure gradients by using specialized organs that can help in their orientation inside the water, in order to feed or escape from predators. The neurological response that these organs give to fish is used in guiding them away from various dangers or in attracting fish on particular routes.

The chapter aims to present an insight regarding several aspects that define a fish guidance system development and operation. Therefore, the chapter is structured in 6 sections, each of them approaching one of the key elements that contribute to the fish behavioral barriers characterization. The first section provides a broader context of the subject addressed by the research and underlines the purpose of the study. The second section approaches some aspects related to the study of water intakes with behavioral barriers. Thus, the physical and nonphysical fish guidance systems, fish hydrodynamics, as well as the main requirements related to the design of ecological river water intakes are discussed. For the experimental study of the behavioral barriers, a bubble curtain operating along with a river water intake scale model was designed, developed, and integrated in an experimental setup. The bubble curtains combine two stimuli (velocity and sound) to deter or guide fish. Hence, sound waves associated to the bubbles propagation and splitting as well as local velocity modification are generated by the curtain in the surrounding water. In addition, the bubbles create a visual barrier for the aquatic fauna. Besides these aspects approached in the third section, the next one presents the developed experimental setup and the methodology applied for the measurement of the velocities induced in the water by the bubbles at different airflow rates. The investigation of the velocities is necessary for evaluating the bubble curtain potential in guiding or deterring fish. In order to assess the water aeration associated to the bubble curtain generation, the experimental setup was provided with the necessary equipment for the dissolved oxygen measurement. The following section summarizes the results and their analysis, as well as the main findings of the study. Since river water intakes can be placed also in remote areas, with no connection to the power grid, the fifth section approaches the analysis of power autonomy solutions for the behavioral barrier based on bubble curtains.

2. General aspects regarding the study of water intakes with behavioral barriers

In order to protect the fish population, various solutions of fish guidance systems have been developed. Thus, several types of physical and nonphysical barriers have been proposed and studied, as reported by Turnpenny [11]. The physical barriers represent direct methods for fish exclusion, while nonphysical barriers stand as indirect solutions, designed to deter fish outside the proximity of water intakes area. The physical barriers consist of specially designed screens or sieves, which can be either fixed or mobile. Bubble curtains, pressure waves, strobe light, sound, electricity, chemical stimuli, magnetic, and velocity fields are nonphysical means of influencing the natural behavior of fish.

Given the fact that the chapter deals with the development and testing of a bubble curtain, the information provided below mainly refers to various research on behavioral barriers.

2.1 Assessment and analysis of behavioral barriers

Behavioral fish guidance barriers represent alternative solutions in comparison to conventional mechanical ones. The main advantages of behavioral systems consist in the fact that these are easy to use, safe, and also need just a low-cost maintenance. They can be applied in locations where grills are difficult to install, being at the same time eco-friendly, not impacting the aquatic fauna by injuring or killing fish. However, behavioral barriers have an important disadvantage considering that they do not create an absolute exclusion barrier and that the exclusion efficiency varies with the species, fish and stage development, environmental conditions (flow and variability), water quality, or lighting. They are also not generally accepted as fish exclusion methods by agencies/fisheries and so on.

In the following paragraphs, the analysis of behavioral barriers is presented [12].

- *Electrical barriers* consist of a series of metal anodes and cathodes placed in the water body. Thus, an electric field is created in the barrier proximity, causing fish to avoid the area [13]. The main limitations of electric barriers are represented by deactivation in the event of power outages and low efficiency for small fish, which are not affected by electric fields.

- It is to be mentioned that when using *strobe lights barriers*, different light levels influence fish behavior. Thus, fish orientation and other survival behavior in the aquatic environment can be affected. The efficiency of these deterrent systems is influenced by the life history of the target species, the lights brightness and design, the turbidity, and the ambient light level [14, 15].

- *Acoustic fish guidance barriers* are suitable when *in situ* conditions do not allow the use of visual stimuli, being based on sound and pressure waves, which can be detected by the inner ear and fish lateral line system [16]. Fish have been broadly classified [17] as either "hearing generalists" or "hearing specialists". While the generalist fish detect sounds with a frequency below 1kHz, specialists can detect frequencies up to several kHz. The carp lateral lines [18] are similar to most of the Romania native species, for example, trout. The anatomical characteristics of the carp (Weberian ossicles) allow the swim bladder to act as an acoustic pressure transducer in addition to the inner ear, thereby increasing sensitivity to environmental sound levels. The best sound detection of the carp corresponds to a frequency in the range of 100–500 Hz and 60–90 dB sound pressure level. One of the main disadvantages of acoustic barriers is that low-frequency sound waves propagate better in deep water and on softer substrates. Thus, they have limited efficiency in the case of shallow water bodies. Usually, acoustic systems are used to repel fish from hydropower facilities area. For example, high-frequency sounds in the range of 122–128 kHz demonstrated an efficiency of 87% in repelling a species of herring (*Alosa pseudoharengus*) from the proximity of a water intake placed in Lake Ontario [19].

- *Velocity barriers* are based on a local velocity increase above the target species' swimming speed. This is usually achieved by water flow limitation through a channel, gutter, or culvert, thereby increasing the velocity. When the fish movement in running waters is achieved with difficulty (e.g., upstream, against a current), the local velocity increase can represent a barrier [20]. An experimental barrier used on a Great Lakes tributary that functioned as a spawning ground for invasive sea barbells showed 33% efficiency [21].

- *Chemical barriers* are based on the use of toxic substances in order to create a hypoxia/hypercapnia effect. They necessitate carrying out laboratory experiments in order to establish the tolerance thresholds for each species [22]. Dissolved gas barriers are quasi-efficient as long as a certain concentration is obtained (e.g., approximately 1.5 mg/L DO [12, 23]). Either purified nitrogen or carbon dioxide was injected into the water to determine the feasibility of a gas bubble curtain [24]. Another example of chemical barriers consists in pheromones use, as they are known for their potential on impacting same species individuals [25]. Thus, pheromones that are directly introduced in the water column can deter or attract fish from a certain area. Additional research on the pheromones efficiency and use against invasive species is needed before they can be considered a permanent nonphysical barrier. In any case, alarm pheromones have a more significant potential as a barrier during temporary shutdown of permanent barrier systems. Many widely applicable pesticides and biocides as well as other toxins specific to various species that can be used as deterrents against aquatic organisms are also available [26]. The systems based on fish toxins can generally qualify as nonphysical barriers if their application is limited to a relatively reduced area, necessary for fish passage. In order to be considered a barrier to fish movement, chemical substances should generally be nonpersistent.

- *Magnetic barriers* are based on fish electroreceptive organs (present in the case of some fish—like sturgeons, *Scaphirhynchus platorynchus*, [27]) that can identify the water electromagnetic fields. Thus, fish can be deterred by the use of magnetic fields in specific cases like diminishing their accidental entrapment in commercial fishing activities.

- *Bubble curtain barriers* generate distinct acoustical and hydrodynamical fields inside the water bodies. By their appropriate use, they can be applied in order to guide juvenile fish. The bubble curtain is generated by air diffusers placed along the water bottom, perpendicular to the channel/river. A continuous "screen" of bubbles in the water column provides an unnatural visual obstacle that fish should avoid. Bubble curtains are limited due to the fact that they do not emit own light and cannot be easily observed from a distance. Light penetration, especially due to turbidity, is a major obstacle to their efficiency. Compared to other alternative guidance technologies, the bubble curtain barriers represent a more economically efficient solution that does not impact the river morphology.

A summary of the barrier types and characteristics is provided in **Table 1**.

Bubble barriers for fish guidance have been approached in several laboratory and field studies. Most of them focus on a wide range of species: Asian carp [28], Atlantic salmon, European flounder, mackerel, and herring [29]. The use of a bubble curtain along with a sound source was studied in [28, 30]. The bubble curtain was generated by injecting airflows with 0.1–1 L/s·m rates through a perforated polyvinyl chloride (PVC) pipe. This combined bubble and sound system is developed by Fish Guidance Systems Ltd. in UK. The experimental results shown in [28] indicate a decrease in Asian carp migration by 95%, while in [30], a smolt-guiding efficiency (young catfish (*Silurus glanis*)) of 20–40% per day is reported, respectively larger than 70% during nighttime. These studies did not consider the bubble curtain as the main guidance solution and therefore did not analyze in detail the generated physical fields. In [31],

Barrier type	**Implementation conditions**	**Description**	**Advantages**	**Disadvantages**
Electric	Site with suitable power supply; suitable water conductivity	Electric current produced by electrodes located on the channel perimeter	It has proven highly effective against upstream migrating fish	Expensive equipment and maintenance; risk to human and animal health; it is not specific to a species
Visual-strobe lights	Low water turbidity	Strobe light with set water flashing frequency	Relatively small influence on water flow; adaptable to flow; in some species, very effective in combination with bubble curtains	It depends very much on the natural light level; cloudy water reduces its efficiency, and this is not proven in all species
Acoustics	Site with suitable acoustic characteristics	Underwater speaker with specified signal	Flexible to different flow conditions; species specific potential; wide range of sounds	Variable efficiency; the frequencies must be selected according to species; expensive equipment that can clog
Bubble curtains	Low water turbidity, relatively shallow water	Compressed air emitted as bubbles through diffusers	Flexible in different flow conditions; multiple physical stimuli; relatively simple construction; reduced cost. The efficiency increases in combination with other types of barriers	The bubbles can be "washed" at high flow rates; it cannot work in all conditions

Barrier type	Implementation conditions	Description	Advantages	Disadvantages
Water velocity	Target species (a weak swimmer); narrow channel, adequate water flow	Based on local change in water flow velocity	Selectively excludes disturbing species	Major channel change; few sites meet the criteria
Chemical-Hypoxia and hypercapnia	Relatively shallow water, space needed for bulk gas storage	Based on chemicals that decrease or increase the water oxygen content	It is possible to exclude all fish	Large capital investments and research period
Chemical-Chlorine	Very limited implementation area	Based on the chlorine entry in water	May selectively exclude all fish	Dangerous to almost all aquatic wildlife; negative public perception
Pheromones	Small areas and/or short-term applications	Based on the entry of pheromones in water	May selectively exclude certain fish	Time and effort to purchase pheromones in bulk quantity
Electromagnetism	Narrow areas, narrow spots	Based on the use of electromagnetism	Cost-effective, low environmental impact	Might not work on all teleost fish

Table 1.
Types of behavioral barriers and their characteristics [12].

the efficiency of a bubble curtain and of an electric barrier, respectively, was studied separately. A reduced attraction of the eurasion ruffe (*Gymnocephalus cernuus*) was observed instead of its repellence when a bubble diffuser with orifices between 0.4 and 1 mm, spaced equidistantly at 6.25 or 12.5 mm, was used. Following the assessment of a bubble curtain in Great Lakes USA, according to [32], it is reported that it is efficient only when used with other guidance systems such as sound or light. One other study that was conducted at a hydroelectric plant site in Michigan, USA [33], demonstrated that choosing a mixed method (by combining bubble curtains with strobe lights) can increase the fish exclusion rate and thus the behavioral barrier efficiency. Therefore, other associations between exclusion or guidance systems based on fish behavior are expected to improve their overall efficiency.

Like other types of behavioral barriers, bubble curtains are most effective when used as a component of an integrated deterrent system. For this purpose, local conditions such as turbidity or river depth must be taken into account. One major disadvantage of bubble curtains regards their ineffectiveness to maintain equal air pressure at different water depths. As mentioned before, studies indicate higher deterrence rates when bubble barriers are associated with an additional source of illumination such as strobe lights [29] or sound barriers [30]. According to [34], an experimental barrier consisting of a bubble curtain and acoustic deterrents is reported to be 95% effective in limiting Asian carp movement.

Turnpenny et al. [35] and Jones et al. [36] also demonstrated that combined behavioral barriers are more efficient than independent ones. Thus, the use of the stroboscopic light or sound systems along with bubble curtains can enhance the effect on fish as well as the barrier efficiency. Frizell and Arndt [37] reported that fish behavior is influenced by ambient light.

Since the bubble curtain can operate as a behavioral barrier to some fish movement, its associated velocity field should be determined and analyzed in order to assess its potential effectiveness for some particular fish species or for particular fish size. To create a velocity barrier, the flow field should be modified as to obtain water velocities superior to the envisaged species swimming capacity [24].

In the design and construction of fish guidance systems, whether attracting or deterring fish, consideration has to be given both to the water body characteristics (flow, depth, seasonal changes, etc.) and to the fish species that populate that water and their swimming characteristics (both anaerobic and aerobic swimming capabilities).

As shown in the following paragraphs, many studies analyze the fish swimming ability and their changing behavior when they encounter barriers (physical or nonphysical). Velocity barriers can be used to attract or deter fish from hydrotechnical facilities in order to avoid their catchment. When the local velocities created by the barriers' presence on rivers are very high, they can negatively influence the fish movement, affecting the species continuity along the river. Regardless of the species, according to the study conducted by Castro-Santos and Haro [38], the performance of a fish passage "is the product of locomotor behavior generally, including guidance, attraction, and ascent or descent through the fishway" and is not only given by the swimming capacity. Also, the motivation of the fish is important, and the necessity to integrate the "knowledge on the interaction between physical environment, physiology and behavior" is highlighted.

In Sanz-Ronda et al. [39], the efficiency of a velocity barrier consisting of a Flat-V Gauging Weir, characterized by the velocity field variation along the barrier for the Iberian barbel (*Luciobarbus bocagei*), is analyzed. Given the need to ensure flow continuity and to control the access or removal of fish to and from various hydrotechnical constructions, Kapitze [40] studied barriers and fishway types and characteristics for road crossings. Both the hydraulic characteristics associated to different barriers types (total, partial, or temporal barriers) and the fish swimming ability and speed have been analyzed. Ensuring the fish population continuity has also been intensively studied. Sanchez-Gonzalez et al. [41] analyzed the fish passage through velocity barriers, in particular the northern straight-mouth nase (*Pseudochondrostoma duriense*), which populates the Portugal and Spain rivers. Experiments conducted in a free surface channel showed that swimming capacity is mainly influenced by the fish size, the larger individuals swimming better; at high velocity, the fish shape has a significant influence. Regarding flow velocity preference, Liang et al. [42] analyzed *Schizothorax oconnori* Lloyd behavior, a local endemic species in China, by performing experiments in four similar channels, with different velocities up to 0.75 m/s. The velocity preference varied by season and between day and night, indicating that the flow field should not be uniform and various depths and light levels should be provided in the water body.

Besides the usual water velocity measuring methods (like the use of Pitot-Prandtl tube), another reliable method suitable to single- or two-phase flows is Particle Image Velocimetry (PIV). Regarding two-phase flows, PIV can be used for various purposes, in both static and dynamic regime of the liquid phase, to determine the mixing and homogenization, the mass transfer, to characterize the bubble size, their ascending velocity and flow regime, and so on. Thus, in Kovats et al. [43], the method was applied in a bubble column to determine the continuous phase (water) velocities, in order to better characterize the flow and to understand and improve the mass transfer mechanism. In Murgan et al. [44], PIV and LIF were applied to determine the continuous phase velocity field induced by the operation of a sparger in a rectangular column with water at rest. Laakkonen et al. [45] used PIV to determine bubble size distribution and gas holdup in the flowing water inside a stirred vessel.

As regarding the use of PIV technique in free surface flows, literature shows there are different particular cases where this measurement method can provide valuable results. Yao et al. [46] applied PIV to determine the turbulence created by a grid placed inside a free surface channel. Seol et al. [47] used towed underwater PIV to determine the water vertical and horizontal velocity fields behind a floating body, in order to characterize the turbulent wake and the influence that free surface has on it. To demonstrate that image velocimetry can be successfully used to characterize relatively large free surface flows (like rivers), Muste et al. [48] performed laboratory measurements on a 3.4 m wide and approximately 23.5 m long free surface channel combining PIV with controlled surface wave image velocimetry in order to adapt the management of sediments at a river water intake. In this scope, velocity fields, streamlines, and vorticity fields were determined. Lindmark [49] used Laser Doppler Velocimetry and PIV to determine the velocity and the flow field in a channel to study both attraction and guidance methods to direct fish upstream and to guide them away from a turbine inlet.

2.2 Water intakes and ecological design requirements

As the conditions impacting the fish swimming ability are variable and complex, the design criteria of water diversion or intake facilities must be expressed in general terms, while their characteristics must respond to the local conditions. Water intakes represent all the constructions and installations (works) that are used to divert and capture water for various aims such as thermoelectric, hydroelectric, and micro hydroelectric plants producing "green" electric power, pumping stations for irrigation systems, water supply for industrial or domestic end users, navigation facilities, fish farms, tourism, and so forth. In the following paragraph, only river intakes will be analyzed.

During their operation, river intake facilities must be protected by the accidental blocking by solid bodies carried along the water. Any solid body caught in the water intake can cause its obstruction or even the temporary removal of service. One constructive recommendation regards the placement of the intake near the river bend. The most suitable location is toward the exterior, downstream of the maximum curvature point of about (0.66–0.9)·R, where R represents the radius of the meander curvature. Thus, mainly the surface currents will be captured [50].

Hydrotechnical facilities for river water diverting or capturing have proven to have a high impact on fish habitat over time [51]. Therefore, it is important to optimally design systems in order to retain and/or guide fish to/from water intakes, taking into account the size and behavior of local species [33].

According to European and national legislation, the design of an ecological water intake must take into account the following characteristics of the fish habitat:

- The presence of migratory species over medium/long distances on the river sector of interest,
- Duration of fish migration,
- The migration period,

- The swimming capacities of each analyzed species,
- Fish behavior to various external stimuli.

In order to improve the efficiency of the fish guidance system in real operating conditions, the authors elaborated a patent application [52] that proposes, in addition to the bubble curtain, the use of an optoelectronic device to generate strobe light. These two systems create a behavioral barrier that prevents the accidental entry of small fish or juveniles through the intake holes.

2.3 Aspects related to fish hydrodynamics

Considering that the chapter is dedicated to the analysis of a fish guiding solution, in this section, a few aspects regarding the fish hydrodynamics are provided.

The fish swimming speed varies according to several physiological, biological, and aquatic environmental factors such as species, musculature system, age, stage of physiological development, season, required swimming time, behavioral aspects, flow speed, or water stream depth.

Fish swimming speed is usually defined in fish length and can be classified as:

- *Cruising speed* (or *sustained speed*) is the fish swimming speed used for long periods of time (2–3 hours) without them getting tired. This speed is characteristic to migration or to maintaining the position in water. It also depends on environmental conditions and biological constraints.
- *Prolonged swimming speed* is the maximum speed that can be maintained for medium periods of time (20–200 s), at the end of which the fish gets tired but recovers by resting [53]. However, since this speed does not lead to fish stress, it is used in the design of fish protection structures. One special category of prolonged speed is represented by the so-called *prolonged critical speed* that fish can maintain for a specified time period (usually 1 hour) but at the end of which fatigue occurs. According to [54], it was identified that the critical speed for a water temperature of 2–10°C is 1.35–1.55 m/s for trout and 1.3–1.4 m/s for grayling. Therefore, an increasing trend of speed along with temperature is observed.
- *Burst speed* is the highest speed fish can reach for short periods of time (<20 s), after which it returns to a much lower speed for rest or recovery. This speed can be maintained for a short period and is only used in panic situations, as the fish becomes tired and stressed due to the additional energy use. Fish often use this speed in order to burst through fast streams areas, for example, when entering or exiting a water intake. After 20 s of swimming, fish can no longer maintain the swimming direction. The maximum burst speed is usually the initial one. The trout burst speed is about 9 lengths/s (in the range of 2–3 m/s), for 10 to 19°C water temperature.

In the case of fish ladders, when fish have to move at high speeds through fast streams areas, it is important that they recover in order to repeat these short periods of intense effort. The ability of fish to swim at burst speed is limited to a few hours per day, as some species require relatively a long time to recover. As a result of such efforts, some fish may die. The mortality rate of a trout that maintains this speed for 6 minutes is 40% at about 4–8 hours later.

Migratory fish species are attracted by specific water flow rates. These currents are called attraction or guiding flows and are considered criteria in the design of fish passage constructions (fish ladders). Design guidelines recommend specific guiding flow velocities at the fish ladder entrance of 1–2 m/s for trout, 1–1.7 m/s for grayling, and 1–1.5 m/s for barbel and barberry (*Barbus petenyi*), respectively. At the same time, it must be taken into account that fish need enough water depth when they "jump" within waterfalls, thresholds, or fords. For example, trout can jump up to 1 m. The ecological intake design and development for the experimental setup dedicated to the characterization of the bubble curtain envisaged in this chapter was carried out based on the trout behavior and characteristics. It is worth mentioning that the trout is the most well-known and common fish in the hill and river areas from Romania.

This chapter presents details on the experimental setup developed for the measurement of the velocities induced in water by the bubble curtain, as well as for the assessment of the respective dissolved oxygen variation. The methodology followed in each of the two test campaigns is also provided.

3. Experimental testing of a bubble curtain barrier for ecological water intakes

3.1 Experimental setup and methodology for induced velocities measurement

The aim of the experimental testing consists in characterizing the operation of a bubble curtain barrier by analyzing its associated induced velocities. The behavioral barrier was placed in the proximity of a scale model of river water intake. In this regard, an experimental setup has been designed and developed. The setup includes two main components: the water intake model and the bubble curtain behavioral barrier (**Figure 1**). The entire design has been developed in order to be fitted in a closed-circuit hydraulic stand.

As described in detail in [12], the first component is represented by an intake chamber with perforated orifices (**Figures 2** and **3**) communicating through an inlet with a lower tank dedicated to discharging the water from the main channel of the test stand; the lower tank is located beneath the intake chamber, with water storage purpose. It allows the gravitational flow of the water drawn through an upper hole connected to the intake chamber. This water is recirculated by a pump and returned to the main channel. The captured water flow rate is determined by using a flow meter [12]. **Figure 1** shows the design of the experimental setup.

The water enters the intake model through 440 orifices with 4 mm diameter that are perforated on a 55 × 400 mm plexiglass plate as shown in **Figure 2**. Only 80 orifices have been used during the testing in order to prevent the flooding of the tank placed below (**Figure 4**) and the suction operating conditions. This way, the gravitational flow could be maintained.

To obtain the bubble curtain, a porous hose designed for aerating fishponds was placed at the bottom of the intake chamber. The bubble generator is supplied with compressed air through a dedicated circuit, provided with a precision valve for regulating the flow rate and a flow meter for measuring its value.

The water intake scale model and the behavioral barrier have been fitted in the closed-circuit hydraulic stand. The two components were placed in the 375 × 300 × 1015 mm

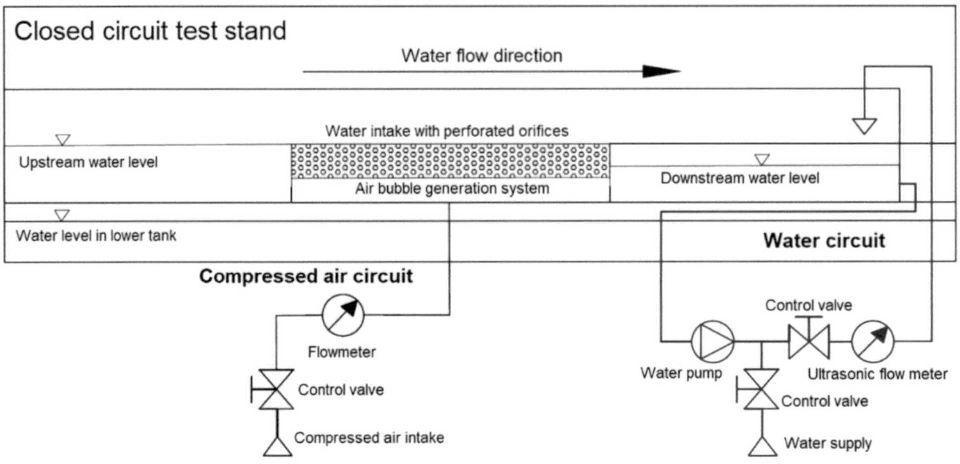

Figure 1.
The design of the water intake and behavioral barrier experimental setup [12].

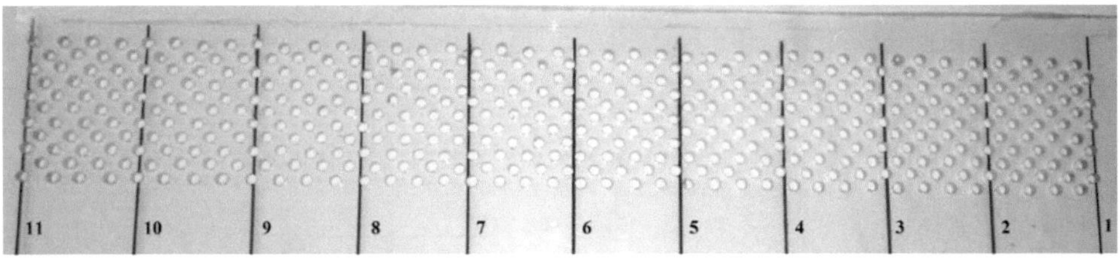

Figure 2.
Perforated plate of the water intake model [12].

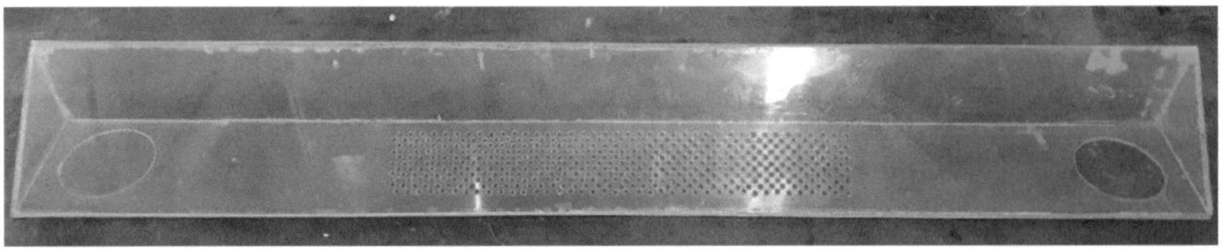

Figure 3.
The water intake chamber [55].

transparent section of the stand. Tap water was used for performing the tests. The testing stand is provided with recirculation pumps and variable speed motors, which ensure the variation of the water flow rate and of the velocities between 0.05 and 1 m/s (**Figure 5**) [12].

The stand is also provided with an automatic system dedicated to flow and velocity parameters control and acquisition. A Pitot-Prandtl tube has been used to measure the water velocity, as indicated in **Figure 6**.

In order to perform a better characterization of the flow, a dedicated device was used, thus ensuring the velocity measurement in certain positions inside the water as indicated in **Figure 6**. The vertical and horizontal coordinates of the velocity measuring points are detailed in **Figure 7**.

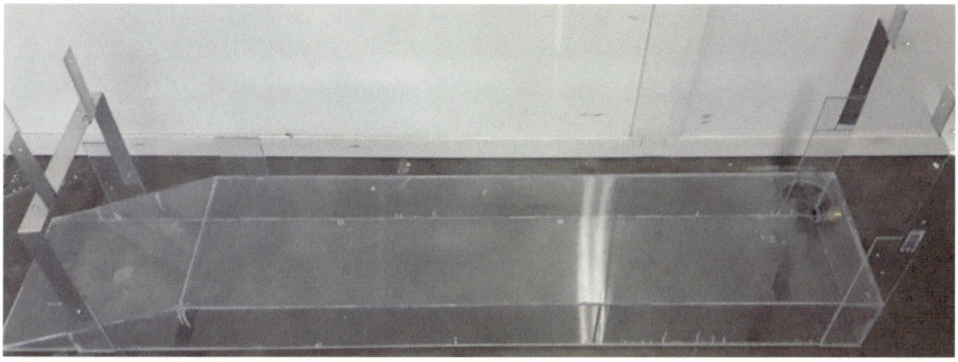

Figure 4.
The lower reservoir, with the role of collecting the water drained through the intake orifices [55].

Figure 5.
The experimental facility of the closed-circuit hydraulic stand [12].

The following paragraph describes the methodology applied for carrying out the experimental testing of the bubble curtain barrier for the ecological water intake scale model. A certain water velocity inside the experimental stand channel was set by varying the pump group rotational speed. To assess the tandem operation of the bubble curtain and water intake, the water levels have to be determined both upstream and downstream the intake chamber upstream, as well as in the drainage area and lower reservoir. A Qalcosonic W1 Axioma smart ultrasonic flow meter was used for measuring the water flow rate (**Figure 8**) and a Cole Parmer flow meter for determining the air flow rate, respectively. The Cole Parmer flow meter scale ranges from 0 to 20 LPM, with a ± 5% accuracy. The differential pressure transducer model that was used for the water velocity measurement is AppliSens APRE-2000, having an accuracy of ±0.1% for the respective calibrated range of −5 to 70 mbar [12]. In order to perform the envisaged measurements, two water depths were considered: 92 and 119 mm, respectively. A deeper water allows more measurement point in the vertical direction. Thus, 36 points were selected for a water depth of 92 mm and 42 points

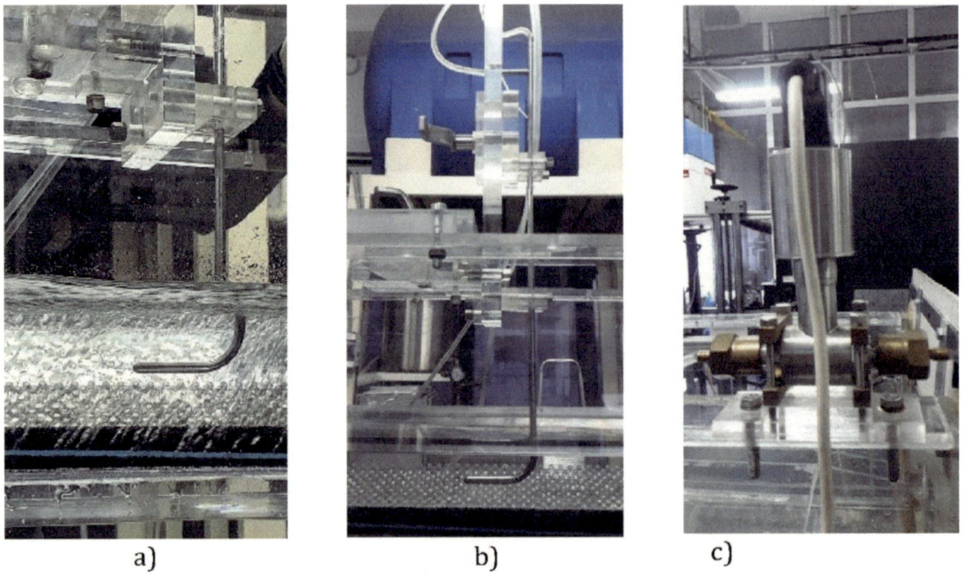

Figure 6.
Pitot-Prandtl tube placed in water parallel to the flow direction for measuring the water velocity with (a) and without (b) air bubbles injection and the differential pressure transducer (c).

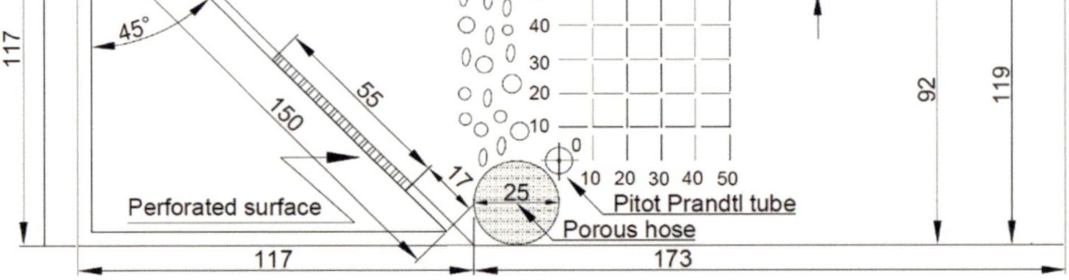

Figure 7.
Velocity measuring points for two different water depths in the channel [12].

for the 119 mm water depth. The positions of the measuring points for both cases are indicated in **Figure 7**.

Figure 9 shows two sides of the experimental setup. These images have been selected for providing a better view of the gravitational flow from the water intake chamber to the lower reservoir.

Table 2 shows the conditions in which the behavior barrier operation was investigated. The registered induced velocities were compared with no air injection case (0 LPM). The two values of the water velocity inside the transparent channel of the

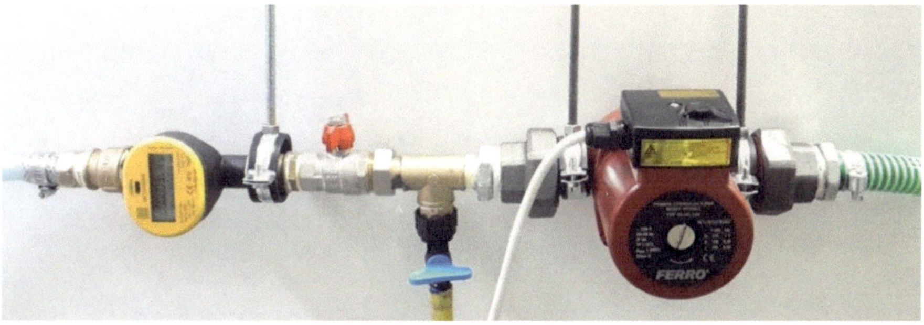

Figure 8.
Components of the water circuit, which ensure the circulation of the flow captured through the water intake model.

experimental setup were either 0.33 or 0.535 m/s having in mind the most common water velocities of the Romanian mountain rivers.

3.2 Experimental setup and methodology for the dissolved oxygen assessment

To evaluate the mass transfer (oxygen transfer) induced by the behavioral barrier, measurements were made on the experimental model presented in chapter 3.1, to which an oximeter was added (**Figure 10**). This measuring device is provided with a membrane sensor. It is worth mentioning that its measuring accuracy does not depend on the recirculated water flow rate. Thus, in order to analyze the mass transfer associated to the bubble curtain, the dissolved oxygen (DO) concentration in the water was measured.

Although the main purpose of the behavioral barrier is to guide fish to the downstream area so that they are not captured by the water intake, the dispersed air injection into the water can lead to an increase in the DO content. Thus, the use of a bubble curtain can improve the living conditions of the fish.

The analysis of the mass transfer induced by the behavioral barrier aimed to determine if the bubble curtain aerates the water and to what extent, as it is known that the water aeration can be achieved by both submerged and surface transfer.

Thus, the DO concentration measurement was carried out with the WTW Oxi 315i oximeter in nonpermanent regime, by determining the dissolved oxygen value at different periods. The oximeter accuracy is $\leq 0.5\%$ of the measured DO value and ≤ 0.1 K of the measured temperature value. This measuring device is a portable instrument that has embedded a microprocessor and a rapid pressure and temperature correction system, while the DO concentration measurement range is between 0.00 and 90.00 mg/L. The oximeter calibration was performed in water vapor saturated air, using an OxiCal-SL vessel for calibration in air.

The oxygen sensor was placed downstream the ecological water intake (as shown in **Figure 10**). The mass transfer generated by the bubble curtain operation along with the water intake scale model was evaluated by measuring the DO concentration variation with time and water temperature. The mass transfer assessment was carried out in two experimental cases, namely, the water intake operation with and without air injection through the porous hose in order to generate the bubble curtain. Thus, to determine the effect of dispersed air introduction into the water, it was necessary to measure the initial DO concentration in the case of water intake single operation. After this initial phase, the measurements were repeated for the case of water intake

a)

b)

Figure 9.
Experimental setup used for testing the bubble curtain barrier for ecological water intakes from both sides (a) and (b) [12].

and bubble curtain tandem operation. The measuring position of the oximeter was kept and the DO values were recorded for two airflow rates injected through the porous hose: 10.5 and 15 LPM, respectively.

4. Results and discussion

Given that a bubble curtain generates changes in the hydrodynamic field, the measurements focused on determining the velocities induced by the bubbles at different airflow rates injected through the porous hose. The field of induced velocities is characterized by the fact that in the far field, the maximum velocity is in the horizontal plane (along the flow), while in the near field, the maximum velocity is in the vertical plane (perpendicular to the flow). Thus, a sharp velocity gradient is created.

$v_{channel}$ [m/s]	H [mm]	Q [LPM]	Figure
0.33	92	0	11
		10.5	12
		15	17
0.535	92	0	13
		10.5	14
		15	18
	119	0	15
		10.5	16
		15	19

Table 2.
Water depths, velocities, and airflow rates used for testing the scale model [12].

Figure 10.
Measurement of the DO concentration in the water circulated through the experimental setup.

The horizontal velocity variation is very low as bubbles rise vertically to the free surface. Since the scope of the behavioral barrier is to repel fish from the water intake area, the vertical movement of the air bubbles was envisaged. Thus, in order to evaluate the impact of the bubble curtain on the channel water flow, only the vertical velocity variation was studied. Even though the bubbles are injected in flowing water, they still rise vertically to the free surface, while the main water flow influence is insignificant.

To analyze the bubble curtain impact on the water velocity inside the channel, the velocity variation curves were plotted for each studied case (**Figures 11–19**, [12]), as shown in **Table 1**. The velocity induced in water by the bubble column was denoted by v, the initial velocity in the channel by $v_{channel}$, the air flow rate by Q, and the vertical and horizontal distance from the origin of the chosen coordinate system to the velocity measuring point by D_v and D_h, respectively [12]. During the experimental testing, the water flow rate of the intake scale model was of 2.04 m^3/h for 92 mm water level and of 2.15 m^3/h for a level of 119 mm.

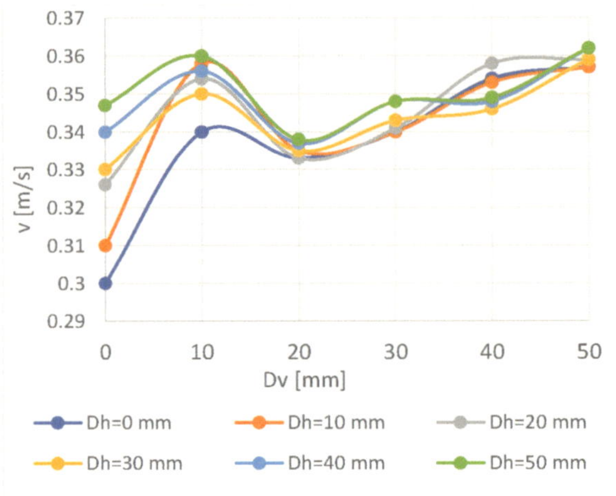

Figure 11.
Induced velocity variation for $v_{channel}$ = 0.33 m/s, Q = 0 LPM, H = 92 mm.

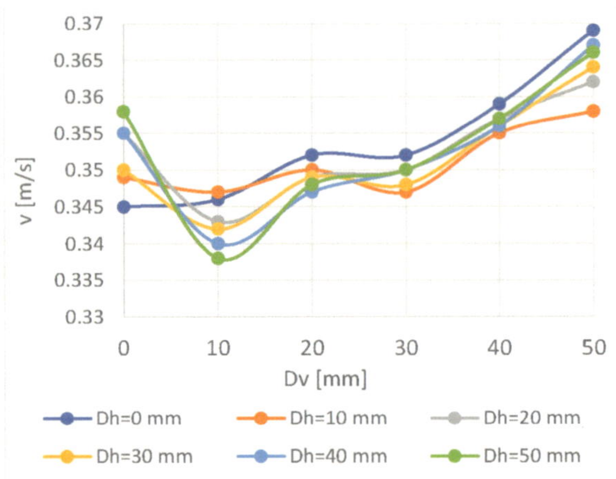

Figure 12.
Induced velocity variation for $v_{channel}$ = 0.33 m/s, Q = 10.5 LPM, H = 92 mm.

Figures 11, 12, and **17** show that for 92 mm water depth, the difference between the curves showing the velocity variation on each vertical is smaller for 0.33 m/s water velocity than for 0.535 m/s. As we approach the free surface, an increasing difference between the curves can be noticed. Moreover, the velocity gradients for the 0, 10, 20, and 30 mm verticals are positively influenced by the air flow rate increase.

For the same water depth of 92 mm but a higher water velocity inside the main channel, namely, 0.535 m/s, **Figures 13, 14**, and **18** show that a higher turbulence degree is developed, mainly in the free surface area. When the velocity measuring points were near the free surface (17 mm, like the case of 40 and 50 mm measuring verticals), the velocities measured by the Pitot-Prandtl tube were lower. For a water depth of 119 mm, the closest measuring point to the free surface is the one located in vertical direction at a distance of 70 mm from the porous hose [12], as shown in **Figure 7**. In this case, the tube measures at 24 mm below the free surface, and thus, the turbulence influence is diminished.

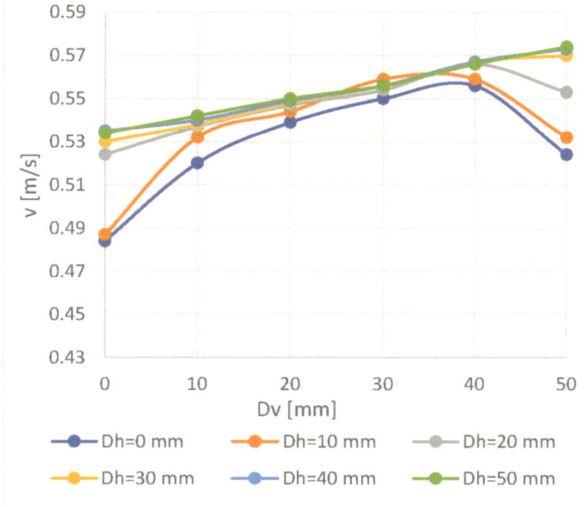

Figure 13.
Induced velocity variation for $v_{channel}$ = 0.535 m/s, Q = 0 LPM, H = 92 mm.

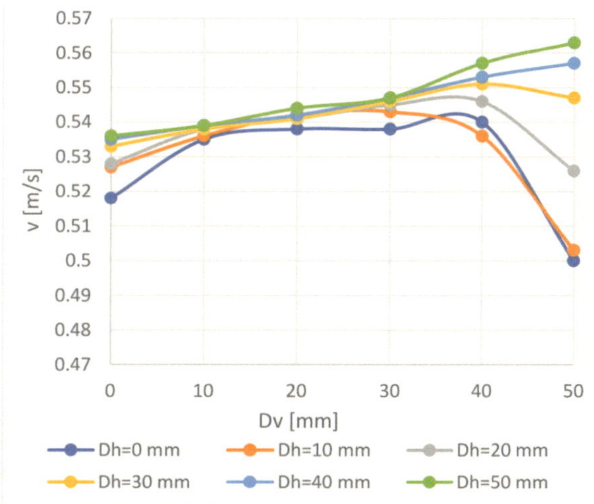

Figure 14.
Induced velocity variation for $v_{channel}$ = 0.535 m/s, Q = 10.5 LPM, H = 92 mm.

Figure 15 shows the water velocity curves variation for v_{channel} = 0.535 m/s, H = 119 mm water depth, and no air injection. As a result of the disturbed water flow in the catchment area of the intake, a rapid increase in the velocity is noticed starting with the 10 to 30 mm verticals. On the other hand, **Figures 16**, **17**, and **19** indicate significant velocity oscillations when the air is dispersed through the porous hose with 10.5 and 15 LPM flow rates. The most substantial oscillations were registered in the vertical measuring sections located in the bubble curtain proximity: 0, 10, and 20 mm, respectively. Starting with the vertical located at 30 mm, the variation becomes uniform. This is due to the increasing distance from the bubble curtain, which leads to a diminished influence on the flow.

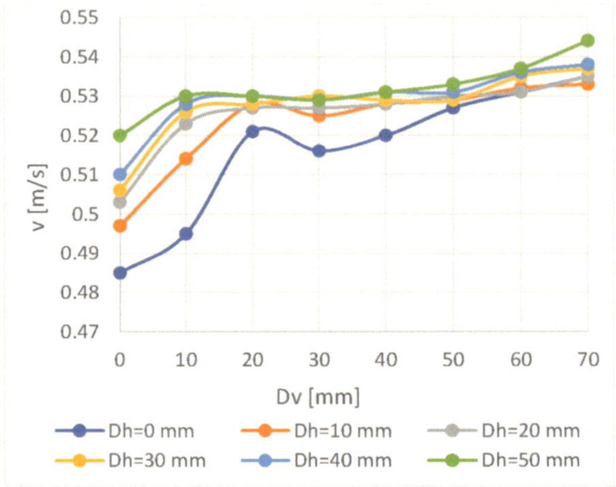

Figure 15.
Induced velocity variation for $v_{channel}$ = *0.535 m/s, Q = 0 LPM, H = 119 mm.*

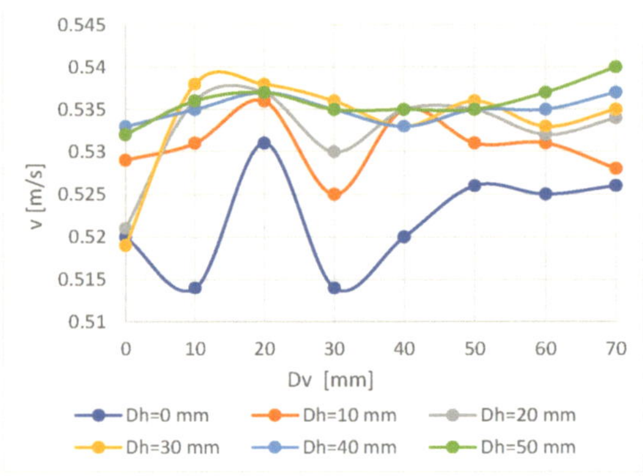

Figure 16.
Induced velocity variation for $v_{channel}$ = *0.535 m/s, Q = 10.5 LPM, H = 119 mm.*

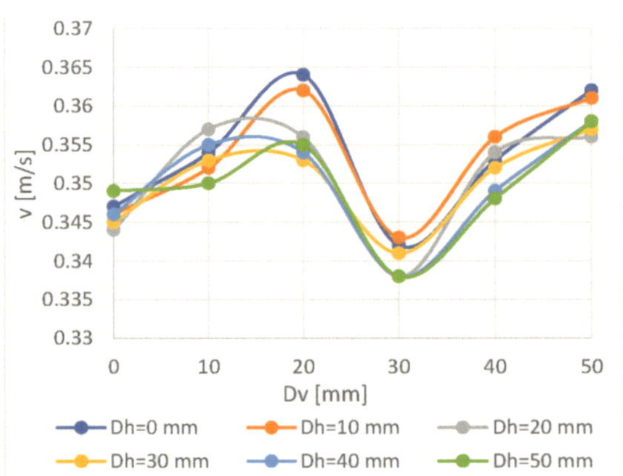

Figure 17.
Induced velocity variation for $v_{channel}$ = *0.33 m/s, Q = 15 LPM, H = 92 mm.*

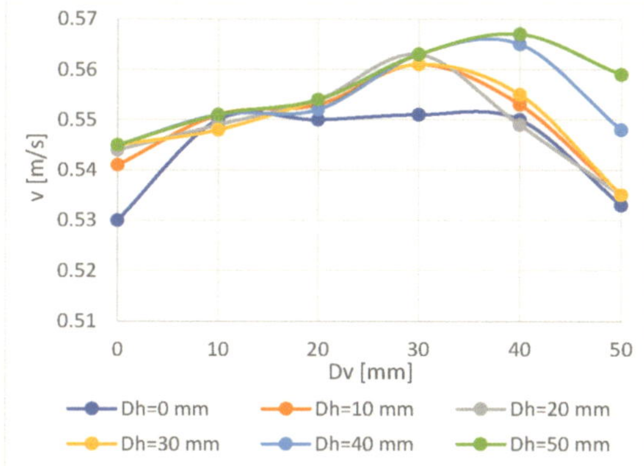

Figure 18.
Induced velocity variation for $v_{channel}$ *= 0.535 m/s, Q = 15 LPM, H = 92 mm.*

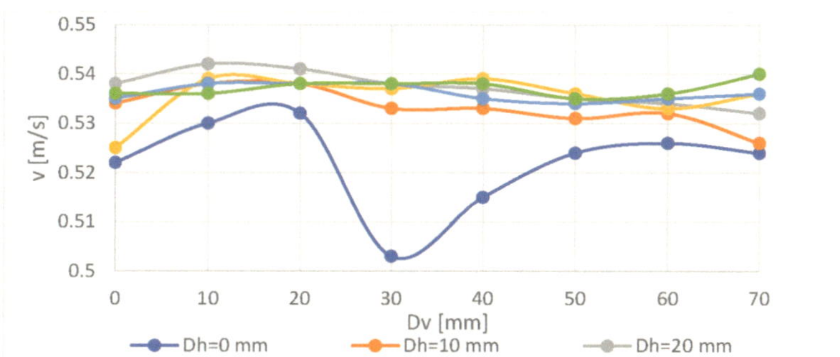

Figure 19.
Induced velocity variation for $v_{channel}$ *= 0.535 m/s, Q = 15 LPM, H = 119 mm.*

The water intake operation also influences the water flow inside the main channel. In the proximity of the intake orifices (at a vertical distance of 20–30 mm from the origin of the chosen coordinate system, [12]), the velocity field is modified. The velocity field in longitudinal direction registers a decrease due to the velocity transverse component increase, as a result of the water lateral catchment trough the orifices. As reported in [12], **Figures 11** and **15** show that when no air is injected, this influence is present in both the low water depth (92 mm) and low velocity (0.33 m/s) conditions, as well as in the high depth (119 mm) and high velocity (0.535 m/s) conditions. From a relative depth of 0.33–0.35 (a distance of 30–40 mm away from the intake area) upwards, the velocities in longitudinal direction are no longer influenced by the intake presence. When air is injected in reduced water depth and low velocity conditions (92 mm and 0.33 m/s, respectively), it determines an increase in the longitudinal velocities field in the intake inlet vicinity area (15–30 mm) until a relative depth of 0.33. As the airflow rate increases, the minimum longitudinal velocities' rise toward the free surface is noticed.

Considering all the above findings of the carried out experimental research, it was concluded that the behavioral barrier causes a local modification in the water velocity field. This effect could be noticed even for low air flow rates (10.5 LPM), indicating that the bubble curtain can be effective also when using reduced airflow rates for its generation. In addition, this effect was noticed for the both tested water velocities in the main channel. In the case of 0.535 m/s, the bubbles are being carried away downstream. As indicated in [12], the position of the porous hose should be reconsidered. Thus, for increased velocities, it has to be placed with few centimeters offset upstream the river water intake model.

The carried-out tests indicated that the environmental barrier modifies the velocity field in the surrounding water, creating recirculation. This finding is motivating and highlights the conclusions reported in [48], which refer to the fact that the fish behavior can be influenced by a sharp velocity gradient.

In contrast to the studies reported in the literature, the present research has analyzed the velocity field under different conditions: two water depths, two velocities in the main testing channel, and two airflow rates for the bubble curtain generation. The results were compared to the case when no air was injected in the water. Furthermore, the experimental analysis was performed on a dedicated experimental setup. It consisted of a hybrid model that comprised both an ecological river water intake that does not influence river morphology and a behavioral barrier based on bubble curtains. The components' individual operation as well as their tandem operation have been investigated. A special interest presents the impact that the bubble curtain has on the surrounding water velocity. This interest is justified by the presence of velocity and pressure gradient detectors in the fish sensory system as reported by Mogdans in [10]. These detectors make fish react to hydrodynamic stimuli such as sounds from air bubble rise and breakup. In the case of the present study, with the injected airflow increase, there are registered variations in the velocity magnitude in longitudinal direction for the same depth and different cross-channel distances. These variations can be noticed by analyzing **Figures 11–19**. Popper and Schilt [56] reported that besides the velocity gradient detectors, fish are also sensitive to acoustic underwater sounds of up to 100–200 Hz [12]. As reported by Braun et al. in [57], some fish have an increased sensitivity up to 1000 Hz. Frizell, and Arndt [37] showed that this is in the noise range that bubbles make when splitting. All these findings show that both hydrodynamic and sound stimuli associated to the bubble curtain operation influence the fish response and may be used in repelling fish solutions and strategies.

Taking into account the findings above (especially those reported in [10, 37, 56, 57]), we ascertain that the current research may bring new insights on the fish behavior when encountering bubble curtains.

Another aim of the carried-out research consisted in assessing the dissolved oxygen induced in the water by the bubble curtain. Therefore, **Table 3** shows the DO values registered in the following operating conditions: with and without air injection, for 92 mm water depth of 0.33 m/s water velocity inside the main channel and different water temperatures, t.

The carried-out experiments showed that the water intake operation when no air was introduced through the porous hose led to an increase in the DO concentration inside the circulated water through the experimental setup. This is explained by the fact that the water flows in a turbulent regime in a free surface channel. Thus, surface mass transfer occurs between the water free surface and the surrounding air.

Figure 20 shows the dissolved oxygen concentration values in three experimental conditions: for Q = 0 LPM, Q = 10.5 LPM, and Q = 15 LPM, respectively.

For a 50-minute operation of the water intake model and no air injected in the water, the experimental results indicated that an increase in the DO concentration of 4% is obtained compared to the initial value (DO increases from 7.42 to 7.72 mg/L). This demonstrates that the simple recirculation of water through the free surface channel leads to a surface mass transfer.

Q = 0 LPM, t = 15.9 °C			Q = 10.5 LPM, t = 16°C		Q = 15 LPM, t = 16.1 °C	
No.	**Time [min.]**	**DO [mg/L]**	**Time [min.]**	**DO [mg/L]**	**Time [min.]**	**DO [mg/L]**
1	0	7.42	0	7.79	0	8.19
2	2	7.45	5	7.88	2	8.22
3	4	7.47	6	7.89	3	8.24
4	5	7.47	7	7.94	4	8.26
5	10	7.48	8	7.98	5	8.28
6	12	7.52	11	8.00	6	8.29
7	15	7.53	12	8.02	7	8.3
8	17	7.54	13	8.03	8	8.31
9	20	7.56	14	8.05	9	8.34
10	25	7.57	15	8.06	10	8.36
11	28	7.58	16	8.07	11	8.38
12	31	7.63	17	8.06	12	8.39
13	33	7.64	18	8.08	13	8.48
14	36	7.64	19	8.12	14	8.5
15	39	7.68	20	8.14	15	8.52
16	42	7.7	—	—	16	8.51
17	45	7.7	—	—	17	8.54
18	48	7.72	—	—	18	8.53
19	50	7.72	—	—	19	8.55
—	—	—	—	—	20	8.59

Table 3.
DO concentrations for $v_{channel}$ = 0.33 m/s, H = 92 mm, and different airflow rates injected for bubble curtain generation.

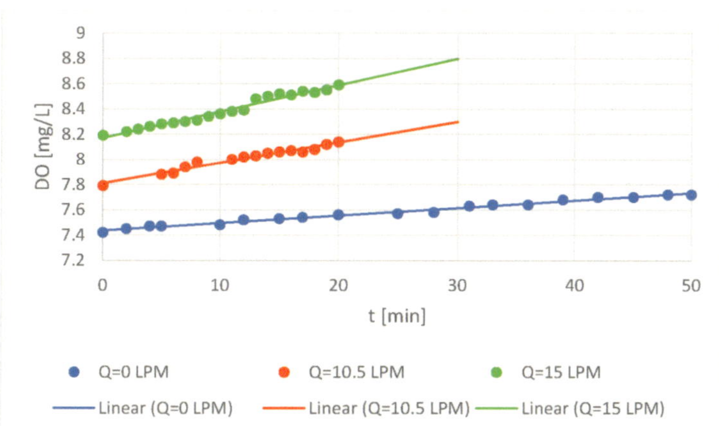

Figure 20.
DO variation trend for different airflow rates injected in the water by the porous hose.

The injection of 10.5 LPM through the porous hose for 20 minutes led to a 5% increase (from 7.72 to 8.14 mg/L) in the DO concentration compared to the value obtained at the end of the measurements with zero airflow feed through the porous hose.

When introducing in the water an air flow rate of 15 LPM for 20 minutes, the bubble curtain generation led to an 11% increase in the DO concentration compared to the value obtained at the end of the measurements with no air injection. Thus, the DO increased to 8.59 mg/L.

The performed experiments showed that both the water recirculation through the experimental setup and the air dispersion through the porous hose in order to generate the behavioral barrier leads to the DO concentration increase. This contributes to the improvement of the aquatic fauna living conditions, as it is well-known that fish need a certain amount of DO for an appropriate development.

5. Analysis of power autonomy solutions for the behavioral barrier based on bubble curtains

The solutions that ensure the power autonomy of the bubble curtain behavioral barrier are based on the use of available renewable energy sources [58]. They are aimed especially at the conversion of solar and wind energy and their integration within an off-grid system, without any connection to the electric public grid. Considering that the water intake is intended to be used within mountain rivers, ensuring the energy autonomy of the behavioral barrier can be achieved by a microgrid deployment that consists of distributed generation sources (photovoltaic panels, wind turbines, and/or kinetic turbines) as well as a storage system. The microgrid must provide the necessary energy to operate a compressor that supplies compressed air to the bubble generation system and to the porous hose, respectively. Thus, for the case analyzed within the chapter, a 25-liter compressor was used, with a maximum pressure of 8 bar and 1.5 kW motor power, equipped with a pressure regulator. When the energy demand is simultaneous to the energy production, the battery is no longer needed. In order to prevent the overcharge or the complete discharge of the battery, a charge controller is used between the photovoltaic (PV) system and the battery. The charge controller usually includes also a discharge protection diode, which prevents the battery from being discharged at night by the PV or other system.

If higher output levels are required or industrial appliances are used, systems must provide an output voltage of 230 V A.C. In order to generate this output voltage, the system is provided with an inverter that transforms the direct current produced by the PV system or the battery current to alternating current.

Considering the specific electrical parameters of a compressor, the power supply system solution consists in the production and use of electric power generated by the PV system as well as the wind-turbine, with off-grid operation. Thereby, the microgrid includes a charge controller, batteries to ensure autonomy with a charging management system, and an inverter for 230 V AC consumers. A typical configuration of microgrid using a wind turbine, a PV system, and a hydro turbine is shown in **Figure 21**.

In order to supply the compressor used for the current proposal of behavioral barrier, the hydrokinetic turbine-based solution was not taken into account due to the fact that in general, the generated power is low, the water depth varies significantly, and it requires special attention when operating (periodic maintenance). A possible solution for the use of hydraulic turbines regards the placement of low-head bulb

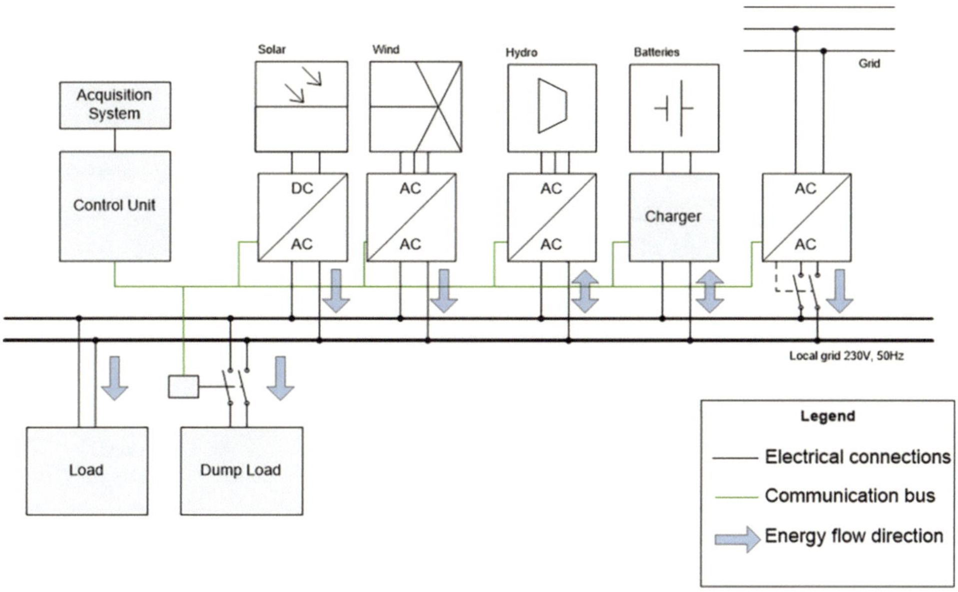

Figure 21.
Typical configuration of a mixed microgrid [57, 59].

turbines, which provide more power, but these require separate hydrotechnical constructions, which affect the morphology of the river and the natural flow conditions.

In order to reduce losses due to processing and conversion of electrical parameters, a potential solution is given by a compressor with direct current drive and low voltage 12/24 V. Therefore, some of the microgrid components can be removed, such as the inverter and the AC voltage stabilizer. When sizing the system, the solar or wind potential of the envisaged location for the deployment of the water intake must be taken into account. It is also necessary to identify appropriate solutions in case certain renewable energy sources are not available (the presence of trees causes a significant shaded area, which impacts the production of solar energy and also disrupts the wind speed, leading to a low production of wind power).

In order to establish the power autonomy solution for the behavioral barrier based on bubble curtains, an important data is represented by the compressor's consumed power ensuring its operation at different flow rates. Thus, the time related to the air compressor's electrical energy supply was determined experimentally, at the following airflow rates injected through the porous hose: 8, 10.5, and 15 LPM. Also, in order to determine the power consumed by the compressor, a FLUKE 434 three-phase power analyzer was used (**Figure 22**).

The designed system supplies compressed air with flow rates between 8 and 15 L/min/m to the porous hose. In the tested case, the length of the porous hose was of 1 m. Therefore, in order to provide general applicable results and indications, all the calculations are related to the reference length of the porous hose. Consequently, for longer lengths, the obtained result in terms of power or energy is multiplied by the length required for the considered river water intake (**Figure 23**).

In order to perform the analysis of the required energy, the electrical parameters of the compressor were also measured during tests using the Fluke 434 type power analyzer (**Figure 24**). This measuring device helps predict, prevent, and solve problems in electricity production and distribution systems.

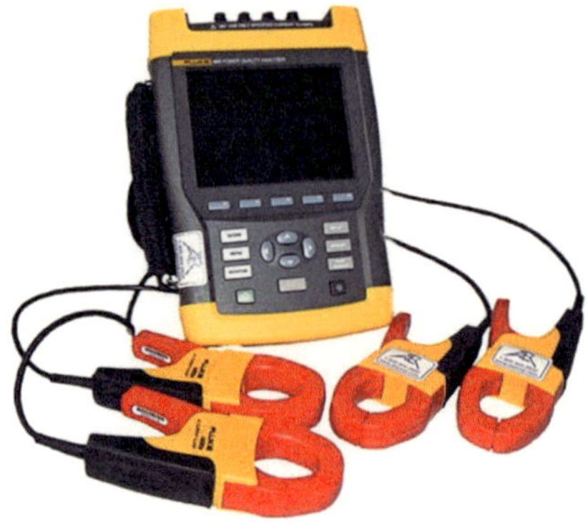

Figure 22.
FLUKE 434 three-phase power analyzer.

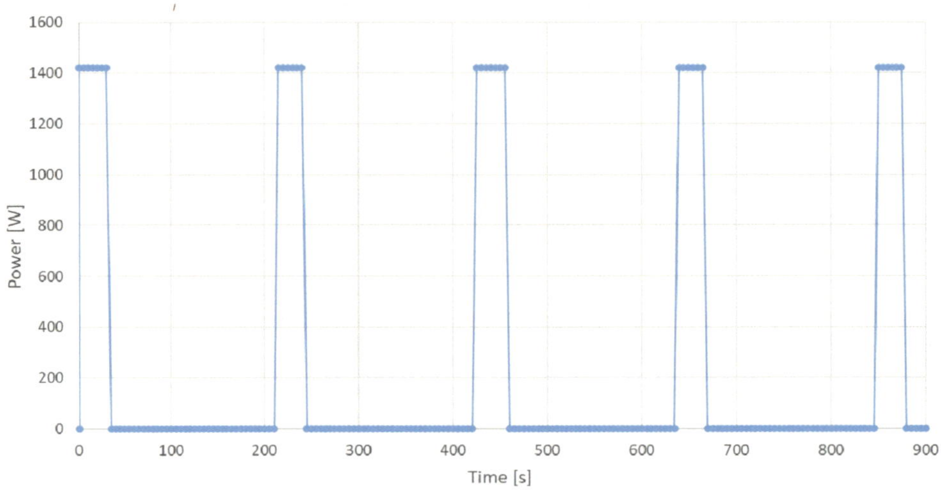

Figure 23.
The compressor operation diagram for 15 minutes highlighting the input power—Functioning at 15 LPM airflow.

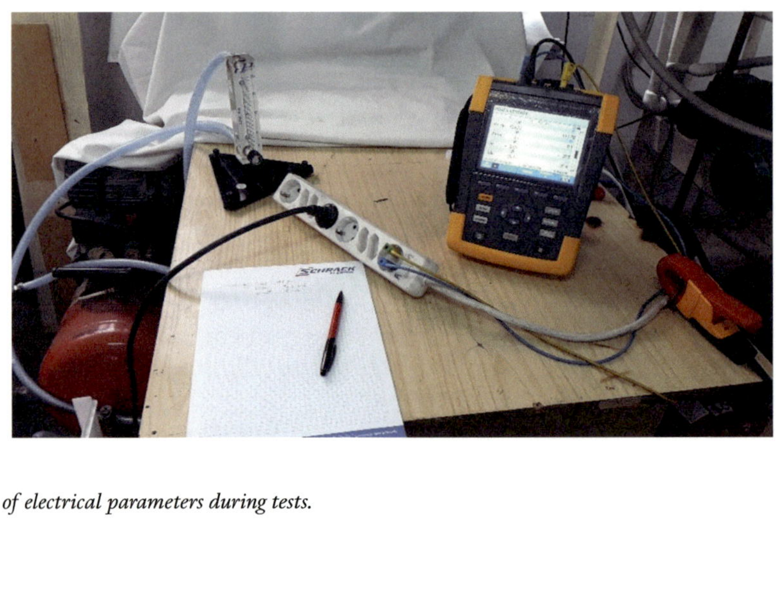

Figure 24.
Acquisition of electrical parameters during tests.

The characteristics of the compressor used for carrying out the tests, according to its data sheet, are as follows:

- Supply voltage: 230 V,
- Nominal power: 1.1 kW,
- Tank capacity: 24 liters,
- Nominal pressure: 8 bar.

The tests aimed to determine the following parameters:

- instant power,
- energy required for the bubble curtain operation for 1 hour,
- operating time of the compressor, respectively the time between two successive drives.

Figure 24 shows an image captured during the electrical parameters measurement.

In the following paragraphs, the results of the performed tests regarding the electricity consumption are presented.

Thus, to initially charge the compressor from 0 to 7.5 bar, it is necessary to power the compressor for 83 seconds. The necessary energy supply is of 30.5 Wh, with 1420 W active absorbed power, 1550 VA apparent power, and a power factor of 0.92. For the proper operation of the bubble curtain (a uniform generation of bubbles over the entire surface of the porous hose), the compressor pressure was adjusted between the minimum threshold of 5 bar and the maximum of 7.5 bar. Thus, after the initial charge, during the bubble curtain generation, the compressor came into operation only to raise the pressure in the tank from 5 bar to 7.5 bar.

During the compressed air injection through the porous hose, respectively the generation of the bubble curtain using different airflow rates, the following values of the envisaged parameters were registered:

a. for the airflow rate of 8 LPM, the compressor was supplied for 28 seconds in order to raise the pressure from 5 to 7.5 bar, with 5-minute and 49-second breaks. The total energy consumption during 1 hour of operation was of 106.2 Wh; each 30-second operation period required the consumption of 10.9 Wh.

b. for the airflow rate of 10.5 LPM, the compressor was supplied for 28 seconds in order to raise the pressure from 5 to 7.5 bar, with 4-minute and 15-second breaks. The total energy consumption during 1 hour of operation was of 138.8 Wh, while each 28-second operation period required a consumption between 10.8 and 11.1 Wh.

c. for the airflow rate of 15 LMP, the compressor operated for 30 seconds to in order raise the pressure from 5 to 7.5 bar, with 3-minute breaks. The total energy consumption over an hour was of 200.6 Wh; each 30-second operating period required an energy of 11.8 Wh.

Consequently, in order to supply compressed air for a 24-hour period, the power required to operate the compressor and to generate the bubble curtain respectively is between 2.54 and 4.81 kWh. An additional 30.5 Wh are also necessary apart to the initial power consumption for loading the compressor. The instantaneous power that has to be ensured is approximately of 1500 W.

It is worth specifying that the values measured and presented above are only valid for one linear meter of porous hose in order to generate the bubble curtain.

Based on the established necessary power, a photovoltaic system can be used, consisting of a string of PV panels, an off-grid controller, and storage batteries (as shown in **Figure 25**). If, in case of a specific application, an increased airflow rate, or a longer bubble curtain length is expected, then a wind turbine can also be integrated within the microgrid. An example of a microgrid suitable to isolated locations based on 2 or 3 polycrystalline photovoltaic panels, with a nominal maximum power of 450 Wp each, an off-grid controller of 3 kVA, which stores the surplus energy in two 12 V batteries connected in series, with a capacity of 67 Ah.

This system provides the pure sine wave needed by the compressor's electric motor. The resulting microgrid will include a controller with battery charging function, inverter for AC consumers, and Maximum Power Point Tracking system (MPPT), which optimizes the operation of the photovoltaic panels.

If other consumers are identified such as lighting, data acquisition and transmission, and GSM/4G communications, the system can be adapted with panels and batteries, according to any requirement. The proposed configuration is suitable to low energy consumption and does not require significant acquisition, installation,

Figure 25.
Example of a 3 kVA photovoltaic system ensuring the power autonomy for the behavioral barrier based on bubble curtains.

and maintenance costs. The power supply of the hybrid water intake also allows the addition of many useful facilities to its operation, such as reporting of extracted water flow, incidents or failures warning, video monitoring, and data transmission. Also, there can be achieved the measurement of electrical, hydraulic, as well as meteorological parameters and even remote actuation of certain valves or cleaning installations when required, without the need to travel to the intervention site.

6. Conclusion

The chapter approached the study of fish behavioral barriers for hydrotechnical facilities. In order to better understand their characteristics and implementation possibilities, some general aspects regarding the study of water intakes with behavioral barriers were presented. Thus, the study includes the assessment and analysis of behavioral barriers, elements regarding river water intakes and ecological requirements related to their design, as well as characteristics of fish hydrodynamics. A fish behavioral barrier based on bubble curtains for a river water intake was experimentally investigated. The experimental setup and the methodology used for measuring the velocities induced by the bubble curtain operation and for assessing its associated contribution to water aeration were also detailed.

The impact of the bubble curtain on the surrounding water velocity was determined by measuring the water-induced velocities in different points located in the experimental channel cross section. The barrier was placed in the proximity of the water intake model, where the flow is influenced by the water admission area. This contributed to the modification of the streamlines, thus helping in diverting the directional flow and deterring the fish from the intake area. Therefore, it can be concluded that the single or combined operation of the bubble curtain may safely divert the fish to a particular route, diminishing their accidental intrusion into the river intake.

The experiments showed a sharp water velocity gradient for almost all of the analyzed cases. This finding demonstrates that the bubble curtain determines the local velocity modification, creating the premises of influencing fish behavior. Furthermore, by modifying the injected airflow rate or the type, number, or placement of bubble generating sources, the curtain can adapt to different river flow conditions. Regarding the on-site operation of the behavioral barrier, its flow regulation can be performed from the river bank. Also, by considering the characteristics of the envisaged fish species, the minimum dispersed airflow rate that ensures an increase in the local velocity has to be identified. Thus, the fish guidance solution may be adapted to different operating conditions.

When operating in remote areas, without access to the electric public grid, a microgrid deployment is considered suitable. Thereby, a PV-based microgrid provided with a storage system was proposed.

The studies presented in this chapter did not include fish in the behavioral barrier testing. This is due to the fact that the current research was carried out in the frame of a national project focused on other scientific means that demonstrate the solution feasibility, namely, the induced velocities determination. The research activity did not consider animal testing for professional ethics reasons and was conducted on a state-of-the-art dedicated hydraulic stand that can accurately measure velocity fields and other parameters. Thus, reliable results can be provided for the preliminary experimental validation of the behavioral barrier. The inclusion of a biotic model is envisaged in the future research, in order to validate the proposed fish guidance solution.

An in-depth insight of the complex flow phenomena can be provided with the help of the Particle Image Velocimetry (PIV) measuring technique. PIV can be easily applied on the experimental setup developed in the current research and can be used to determine the velocity vectors direction in the proximity of the bubble curtain and in the water intake area as well.

After performing this laboratory testing, a combined water intake-behavioral barrier fish guidance solution can be deployed on-site to study the fish interaction with the bubble curtain in a real water body. Since the bubble plume better develops in higher water depth, a general recommendation for real operating conditions would refer to placing the diffuser as low as possible in the riverbed.

Acknowledgements

This work was supported by the Romanian Ministry of Education, Research and Digitalization, UEFISCDI, project number PN-III-P2-2.1-PED-2019-1444–Eco-hybrid Water Intake with Behavioral Barrier to Reduce the Impact on Fish Fauna and River Morphology, project number PN-III-P2-2.1-PTE-2021-0075 within PNCDI III as well as Nucleu contract no. 42 N/2023, project no. PN 23140101/2023.

Conflict of interest

The authors declare no conflict of interest.

Author details

Gabriela Circiumaru*, Rares-Andrei Chihaia, Lucia-Andreea El-Leathey and Andreea Voina
Electrical Engineering ICPE-CA, National Institute for Research and Development, Bucharest, Romania

*Address all correspondence to: gabriela.circiumaru@icpe-ca.ro

References

[1] World Wide Fund for Nature [Internet]. 2016. Available from: https://wwfint.awsassets.panda.org/downloads/lpi_technical_supplement_2016.pdf [Accessed: October 26, 2023]

[2] Brasovanu L, Gogoașe Nistoran DE, Armaș I. The hydrological regime of the Prahova River along Carpathian and Subcarpathian stretch. Gtaylorruebuseo Patterns. 2018;**3**(1):32-42. DOI: 10.5719/GeoP.3.1/5

[3] Water Framework Directive 2000/60/EC - Directive 2000/60/EC of the European Parliament and of the Council of 23 October 2000 establishing a framework for Community action in the field of water policy, Official Journal of the European Communities. 22.12.2000, L 327/1-L 327/72. Available from: https://eur-lex.europa.eu/resource.html?uri=cellar:5c835afb-2ec6-4577-bdf8-756d3d694eeb.0004.02/DOC_1&format=PDF [Accessed: October 26, 2023]

[4] Bunt CM, Castro-Santos T, Haro A. Performance of fish passage structures at upstream barriers to migration. River Research and Applications. 2012;**28**:457-478. DOI: 10.1002/rra.1565

[5] WWF-România [Internet]. Analysis of Specific Legislation Related to Planning and Issuance of Regulatory Acts Related to the Construction and Operation of Micro Hydropower Plants in Romania (Analiza Legislației Specifice Din Domeniul Planificării şi Emiterii Actelor de Reglementare Aferente Construirii şi Funcționării Microhidrocentralelor în România). 2013. 1-53. WWF-România, Bucharest, Romania. Available from: https://wwfeu.awsassets.panda.org/downloads/wwf_raport_legal_mhc_nov_2013.pdf [Accessed: October 26, 2023]

[6] Linnansaari T, Wallace B, Allen Curry R, Yamazaki G. Fish passage in large rivers: A literature review. Mactaquac Aquatic Ecosystem Study Report Series. 2015:2015-2016. DOI: 10.13140/RG.2.2.32204.41604 [Accessed: October 26, 2023]

[7] Kerr JR, Kemp PS. Masking a fish's detection of environmental stimuli: Application to improving downstream migration at river infrastructure. Journal of Fish Biology. 2019;**95**(1):228-237. DOI: 10.1111/jfb.13812. Available from: https://onlinelibrary.wiley.com/doi/full/10.1111/jfb.13812

[8] Silva AT, Bærum KM, Hedger RD, Baktoft H, Fjeldstad HP, Gjelland KØ, et al. The effects of hydrodynamics on the three-dimensional downstream migratory movement of Atlantic Salmon. Science of the Total Environment. 2020;**705**:1-10. DOI: 10.1016/j.scitotenv.2019.135773

[9] Hockley FA, Wilson CAM, Brew A, Cable J. Fish responses to flow velocity and turbulence in relation to size, sex and parasite load. Journal of the Royal Society Interface. 2014;**11**(91):1-11. DOI: 10.1098/rsif.2013

[10] Mogdans J. Sensory ecology of the fish lateral-line system: Morphological and physiological adaptations for the perception of hydrodynamic stimuli. Journal of Fish Biology. 2019;**95**:53-72. DOI: 10.1111/jfb.13966

[11] Turnpenny AWH, O'Keeffe N. Screening for Intakes and Outfalls: A Best Practice Guide [Science Report SC030231]. Bristol, UK: Environment Agency; 2005

[12] Cîrciumaru G, Chihaia RA, Voina A, Gogoașe Nistoran DE, Simionescu ȘM, El-Leathey LA, et al. Experimental analysis of a fish guidance system for a river water intake. Water. 2022;**14**(3):1-22. DOI: 10.3390/w14030370

[13] Katopodis C, Koon EM, Hanson L. Sea Lamprey Barriers: New Concepts and Research Needs. Ann Arbor, MI. Available from: http://www.glfc.org/pubs/pdfs/research/reports/Katopodisbarriers.pdf: Great Lakes Fishery Commission; 1994 [Accessed: October 26, 2023]

[14] Li L, Maaswinkel H. Visual sensitivity and signal processing in teleosts. Fish Physiology. 2006;**25**:179-241. DOI: 10.1016/S1546-5098(06)25005-2

[15] McLean AR. Strobe lights as a fish deterrent [thesis]. Victoria, BC: Royal Roads University; 2008

[16] Bass AH, Lu Z. Neural and behavioral mechanisms of audition. Fish Physiology. 2006;**25**:377-410. DOI: 10.1016/S1546-5098(06)25009-X

[17] Popper AN, Fay RR. Rethinking sound detection by fishes. Hearing Research. 2011;**273**(1-2):25-36. DOI: 10.1016/j.heares.2009.12.023

[18] Zielinski D. Bubble barrier technologies for common carp [thesis]. USA: University of Minnesota; 2011

[19] Ross WE, Dunning DJ, Thorne R, Menezes JK, Tiller GW, Watson JK. Response of alewives to high-frequency sound at a power plant intake on lake Ontario. North American Journal of Fisheries Management. 1993;**13**(2):291-303. DOI: 10.1577/1548-8675%281993%29013%3C0291%3AROATHF%3E2.3.CO%3B2

[20] Haro A, Castro-Santos T, Noreika J, Odeh M. Swimming performance of upstream migrant fishes in open channel flow: A new approach to predicting passage through velocity barriers. Canadian Journal of Fisheries and Aquatic Sciences. 2004;**61**(9):1590-1601. DOI: 10.1139/f04-093

[21] Heinrich JW, Mullett KM, Hansen MJ, Adams JV, Klar GT, Johnson DA, et al. Sea lamprey abundance and management in Lake Superior, 1957 to 1999. Journal of Great Lakes Research. 2003;**29**(1):566-583. DOI: 10.1016/S0380-1330(03)70517-6

[22] Davis JC. Minimal dissolved oxygen requirements of aquatic life with emphasis on Canadian species: A review. Journal of the Fisheries Board of Canada. 1975;**32**(12):2295-2332. DOI: 10.1139/f75-268

[23] Miranda LE, Hodges KB. Role of aquatic vegetation coverage on hypoxia and sunfish abundance in bays of a eutrophic reservoir. Hydrobiologia. 2000;**427**(1):51-57. DOI: 10.1023/A:1003999929094

[24] Noatch MR, Suski CD. Non-physical barriers to deter fish movements. Environmental Reviews. 2012;**20**:71-82. DOI: 10.1139/a2012-001

[25] Sorensen PW, Stacey NE. Brief review of fish pheromones and discussion of their possible uses in the control of non-indigenous teleost fishes. New Zealand Journal of Marine and Freshwater Research. 2004;**38**(3):399-417. DOI: 10.1080/00288330.2004.9517248

[26] Bettoli PW, Maceina MJ. Sampling with toxicants. In: Murphy BR, Willis DW, editors. Fisheries Techniques. 2nd ed. Bethesda, MD: American Fisheries Society; 1996. pp. 303-333

[27] Gibbs MA, Glenn NR. Development of the lateral line system in the shovelnose sturgeon. Brain, Behavior and Evolution. 2004;**64**(2):70-84. DOI: 10.1159/000079117

[28] Taylor RM, Pegg MA, Chick JH. Response of bighead carp to a bioacoustic behavioural fish guidance system. Fisheries Management and Ecology. 2005;**12**:283-286. DOI: 10.1111/j.1365-2400.2005.00446.x

[29] Patrick PH, Christie AE, Sager D, Hocutt C, Stauffer J Jr. Responses of fish to a strobe light/air-bubble barrier. Fisheries Research. 1985;**3**:157-172. DOI: 10.1016/0165-7836(85)90016-5

[30] Welton JS, Beaumont WRC, Clarke RT. The efficacy of air, sound and acoustic bubble screens in deflecting Atlantic salmon salmo Salar L., smolts in the river Frome. Fisheries Management and Ecology. 2002;**9**(1):11-18. DOI: 10.1046/j.1365-2400.2002.00252.x

[31] Dawson HA, Reinhardt UG, Savino JF. Use of electric or bubble barrier to limit the movement of eurasion ruffe (gymnocephalus cernuus). Journal of Great Lakes Research. 2006;**32**:40-49. DOI: 10.3394/0380-1330(2006)32[40:UOEOBB]2.0.CO;2

[32] Taft EP. Fish protection technologies: A status report. Environmental Science & Policy. 2000;**3**(S.1):349-359. DOI: 10.1016/S1462-9011(00)00038-1

[33] Bureau of Reclamation. Fish Protection at Water Diversions. A Guide for Planning and Designing Fish Exclusion Facilities. Denver, Colorado: Bureau of Reclamation; 2006; 480 p

[34] Pegg MA, Chick JH. Aquatic Nuisance Species: An Evaluation of Barriers for Preventing the Spread of Bighead and Silver Carp to the Great Lakes. Final report for Illinois-Indiana Sea Grant A/SE (ANS)-01-01. Illinois, Indiana; 2018; 20 p

[35] Turnpenny AWH. Review of Potential for Coarse Fish Entrainment at Potable Water Intakes on the Lower Freshwater Thames (Hurley-Teddington). Fawley ARL Ltd. Thames Region, Bristol, UK: Client Report to the Environment Agency; No. FCR 302/99; 1999

[36] Jones PE, Tummers JS, Galib SM, Woodford DJ, Hume JB, Silva LGM, et al. The use of barriers to limit the spread of aquatic invasive animal species: A global review. Frontiers in Ecology and Evolution. 2021;**9**:1-9. DOI: 10.3389/fevo.2021.611631

[37] Frizell WK, Arndt REA. Noise Generation by Air Bubbles in Water: An Experimental Study of Creation and Splitting. Minneapolis, USA: University of Minnesota Digital Conservancy; St. Anthony Falls Hydraulic Laboratory; 1987. [Internet]. Available from: https://www.usbr.gov/tsc/techreferences/hydraulics_lab/pubs/PAP/PAP-1044.pdf [Accessed: October 26, 2023]

[38] Castro-Santos T, Haro A. Fish guidance and passage at barriers. In: Fish Locomotion. New Hampshire, USA: Science Publishers; 2010. pp. 62-89. DOI: 10.1201/b10190-4

[39] Sanz-Ronda FJ, Fuentes-Pérez JF, Bravo-Córdoba FJ, García-Vega A, Ruiz-Legazpi J, Martínez de Azagra A. Estimating fish passage over velocity barriers for non-uniform flow conditions: A case study in flat-V gauging weirs. Biology and Life Sciences Forum. 2022;**13**(1):1-20. DOI: 10.3390/blsf2022013020

[40] Kapitze R. Culvert Fishway Planning and Design Guidelines, Part C – Fish

Migration Barriers and Fish Passage Options for Road Crossings. Townsville, Australia: James Cook University; 2010. [Internet]. Available from: http://www.jcu.edu.au/fishpassagedesign/ [Accessed: October 26, 2023]

[41] Sánchez-González JR, Morcillo F, Ruiz-Legazpi J, Sanz-Ronda FJ. Fish morphology and passage through velocity barriers. Experience with northern straight-mouth nase (*Pseudochondrostoma duriense* Coelho, 1985) in an open channel flume. Hydrobiologia. 2022;**849**:1351-1366. DOI: 10.1007/s10750-021-04712-9

[42] Liang Y, Hou Y, Hu W, Johnson D, Wang J. Flow velocity preference of Schizothorax oconnori Lloyd swimming upstream. Global Ecology and Conservation. 2021;**32**:e01902. DOI: 10.1016/j.gecco.2021.e01902

[43] Kováts P, Thévenin D, Zähringer K. Characterizing fluid dynamics in a bubble column aimed for the determination of reactive mass transfer. Heat and Mass Transfer. 2018;**54**:453-461. DOI: 10.1007/s00231-017-2142-0

[44] Murgan I, Bunea F, Ciocan GD. Experimental PIV and LIF characterization of a bubble column flow. Flow Measurement and Instrumentation. 2017;**54**:224-235. DOI: 10.1016/j.flowmeasinst.2017.02.004

[45] Laakkonen M, Honkanen M, Saarenrinne P, Aittamaa J. Local bubble size distributions, gas–liquid interfacial areas and gas holdups in a stirred vessel with particle image velocimetry. Chemical Engineering Journal. 2005;**109**(1-3):37-47. DOI: 10.1016/j.cej.2005.03.002

[46] Yao H, Cao L, Wu D, Gao Y, Qin S, Yu F. PIV study on grid-generated turbulence in a free surface flow. Water. 2021;**13**(7):1-909. DOI: 10.3390/w13070909

[47] Seol DM, Seo JH, Rhee SH. Towed underwater PIV measurement for free-surface effects on turbulent wake of a surface-piercing body. International Journal of Naval Architecture and Ocean Engineering. 2013;**5**(3):404-413. DOI: 10.3744/JNAOE.2013.5.3.404

[48] Muste M, Xiong Z, Schöne J, Li Z. Validation and extension of image velocimetry capabilities for flow diagnostics in hydraulic modeling. Journal of Hydraulic Engineering. 2004;**130**(3):175-185. DOI: 10.1061/(ASCE)0733-9429(2004)130:3(175)

[49] Lindmark EM. Flow design for migrating fish [thesis]. Sweden: Luleå University of Technology, Department of Applied Physics and Mechanical Engineering, Division of Fluid Mechanics; 2008

[50] Chihaia RA. Innovative Solutions Used for Developing Small Hydropower Plants with Reduced Environmental Impact (Soluții Inovative Utilizate Pentru Amenajarea de Microhidrocentrale cu Impact Redus Asupra Mediului). Olt County, Romania: Hoffman Publisher; 2021. 126 p. ISBN 978-606-46-1483-4

[51] Poletto JB, Cocherell DE, Mussen TD, Ercan A, Bandeh H, Levent Kavvas M, et al. Fish-protection devices at unscreened water diversions can reduce entrainment: Evidence from behavioural laboratory investigations. Conservation Physiology. 2015;**3**(1):1-12. DOI: 10.1093/conphys/cov040

[52] Cîrciumaru G, Chihaia RA, Dancă P, Voina A. Patent application A/00357/23.06.2022. River water intake with behavioral barrier to reduce impact on fish fauna (Priză de apă pentru râuri

cu barieră comportamentală pentru reducerea impactului asupra faunei piscicole). Bucharest, Romania: State Office for Inventions and Trademarks; 2022

[53] Cano-Barbacil C, Radinger J, Argudo M, Rubio-Gracia F, Vila-Gispert A, García-Berthou E. Key factors explaining critical swimming speed in freshwater fish: A review and statistical analysis for Iberian species. Scientific Reports. 2020;**10**(1):1-12. DOI: 10.1038/s41598-020-75974-x

[54] Taugbøl A, Olstad K, Magnus Bærum K, Museth J. Swimming performance of brown trout and grayling show species-specific responses to changes in temperature. Ecology of Freshwater Fish. 2019;**28**(2):241-246. DOI: 10.1111/eff.12447

[55] Project number PN-III-P2-2.1-PED-2019-1444. Eco-Hybrid Water Intake with Behavioral Barrier to Reduce the Impact on Fish Fauna and River Morphology. Acronym Eco-WIBB; 2020-2022

[56] Popper AN, Schilt CR. Hearing and acoustic behavior: Basic and applied considerations. In: Webb FJ, Fay RR, Popper AN, editors. Fish Bioacoustics. New York: NY, USA: Springer; 2008. pp. 17-48

[57] Braun CB, Sand O. Functional overlap and nonoverlap between lateral line and auditory systems. In: Coombs S, Bleckmann H, Fay R, Popper A, editors. The Lateral Line System. Springer Handbook of Auditory Research. Vol. 48. New York, NY: Springer; 2013. DOI: 10.1007/2506_2013_19

[58] Research project financed by ANCS Romania. Increasing the Efficiency of Technological Equipment and Processes for Energy Conversion from Renewable Sources, Objective 5: 1.5 kW Experimental Model of Autonomous Power Supply System Consisting of Integrated Renewable Energy Sources Grid. Contract no. 09350201. Coordinator: INCDIE ICPE-CA (Team Member);

[59] El-Leathey LA. Energy management system designed for the interconnected or islanded operation of a microgrid using LabVIEW software. Chapter 3. In: Nayeripour M, Waffenschmidt E, Khesht M, editors. Smart Microgrids. London, UK: IntechOpen; 2018. pp. 45-64. DOI: 10.5772/intechopen.74856

Chapter 9

Time Trends in Fish Tissue Methylmercury in Northern Watersheds: Implications of Phosphorus Loading and Eutrophication on Subsistence Fisheries

Dean G. Fitzgerald and Lynn S. McCarty

Abstract

Subsistence fisheries for Michipicoten First Nation (MFN) in habitats across an area north of Lake Superior in Ontario, Canada were assessed. This assessment used reports by Ontario, private entities (e.g., mines), and MFN to evaluate contaminant concentrations in fishes from the 1980s to 2021; methylmercury was determined to be the contaminant of primary concern in fish tissues. Methylmercury tissue concentrations for varied fish species from four lakes and one river were used to establish contaminant-fish length relationships. Observed methylmercury tissue concentrations for these fishes allowed for the creation of updated consumption recommendations in MFN's subsistence fisheries. This study recommended updated consumption rates for fish species including Northern Pike (*Esox lucius*), Walleye (*Sander vitreus*), Lake Trout (*Salvelinus namaycush*), Burbot (*Lota lota*), Lake Whitefish (*Coregonus clupeaformis*), White Sucker (*Catostomus commersoni*), Longnose Sucker (*Catostomus catostomus*), and introduced Smallmouth Bass (*Micropterus dolomieu*). Elevated methylmercury concentrations followed increased eutrophication in these naturally oligotrophic watersheds from loading of plant nutrients, from both diffuse and defined regional sources. Nutrient mitigation measures to control *in situ* methylmercury production cannot be implemented as neither the nature or extent of past or current nutrient loading from various sources has been identified or estimated in the region.

Keywords: fish, eutrophication, methylmercury, Northern Ontario, lakes, rivers, subsistence fisheries

1. Introduction

There has been a worldwide increase in the levels of a variety of chemical substances in the natural environment, largely since the early 1900s [1–6]. These increases can be attributed to human activities, including agriculture, road building, dams, natural floods, forest harvest, forest fires, mining, burning of coal and wood, urbanization, exhaust from combustion engines, etc. Such activities have contributed to the widespread dispersal of chemical substances across ecosystems. When ecosystem components are disturbed, the scale and duration of the activity influence how chemical substances are distributed [6]. Disturbances can also result in immediate or delayed environmental effects on organisms ranging in magnitude from small to large and varying across organisms and life stages [6].

Some substances associated with ecosystem disturbances can bioconcentrate in the tissues of both plants and animals over time. Bioconcentration through direct exposure, as well as bioaccumulation via dietary uptake, can result in net bioconcentration/bioaccumulation within food webs, a process formally termed biomagnification. This process transfers substance(s) from water and soil to plants and then to wildlife in terrestrial ecosystems or from water and sediment to plants to fish to predators in aquatic ecosystems [6]. Due to differences across ecosystems habitats and species, and variability in the rate of transfer of substances, monitoring studies are required to manage this risk for humans and wildlife [3, 7, 8].

Such bioaccumulation/biomagnification across species and ecosystems justifies biomonitoring of traditionally harvested items such as berries, plants, fish, and wildlife. A key component is estimating typical consumption patterns for various foodstuffs that may contain substances considered hazardous at elevated concentrations. The most effective process is to measure concentrations in monitoring data for the food, along with samples from air, water, sediment, plants, fish, and wildlife [6].

The nature and degree of risks to human health vary with circumstances such as age, sex, and diet [6]. Although consumers of fish are exposed to increased concentrations of chemicals, Indigenous consumers are a particular concern as they may consume a higher proportion of wild-caught local fish, for cultural or subsistence reasons [6, 9, 10]. Thus, it is important to study the consumption of country foods separately from foods used in urban areas, to ensure consumption risks are understood, prevented, or mitigated.

Canada in general, and Ontario in particular, has been a leader in monitoring domestic freshwater fish contamination and recommending safe consumption rates for habitats ranging from rivers to lakes to reservoirs [3, 5, 6, 11]. The first Guide to Eating Ontario Sport Fish (GEOSF) was published in 1978 [12]. This public information program, with recommended human fish consumption by waterbody and species, continues to this day [13].

The main contaminants of concern (COCs) reported in GEOSF are mercury (bioavailable organic methylmercury in tissue, not inorganic mercury), dioxins, polychlorinated biphenyls (PCBs), and several pesticides (e.g., mirex, DDT, toxaphene). The standard GEOSF approach measures the concentration of chemical substances in a skinless, boneless sample of dorsal fish muscle. This measurement of chemical substances in muscle flesh represents a surrogate for other tissues within the fish and provides a consistent strategy to quantify consumption [6].

The inclusion of the chemical substances of concern within GEOSF reflects findings from studies linking dietary exposure to health impacts in humans and wildlife. The linkage between exposure to methylmercury with a range of health impacts has been documented within various settings across North America (reviewed in [6, 9]). A suite of studies [3, 5, 6, 11] have quantified the linkage between methylmercury in water or sediments of aquatic habitats and corresponding concentrations in different components of the food web (e.g., plants, invertebrates, fish, wildlife). These studies have linked concentrations of methylmercury in sediment with other components of the food web, and demonstrate how the concentration of methylmercury in fish and wildlife tissue reflects existing environmental conditions, especially total phosphorus (TP) [6].

1.1 Observation of eutrophication in northern Ontario forested watersheds

Studies in northern Ontario have quantified the relationship between disturbance within primarily forested watersheds leading to greater eutrophication of lakes and rivers. When these forested watersheds are disturbed, this can trigger episodic algal blooms and elevated growth of macrophytes with increasing frequency [1, 6, 14–16]. It is now understood such disturbance in forested watersheds leading to observation of eutrophication is linked to increases in TP loading to surface waters, particularly from high rates of atmospheric dust along with wood ash deposition [14–17]. Hence, disturbance of forested watersheds, starting with road building, forest harvest, or other anthropogenic activities, leads to the secondary effect of an increase in dust and ash, concomitant with elevated TP deposition to surface waters causing eutrophication. With this understanding, the general pattern of increased TP loading into lakes and rivers results in the unexpected effect of enhanced rates of biotransformation of inorganic mercury to methylmercury [3, 18–20]. However, in the nutrient-poor (i.e., oligotrophic) watersheds common on the Canadian Shield of northern Ontario, even modest increases in TP leads to eutrophication-enhanced increases in the rate of biotransformation of inorganic mercury already present in the watershed to methylmercury. This poses a problem for recreational and subsistence fishers, as the methylmercury in fish tissue has been observed to increase rapidly over time within forested watersheds.

Confirmation of the link between concentration of COCs, such as inorganic mercury and TP in the environment, with eutrophication-enhanced biomethylation and associated food web biomagnification of organic methylmercury into tissues of fishes and wildlife, is both the justification for monitoring ecosystem components and the means to manage risks related to consumption of contaminated fish and wildlife [3, 6, 9, 19].

Such monitoring allows for the identification of risk thresholds for varying consumption rates among species under differing scenarios across ecosystems [3, 6, 18, 21]. For example, how many portions of fish or wildlife can be safely consumed during a period by a person in a country setting compared with periodic consumption by an urban resident? Or, how many portions of fish or wildlife can be consumed by a predator such as Bald Eagle (*Haliateetus leucocephalus*) that consumes fish and wildlife or American Marten (*Martes americana*) that also consumes fish and wildlife from the same habitat as humans.

As circumstances can change over decades, monitoring of time trends in fish tissue methylmercury levels is vital for ongoing risk assessment. Monitoring fish populations provides the basis to update the status of regional fish tissue methylmercury concentrations and ensure that human and wildlife consumption guidance is sufficiently protective.

Unlike typical environmental monitoring programs, that focus on air, soil, water, and sediment concentrations, organism-based dose metrics (i.e., body- or tissue-based concentrations), such as those employed in the GEOSF program, largely bypass complex bioavailability, phase partitioning, and organism toxicokinetics issues and allow more direct residue-based risk assessment approaches of human food consumption pattern [22]. Thus, not only are data less confounded by extraneous modifying factors, the information and measurements are more readily interpretable and thereby easily communicated to non-technical parties. This process also simplifies the calculation of risk to human consumers and adds objectivity to the risk decision-making process [22, 23].

1.2 Management of nutrients that cause eutrophication in northern Ontario

In Ontario, the PWQO for TP addresses the aesthetic, algal bloom issue with 10, 20, 30 µg/l objectives for lakes and rivers, also noting that for all lakes naturally below 10 µg/l, the TP objective should be to remain below 10 µg/l.

Referring to these provisional objectives formulated in 1979, Ontario ([24], pp. 13–14) notes: "*Current scientific evidence is insufficient to develop a firm objective at this time. Accordingly, the following phosphorus concentrations should be considered as general guidelines which should be supplemented by site-specific studies* ". The current site-specific Ontario PWQO for phosphorus is provided in the Lakeshore Capacity Assessment Handbook with both the rationale and application methodology, Ontario ([25], pp. 13–14) stating: "*The revised PWQO for lakes on the Precambrian Shield allows a 50 per cent increase in phosphorus concentration from a modelled baseline of water quality in the absence of human influence*."

Thus, it is clear that the regulatory PWQO for TP in water bodies on the Precambrian Shield according to current Ontario regulations, is not the aesthetic-based objects of 10, 20, or 30 µg/l since the eutrophication problem is mercury methylation not aesthetic. Rather, site-specific objectives for lakes and rivers must be developed using the methods presented in Ontario [25]. Additionally, a key policy objective of the PWQO process stated by Ontario ([24], p. 6) is: "*Policy 2 states, "Water quality which presently does not meet the Provincial Water Quality Objectives shall not be degraded further and all practical measures shall be taken to upgrade the water quality to the Objectives*."

1.3 Problem identification

Studies in northwestern Ontario suggest when lakes are subjected to increased nutrient stress, the eutrophication process appears to be amplifying existing mercury cycling. This multi-trophic level process is evident across lake and river ecosystems, as supported by studies that indicated increased macrophyte and algal production stimulated by elevated nutrient levels. Such increases in plant production contributes to enhanced microbial production of methylmercury over relatively short periods [3, 20, 21]. Hence, with increased TP leading to eutrophication, concentrations of elemental mercury measured within ecosystems are not a key metric for mercury content in freshwater lake and rivers. Rather, the key driver of the rate of methylation of mercury is nutrient loading [18–20]. In this process, modest increases in annual plant nutrient loadings, particularly TP, are adequate to drive increases of *in situ* methylation with existing low inorganic mercury concentrations in water or sediment to produce sufficient amounts of very bioaccumulative methylmercury. A key diagnostic parameter for eutrophication stress in such situations is elevated mercury levels in sport fish.

Hence, an important first step for ecosystem assessment is to assess fish tissue concentrations for chemicals such as methylmercury, to quantify local patterns of risk to consumers of fish within traditional fisheries. The next important step is to reduce local anthropogenic sources of increased nutrient loading into watersheds. This approach is justified, as diets with a focus on country foods such as subsistence fisheries can have high proportions of food types with elevated mercury residues compared with consumers that do not regularly consume country foods.

1.4 Study design

Due to the paucity of recent fish contaminant monitoring information available within Michipicoten First Nation (MFN) territory, a study was recently completed [26]. The report documented verified observations from fish monitoring studies in MFN traditional territory. The habitats studied included primarily forested watersheds with seasonal cottages, roads, and/or campgrounds, a watershed with established large-scale metal mine, and a large river with upstream dams, as:

- Traditional fisheries in Dog Lake, Missinaibi Lake, Wawa Lake, and Michipicoten River; and

- Limited traditional fisheries in Nemegosenda Lake, downstream of Borden Gold Mine (mine opened in 2018, within Borden Lake/River watershed).

Fish species sampled for monitoring from these five habitats include Walleye (*Sander vitreus*), Northern Pike (*Esox lucius*), Lake Trout (*Salvelinus namaycush*), Burbot (*Lota lota*), Lake Whitefish (*Coregonus clupeaformis*), White Sucker (*Catostomus commersoni*), Longnose Sucker (*Catostomus catostomus*), and introduced Smallmouth Bass (*Micropterus dolomieu*). This suite of fish species represents predators that primarily consume fish, like Walleye, Northern Pike, and Burbot, species that consume fish and invertebrates, such as Lake Trout and Smallmouth Bass, and species that primarily consume plants and detritus, such as Lake Whitefish, White Sucker, and Longnose Sucker. Analysis of different fish species in lakes and rivers should identify any habitat-specific and/or species-specific patterns in fish methylmercury tissue concentrations.

1.5 Contaminants of concern

During 2019 and 2020, Staff from the Ontario Ministry of Environment, Conservation, and Parks (MECP) confirmed with MFN that routine fish tissue monitoring includes the following COCs: the metalloids arsenic and selenium, along with the metals: aluminum, cadmium, chromium, cobalt, copper, iron, lead, magnesium, manganese, molybdenum, nickel, tin, and zinc. Of this suite of COCs, mercury was identified as the primary COC for human health stemming from fish tissue consumption in recreational and subsistence fisheries. Thus, the focus of the analyses in this chapter concerns methylmercury. In northern Ontario, both metallic mercury (Hg) and organic methylmercury were historically low in oligotrophic waters across the Lake Superior watershed until the early 1980s [3–5]. During this historic period before the early 1980s, fish tissue concentrations across species and habitats were usually $\leq$0.06 μg/g WW. However, concentrations of methylmercury were observed to increase during recent decades, representing an identified risk to regular consumers of fish tissue at >0.06 μg/g WW [4, 5].

This study proposed to collect fish specimens from areas across MFN traditional territories and carry out fish tissue methylmercury analyses. In addition, the study proposed to search for other studies with verifiable fish tissue methylmercury analyses, completed since 2009, from other private and public sources. This cut-off date was used due to recent advances in analytical measurement of contaminants such as Hg in fish tissues. The private sources included the proponents at mines as well as selected MECP monitoring datasets. The study objectives were as follows:

1. to document past studies of fish in MFN territory, to better establish baseline records for COCs, such as methylmercury in fish tissues;

2. to compare the concentrations of COCs from these fish species to available recent data sets (last 10–15 years) from private and public sources;

3. to update the consumption guidelines for fishes in these habitats, and to share these findings with MFN residents and other stakeholders.

These biomonitoring study objectives were intended to achieve enhanced environmental management of fishes of importance to MFN within their traditional territory.

2. Methods and materials

2.1 Fish sample collections

The study was conducted from 2019 to 2022. It involved members of MFN as well as representatives from government departments of Canada and Ontario, along with representatives from private companies. From 2019 until late 2020, the MFN effort was led by Aaron Bumstead, Director of Lands and Economic Development. For 2021 and 2022, the MFN effort was led by Steven Murphy, Manager of Lands and Environmental Stewardship.

During this effort, traditional knowledge on fishing from MFN residents was collected by Mr. Bumstead, Mr. Murphy, and Dr. Fitzgerald. Information on fish harvest, fishing locations, and other details were also collated.

In addition, various MFN residents participated in fish collections for Northern Pike with this effort led by MFN Elder Paul Jaques. During the summer of 2020, only a small number of specimens were collected due to disruptions associated with the COVID-19 pandemic. Specimens were processed in the field, placed on ice and in bags, and then transported to secure freezers at MFN, awaiting analysis. In 2021, fishing was more extensive across select habitats, with specimens again processed in the field, placed on ice and in bags, and then transported to freezers at MFN. Habitats considered in this study are listed in **Table 1** and include Dog Lake, Michipicoten River, Missinaibi Lake, Nemegosenda Lake, and Wawa Lake. The types of known

Habitat name	Source of samples	Description of types of disturbance
Dog Lake	MFN, Ontario	Traditional MFN fishery, campground, cottages, roads
Michipicoten River	Ontario	Traditional MFN fishery, cottages, upstream dams, roads
Missinaibi Lake	Ontario	Traditional MFN fishery, campground, cottages, roads
Nemegosenda Lake	Ontario, private	Cottages, roads, upstream dam, receives water via Borden River, downstream of Borden Mine
Wawa Lake	Ontario	Traditional MFN fishery, cottages, Town of Wawa, roads

Source of samples as either MFN, Ontario, or private company.

Table 1.
Habitats where fish samples were assessed for this study.

disturbance for each habitat is also listed in **Table 1**. Fish species considered in this study across the habitats are listed in **Table 2** and **Figure 1**.

2.2 Fish sample processing

In the autumn 2021, Dr. Mary-Claire Buell of Collective Environmental Consulting (CEC) trained MFN residents for fish processing, and then supervised fish processing of frozen specimens by community members in Wawa, ON. She also arranged delivery of the processed fish samples to an accredited laboratory (SGS Laboratories, Lakefield, ON, Canada, KOL 2HO) for measurement of COCs in fish tissues. Fish

Habitat name	Fish species assessed in monitoring studies						
	Northern Pike	Walleye	Lake Trout	Smallmouth Bass	Burbot	Lake Whitefish	Longnose Sucker
Dog Lake	X	X	X	X	X	X	X
Michipicoten River	X	X		X		X	X
Missinaibi Lake	X	X	X		X		
Nemegosenda Lake	X	X	X	X			
Wawa Lake		X	X				

Table 2.
Summary of habitats and resident fish species assessed in this study.

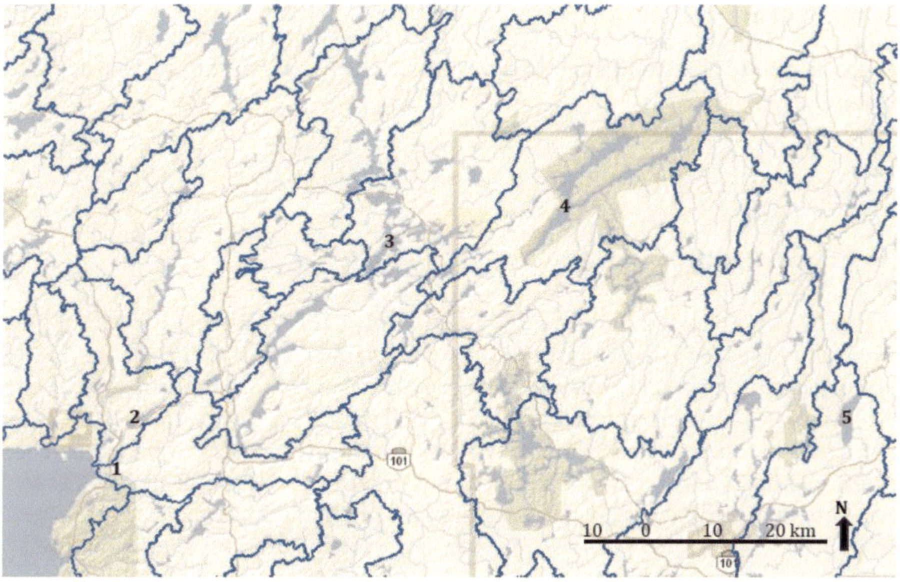

Figure 1.
Map of northern Ontario watershed boundaries (blue lines) with habitats under study identified. These habitats include: 1. Michipicoten River, 2. Wawa Lake, 3. Dog Lake, 4. Missinaibi Lake, and 5. Nemegosenda Lake. (Source of base map polygon: Ontario Watershed Boundaries).

processing options were limited at this time due to the disruptions caused by the COVID-19 pandemic.

All fish muscle tissue samples were processed by SGS Laboratories for a number of COCs: total mercury, arsenic, lead, aluminum, cadmium, chromium, cobalt, copper, iron, magnesium, manganese, molybdenum, nickel, selenium, tin, and zinc. The raw data, along with initial data processing (i.e., conversion to wet weight), were provided to MFN by CEC in February, 2022.

2.3 Statistical analyses

Concentrations of chemical substances in fish muscle tissues were provided to MFN by CEC, or obtained directly from the Ontario monitoring database, or from technical reports provided by private sources (e.g., mines). For these studies, the analytical results were verified as valid before analysis. Following verification, these results were transcribed to spreadsheets followed by statistical analysis including mean, standard deviation, and range. As most sample sizes for the fishes under study were small (<30 specimens), simple Model-1 linear regression was selected to quantify the relationships of mercury in fish tissues and fish length. Such small sample sizes for fish in the habitats under study were common across the Ontario and private data sets, as a goal of routine monitoring is for this effort to not modify the population under study and to avoid over-fishing over the monitoring period.

Availability of at least three sample years for Northern Pike within Dog Lake and Missinaibi Lake with reasonable sample sizes (often ≥10) allows for the opportunity to compare mercury tissue concentration relationship with fish length over time. The Analysis of Covariance (ANCOVA) was used to quantify how mercury tissue concentration changed over time with fish length as a covariate. The significance of the ANCOVA was interpreted at the probability value of $\alpha = 0.05$. In contrast, the analyses of regressions were considered significant if the R^2 (explanation of fit of regression to the data) was equal to or exceeded 0.5. This use of R^2 to assess the significance of the regression is due to the small sample sizes under study that do not justify detailed statistical analyses [27, 28].

3. Results

3.1 Traditional knowledge

At meetings completed between 2018 and 2021, MFN residents identified general fishing preferences for habitats around Wawa, Chapleau, and White River including preferred fish target species, typical fishing methods, and seasonal considerations. MFN residents also communicated that they fish for different species across habitats coinciding with seasonal movement of fishes noting safe periods to complete these activities. Fishing activities include ice fishing in winter with hook and line as well as with gillnets under the ice, and hook and line and gillnets during spring, summer, and autumn. Some fishing also uses fish traps in large lakes and rivers, to allow for undesirable fishes to be released alive.

The residents readily identified Northern Pike and Walleye as the two primary species harvested and captured across the entire calendar year. Other residents identified Lake Trout and Lake Whitefish as the species they usually target in autumn. Other species fished and generally regarded as less popular included: White Sucker,

DOI: http://dx.doi.org/10.5772/intechopen.113368

Fish species	Habitat fishes	Preferred season
Northern Pike	Lakes, rivers, creeks	All seasons
Walleye	Lake, rivers	All seasons
Lake Trout	Lakes	Spring, autumn, winter
Burbot	Lakes	Autumn, winter
Lake Whitefish	Lakes	Spring, autumn, winter
White Sucker	Lakes, rivers	Spring, autumn
Longnose Sucker	Lakes, rivers	Spring, autumn
Smallmouth Bass	Lakes, rivers, creeks	Summer, autumn

The common fishing methods by habitat and species as reported.

Table 3.
Overview of fish species harvested by MFN residents across habitats and the preferred season of harvest.

Longnose Sucker, Yellow Perch (*Perca flavescens*), Smallmouth Bass, and Burbot. Some MFN residents noted that Smallmouth Bass were stocked across MFN territory by Ontario during the early 1900s. Stocking of Smallmouth Bass was often in habitats with limited or no sport fishing opportunities while sometimes included lakes with Lake Trout and/or Walleye populations. A summary of fish captured across habitats, fishing gear, and seasons is presented in **Table 3**.

3.2 Available information from fish monitoring studies

3.2.1 Northern Pike

3.2.1.1 Morphology of Northern Pike samples collected by MFN in Dog Lake

Northern Pike samples were assessed for tissue body residues collected by MFN during 2021 from Dog Lake had a mean total length of 49.8 cm (range: 34.7–77.1 cm; SD = 10.97; n = 22) and mean total weight of 851 g (255–3121; 726; 22). Sex of the specimens was not recorded. A number of specimens were observed to have external sores, external parasites, fin damage, and other evidence of stress. All fish had small external sores with approximately 10–15% with the presence of external parasites and fin damage. Some specimens had healed sores suggesting past encounters with fish-eating birds or fish-eating wildlife; some had evidence of open sores and healed sores with some likely caused during capture. However, no specimens demonstrated major skeletal deformities such as scoliosis or external tumors. This information from MFN for Northern Pike was combined with data sets identified from other sources (Ontario, private companies), to examine patterns across habitats.

3.3 Comparison of fish species across habitats

3.3.1 Northern Pike

Estimates of mercury tissue concentrations in Northern Pike are reported for Dog Lake, Missinaibi Lake, Nemegosenda Lake, and Michipicoten River (**Table 4**). For

Habitat	Northern Pike year of sample: mean mercury tissue Concentrations (μg/g WW) for male and female fish or range		
Dog Lake	**2011: 0.62 (0.5, 0.8)**	**2016: 0.44 (0.43, 0.46)**	**2021: 0.30 (0.11–0.98)**
Missinaibi Lake	2009: 1.14 (0.89, 1.28)	2011: 0.89 (0.61, 1.04)	2016: 0.94 (0.63, 1.18)
Nemegosenda Lake		2015: 1.14 (NF*, 1.14)	
Michipicoten River			2018: 0.26 (0.27, 0.24)

**NF: no male fish captured.*

Table 4.
Comparison of the mercury tissue concentrations for Northern Pike dorsal muscle flesh in male and female fish, available across years in Dog Lake, Missinaibi Lake, Nemegosenda Lake, and Michipicoten River.

Northern Pike sampled in Dog Lake, the 2011 mean mercury tissue concentration was 0.62 μg/g WW with a range of 0.12 μg/g WW for a 33.3 cm specimen to 1.70 μg/g WW for a 74.5 cm specimen. For the 2016 sample from Dog Lake, the mean mercury tissue concentration was 0.44 μg/g WW with a range of 0.11 μg/g WW for a 39.7 cm specimen to 0.81 μg/g WW for a 68.5 cm specimen. For the 2021 sample from Dog Lake, the mean mercury tissue concentration was 0.30 μg/g WW with a range of 0.11 μg/g WW for a 36.8 cm specimen to 0.98 μg/g WW for a 77.1 cm specimen.

For Missinaibi Lake, the 2009 mean mercury tissue concentration was 1.14 μg/g WW with a range of 0.25 μg/g WW for a 41.5 cm specimen to 2.5 μg/g WW for an 87.3 cm specimen. For the 2011 sample from Missinaibi Lake, the mean mercury concentration was 0.89 μg/g WW with a range of 0.24 μg/g WW for a 46.8 cm specimen to 2.2 μg/g WW for a 71.4 cm specimen. For the 2016 sample from Missinaibi Lake, the mean mercury concentration was 0.94 μg/g WW with a range of 0.39 μg/g WW for a 55.6 cm specimen to 1.7 μg/g WW for an 84.9 cm specimen.

For the 2015 sample from Nemegosenda Lake, the mean mercury concentration was 1.14 μg/g WW with a range of 0.88 μg/g WW for a 52.6 cm specimen to 1.6 μg/g WW for a 94.2 cm specimen. For the Michipicoten River, the 2018 mean mercury tissue concentration was 0.26 μg/g WW with a range of 0.24 μg/g WW for a 56.5 cm specimen to 0.3 μg/g WW for a 60.5 cm specimen.

These observations for Northern Pike indicated the mercury tissue concentration in male and female specimens were similar within Michipicoten River and variable between sexes for Dog Lake, Missinaibi Lake, and Nemegosenda Lake.

The relationship between mercury tissue concentration and length for Northern Pike based on the observations for specimens from the Dog Lake and Missinaibi Lake are presented in **Figure 2**. These Northern Pike show linear relationships between mercury muscle tissue concentration and total length:

Dog Lake 2009: $Y = 0.24–0.85$; $R^2 = 0.465$;
Dog Lake, 2016: $Y = 0.018x - 0.60$, $R^2 = 0.806$;
Dog Lake 2021: $Y = 0.015–0.47$, $R^2 = 0.730$;
Missinaibi Lake 2009: $Y = 0.047x - 1.97$, $R^2 = 0.870$;
Missinaibi Lake 2011: $Y = 0.032x - 1.13$, $R^2 = 0.578$; and
Missinaibi Lake 2016: $Y = 0.040x - 1.73$, $R^2 = 0.941$.

The exception to the observation of linear relationships was in the 2009 Northern Pike sample from Dog Lake (i.e., $Y = 0.24–0.85$; $R^2 = 0.465$). Two fish with a total length of 66.8 cm had mercury tissue concentrations of 0.49 and 1.5 μg/g WW

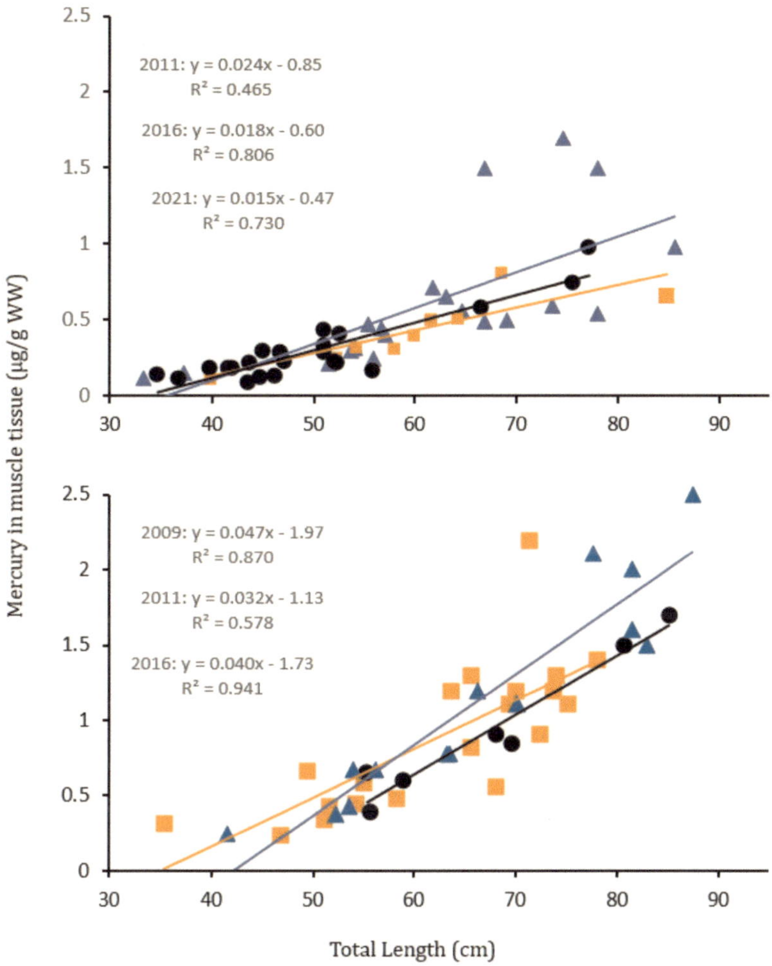

Figure 2.
Upper panel: the relationship between total length and mercury in muscle tissue for Northern Pike sampled within Dog Lake during 2011 (▲), 2018 (■) and 2021 (●). The lower panel is similar except for the Northern Pike sampled from Missinaibi Lake in 2009 (▲), 2011 (■), and 2016 (●). For the Northern Pike in Dog Lake, the methylmercury fish tissue concentration in muscle equals ~0.06 μg/g WW (Ontario recommendation for sensitive population wishing to consume 32–16 meals per month) is near 30 cm and 0.5 μg/g (Ontario "do not eat" recommendation for sensitive populations) occurs with fish lengths near 50 cm. For Northern Pike in Missinaibi Lake, fish tissue concentration in muscle equals ~0.06 μg/g WW (Ontario recommendation for sensitive population wishing to consume 32–16 meals per month) near 35 cm and 0.5 μg/g (Ontario "do not eat" recommendation for sensitive populations) occurs with fish lengths near 55 cm.

whereas two other fish with a total length of 77.9 cm had mercury tissue concentrations of 0.54 and 1.5 μg/g WW. The disparity in the observed mercury tissue concentrations in these four specimens represents natural variability in the samples and is why the relationship does not appear to be linear. For the other five samples from Dog Lake and Missinaibi Lake, the observations for Northern Pike mercury tissue concentrations support the hypothesis that the tissue concentration increases with age, as represented by the length of the specimen. Thus, the Northern Pike total length represents a reasonable surrogate for expected mercury tissue concentration.

The Dog Lake samples from 2011, 2016, and 2021 (**Figure 2**) compared by ANCOVA, included homogeneous slopes and confirmed the comparison was valid.

Then the ANCOVA was identified as non-significant ($F_{2,48}$ = 1.602, P = 0.212). This comparison indicates the mean mercury tissue concentrations did not differ significantly over the 3 years. For the Missinaibi Lake samples from 2009, 2011, and 2016 compared by ANCOVA, the slopes were homogeneous and confirmed the comparison was valid. Then the ANCOVA was found to be non-significant ($F_{2,37}$ = 1.619, P = 0.212). These results indicate the methylmercury: total length relationship remained similar across the 3 sample years for both lakes during the study period.

3.3.2 Walleye

Estimates of mercury tissue concentrations in Walleye are reported for Dog Lake, Missinaibi Lake, Michipicoten River, Nemegosenda Lake, and Wawa Lake (**Table 5**). For Walleye sampled in Dog Lake, the 2011 mean mercury tissue concentration was 0.56 µg/g WW with a range of 0.24 µg/g WW for a 38.8 cm specimen to 1.80 µg/g WW for a 71.4 cm specimen. For the 2016 sample from Dog Lake, the mean mercury tissue concentration was 0.70 µg/g WW with a range of 0.23 µg/g WW for a 36 cm specimen to 1.7 µg/g WW for a 60.5 cm specimen.

For Missinaibi Lake, the 2011 mean mercury tissue concentration was 1.24 µg/g WW with a range of 0.35 µg/g WW for a 30 cm specimen to 5.1 µg/g WW for a 75 cm specimen. For the 2016 sample from Missinaibi Lake, the mean mercury concentration was 1.41 µg/g WW with a range of 1.0 µg/g WW for a 31.6 cm specimen to 2.1 µg/g WW for a 71.6 cm specimen. For the Michipicoten River, the 2017 mean mercury tissue concentration was 0.98 µg/g WW with a range of 0.27 µg/g WW for a 35.3 cm specimen to 1.40 µg/g WW for a 44.8 cm specimen.

Nemegosenda Lake, the 2010 mean mercury tissue concentration was 0.79 µg/g WW with a range of 0.45 µg/g WW for a 32.2 cm specimen to 1.40 µg/g WW for a 45.4 cm specimen. For the 2015 sample from Nemegosenda Lake, the 2012 mean mercury tissue concentration was 1.96 µg/g WW with a range of 0.89 µg/g WW for a 36.6 cm specimen to 2.90 µg/g WW for a 55.8 cm specimen. For Wawa Lake, the 2012 mean mercury tissue concentration was 0.36 µg/g WW with a range of 0.10 µg/g WW for a 20.9 cm specimen to 0.99 µg/g WW for a 66.3 cm specimen. For the 2017 sample from Wawa Lake, the mean mercury tissue concentration was 0.70 µg/g WW with a range of 0.45 µg/g WW for a 53 cm specimen to 0.89 µg/g WW for a 63.4 cm specimen.

The above observations for Walleye indicate that mercury tissue concentrations in male and female specimens were generally similar within Michipicoten River and variable for Dog Lake, Missinaibi Lake, Nemegosenda Lake, and Wawa Lake.

Habitat	**Walleye year of sample: mean mercury tissue concentrations from muscle flesh (µg/g WW) for male and female fish**	
Dog Lake	**2011: 0.56 (0.47, 0.67)**	**2016: 0.70 (0.98, 0.65)**
Missinaibi Lake	2011: 1.24 (0.65, 1.62)	2016: 1.41 (1.37, 1.44)
Michipicoten River		2017: 0.98 (0.97, 1.0)
Nemegosenda Lake	2010: 0.79 (0.69, 0.94)	2015: 1.96 (1.3, 2.18)
Wawa Lake	2012: 0.36 (0.18, 0.44)	2017: 0.70 (0.64, 0.84)

Table 5.
Comparison of the mercury tissue concentrations for Walleye dorsal muscle flesh in male and female fish, available across years in Dog Lake, Missinaibi Lake, Nemegosenda Lake, Wawa Lake, and Michipicoten River.

The relationship between mercury tissue concentration and length for Walleye, for specimens from 2011 and 2016 samples from Dog Lake and Missinaibi Lake are presented in **Figure 3**. These Walleye all show linear relationships between mercury muscle tissue concentration and total length for Dog Lake (2011: $Y = 0.035x - 1.08$, $R^2 = 0.849$; 2016: $Y = 0.039–1.18$, $R^2 = 0.713$) and Missinaibi Lake (2011: $Y = 0.064x - 1.74$, $R^2 = 0.720$; 2016: $Y = 0.027x - 0.13$, $R^2 = 0.752$).

These observations for Walleye indicate mercury tissue concentration follows the hypothesis that the tissue concentration increases with age, as represented by the length of the specimen. Thus, Walleye total length is a reasonable surrogate for expected mercury tissue concentration.

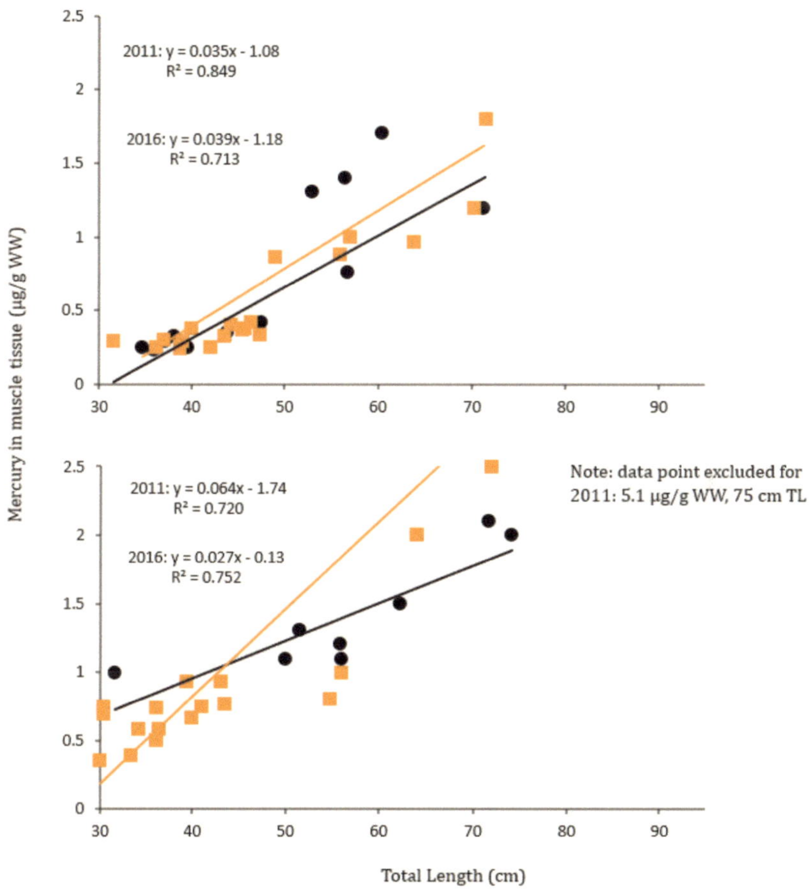

Figure 3.
Upper panel: the relationship between total length and mercury in muscle tissue for Walleye sampled within Dog Lake during 2011 (■) and 2016 (●). The lower panel is similar except for Walleye sampled from Missinaibi Lake during 2011 (■), and 2016 (●). One data point was excluded from the Missinaibi Lake sample from 2011 for a 75 cm specimen with 5.1 µg/g WW mercury in tissue. For the Walleye in Dog Lake, the methylmercury fish tissue concentration in muscle equals ~0.06 µg/g WW (Ontario recommendation for sensitive population wishing to consume 32–16 meals per month) is near 35 cm and 0.5 µg/g (Ontario "do not eat" recommendation for sensitive populations) occurs with fish lengths near 55 cm. For Walleye in Missinaibi Lake, fish tissue concentration in muscle equals ~0.06 µg/g WW (Ontario recommendation for sensitive population wishing to consume 32–16 meals per month) near 25 cm and 0.5 µg/g (Ontario "do not eat" recommendation for sensitive populations) occurs with fish lengths near 35 cm.

3.3.3 Lake Trout

Estimates of mercury tissue concentrations in Lake Trout are reported for Dog Lake, Missinaibi Lake, Nemegosenda Lake, and Wawa Lake (**Table 6**). For Lake Trout sampled in Dog Lake, the 2011 mean mercury tissue concentration was 0.49 μg/g WW with a range of 0.16 μg/g WW for a 34 cm specimen to 1.50 μg/g WW for a 92 cm specimen. For the 2016 sample from Dog Lake, the mean mercury tissue concentration was 0.88 μg/g WW with a range of 0.43 μg/g WW for a 56.6 cm specimen to 1.8 μg/g WW for an 83.6 cm specimen.

For Missinaibi Lake, the 2011 mean mercury tissue concentration was 0.63 μg/g WW with a range of 0.32 μg/g WW for a 33.8 cm specimen to 1.4 μg/g WW for a 90.4 cm specimen. For the 2016 sample from Missinaibi Lake, the mean mercury concentration was 0.94 μg/g WW with a range of 0.60 μg/g WW for a 64.9 cm specimen to 1.7 μg/g WW for a 75.8 cm specimen.

For Nemegosenda Lake, the 2010 mean mercury tissue concentration was 1.23 μg/g WW with a range of 0.96 μg/g WW for a 53 cm specimen to 1.60 μg/g WW for a 63.5 cm specimen. For the 2015 sample from Nemegosenda Lake, the mean mercury tissue concentration was 1.62 μg/g WW with a range of 1.20 μg/g WW for a 63.6 cm specimen to 2.10 μg/g WW for a 72.6 cm specimen. For Wawa Lake, the 2012 mean mercury tissue concentration was 0.40 μg/g WW with a range of 0.19 μg/g WW for a 21.5 cm specimen to 0.91 μg/g WW for a 78.1 cm specimen. For the 2017 sample from Wawa Lake, the mean mercury tissue concentration was 0.42 μg/g WW with a range of 0.18 μg/g WW for a 32.7 cm specimen to 0.87 μg/g WW for a 76.5 cm specimen.

The above observations for Lake Trout indicate the mercury tissue concentration in male and female specimens was generally similar in Nemegosenda Lake and Wawa Lake and sometimes different for Dog Lake and Missinaibi Lake.

The relationship between mercury tissue concentration and length of Lake Trout for specimens from 2011 and 2016 samples from Dog Lake and Missinaibi Lake are presented in **Figure 4**. These Lake Trout all show linear relationships between mercury muscle tissue concentration and total length for Dog Lake (2011: $Y = 0.017x - 0.046$, $R^2 = 0.867$; 2016: $Y = 0.030–1.03$, $R^2 = 0.641$) and Missinaibi Lake (2011: $Y = 0.16x - 0.35$, $R^2 = 0.767$; 2016: $Y = 0.044x - 1.94$, $R^2 = 0.625$).

The above observations for Lake Trout indicated mercury tissue concentration follows the hypothesis that the tissue concentration increases with age, as represented by the length of the specimen. Thus, Lake Trout total length is a reasonable surrogate for expected mercury tissue concentration.

Habitat	**Lake trout year of sample: mean mercury tissue concentrations from muscle flesh (μg/g WW) for male and female fish or range**	
Dog Lake	**2011: 0.49 (0.59, 0.43)**	**2016: 0.88 (1.11, 0.75)**
Missinaibi Lake	2011: 0.63 (0.55, 0.69)	2016: 0.94 (0.60, 1.02)
Nemegosenda Lake	2010: 1.23 (1.13, 1.3)	2015: 1.62 (1.47, 1.62)
Wawa Lake	2012: 0.40 (0.39, 0.41)	2017: 0.42 (0.42, 0.43)

Table 6.
Comparison of the mercury tissue concentrations for Lake Trout dorsal muscle flesh in male and female fish, available across years in Dog Lake, Missinaibi Lake, Nemegosenda Lake, and Wawa Lake.

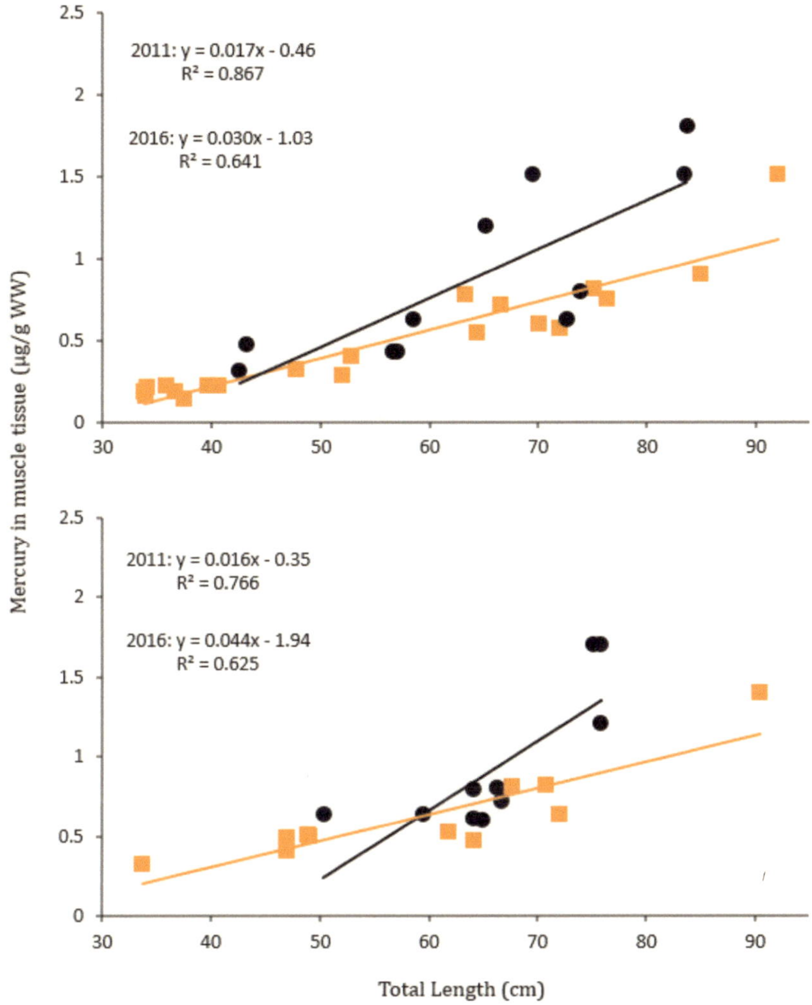

Figure 4.
Upper panel: relationship between total length and mercury in muscle tissue for Lake Trout sampled within Dog Lake during 2011 (■) and 2016 (●). Lower panel is similar except for Lake Trout sampled from Missinaibi Lake during 2011 (■), and 2016 (●). For the Lake Trout in Dog Lake, the methylmercury fish tissue concentration in muscle equals ~0.06 µg/g WW (Ontario recommendation for sensitive population wishing to consume 32–16 meals per month) is near 35 cm and 0.5 µg/g (Ontario "do not eat" recommendation for sensitive populations) occurs with fish lengths near 60 cm. For the Lake Trout in Missinaibi Lake, fish tissue concentration in muscle equals ~0.06 µg/g WW (Ontario recommendation for sensitive population wishing to consume 32–16 meals per month) near 30 cm and 0.5 µg/g (Ontario "do not eat" recommendation for sensitive populations) occurs with fish lengths near 55 cm.

3.3.4 Smallmouth Bass

Estimates of mercury tissue concentrations in Smallmouth Bass are reported for Dog Lake, Missinaibi Lake, Nemegosenda Lake, and Michipicoten River (**Table 7**). For Smallmouth Bass sampled in 2016 in Dog Lake, the mean mercury tissue concentration was 0.40 µg/g WW with a range of 0.25 µg/g WW for a 37.6 cm specimen to 0.7 µg/g WW for a 43.6 cm specimen.

Habitat	Smallmouth Bass year of sample: mean mercury tissue concentrations from muscle flesh (μg/g WW) for male and female fish or range
Dog Lake	2016: 0.40 (0.28, 0.52)
Missinaibi Lake	2016: 0.57 (range: 0.43–0.65)*
Nemegosenda Lake	2015: 0.60 (0.55–0.63)*
Michipicoten River	2017: 0.29 (0.30, 0.27)

**A total of three specimens for these habitats.*

Table 7.
Comparison of the mercury tissue concentrations for Smallmouth Bass dorsal muscle flesh in male and female fish, available across years in Dog Lake, Missinaibi Lake, Nemegosenda Lake, and Michipicoten River.

For the 2016 sample from Missinaibi Lake, the mean mercury tissue concentration was 0.57 μg/g WW with a range of 0.43 μg/g WW for a 36.2 cm specimen to 0.65 μg/g WW for a 39 cm specimen representing three females.

For the 2015 sample from Nemegosenda Lake, the mean mercury tissue concentration was 0.60 μg/g WW with a range of 0.55 μg/g WW for a 48.1 cm specimen to 0.63 μg/g WW for a 34.5 cm specimen representing three females.

For the 2017 sample from the Michipicoten River, the mean mercury tissue concentration was 0.29 μg/g WW with a range of 0.17 μg/g WW for a 28.7 cm specimen to 0.48 μg/g WW for a 36.9 cm.

These observations for Smallmouth Bass indicate mercury tissue concentrations follow the hypothesis that the tissue concentration increases with age, as represented by the length of the specimen. Thus, Smallmouth Bass total length is a reasonable surrogate for expected mercury tissue concentration.

3.3.5 Burbot

Estimates of mercury tissue concentrations in Burbot are reported for Dog Lake and Missinaibi Lake (**Table 8**). For Dog Lake, Burbot sampled during 2016 had a mean mercury tissue concentration of 0.43 μg/g WW with a range of 0.32 μg/g WW for a 40 cm specimen to 0.52 μg/g WW for a 58.5 cm specimen.

For the 2009 sample from Missinaibi Lake, the mean mercury tissue concentration was 0.78 μg/g WW with a range of 0.37 μg/g WW for a 40.8 cm specimen to 1.2 μg/g WW for a 73 cm specimen. These observations indicate the mercury tissue concentration in male and female specimens was generally similar in Dog Lake and Missinaibi Lake.

These observations for Burbot indicate the mercury tissue concentration follows the hypothesis that the tissue concentration increases with age, as represented by the length of the specimen. Thus, Burbot total length represents a reasonable surrogate for expected mercury tissue concentration.

Habitat	Burbot mean mercury tissue concentrations from muscle flesh (μg/g WW) for male and female fish
Dog Lake	2016: 0.43 (0.51, 0.40)
Missinaibi Lake	2009: 0.78 (0.70, 0.82)

Table 8.
Comparison of the mercury tissue concentrations for Burbot dorsal muscle flesh in male and female fish, available across years in Dog Lake and Missinaibi Lake.

Habitat	Mean mercury tissue concentrations from muscle flesh (µg/g WW) for male and female fish		
	Lake Whitefish	White Sucker	Longnose Sucker
Dog Lake	2016: 0.15 (0.15, 0.15)	2016: 0.18 (0.12, 0.19)	2016: 0.26 (0.30, 0.20)
Michipicoten River		2017: 0.22* (0.22; *NF)	2018: 0.18 (0.17, 0.24)

**Sample based on two specimens; NF: no female fish sampled.*

Table 9.
Comparison of the mercury tissue concentrations for Lake Whitefish, White Sucker, and Longnose Sucker dorsal muscle flesh in male and female fish, available across years in Dog Lake and Michipicoten River.

3.3.6 Lake Whitefish, Common Sucker, and Longnose Sucker

Estimates of mercury muscle tissue concentrations for Lake Whitefish, White Sucker, and Longnose Sucker are reported for Dog Lake and Michipicoten River (**Table 9**). For Lake Whitefish sampled in 2016 from Dog Lake, the mean mercury muscle concentration was 0.15 µg/g WW with a range of 0.1 µg/g for a 38 cm specimen to 0.19 µg/g WW for a 50.4 cm specimen. For White Sucker in 2016 from Dog Lake, the mean mercury concentration was 0.18 µg/g WW with a range of 0.07 µg/g WW for a 49.4 cm specimen to 0.34 µg/g WW for a 46.2 cm specimen.

The 2017 White Sucker from Michipicoten River had a mean of 0.22 µg/g WW with a range of 0.22 µg/g WW for a 45.6 cm specimen to 0.23 µg/g WW using a sample of $n = 2$ males. For Longnose Sucker sampled in 2016 from Dog Lake, the mean mercury tissue concentration was 0.26 µg/g WW with a range of 0.11 µg/g WW for a 53.4 cm specimen to 0.39 µg/g WW for a 55.4 cm specimen.

The 2018 Longnose Sucker from Michipicoten River had a mean mercury tissue concentration of 0.18 µg/g WW with a range of 0.06 µg/g WW for a 38.4 cm specimen to 0.36 µg/g WW for a 41.8 cm specimen. For all three benthic species, the male and female mercury tissue concentrations were generally similar (**Table 9**).

For these three fish species, the Dog Lake mean mercury muscle concentrations were consistently higher than samples sampled from the Michipicoten River. In addition, for Lake Whitefish, White Sucker, and Longnose Sucker, these observations indicate the mercury tissue concentration follows the hypothesis that the tissue concentration increases with age, as represented by larger lengths of the specimens. For all three species, total length appears to be an adequate surrogate for expected mercury tissue concentration. Also, for these three species, the mercury concentrations increased with length but more gradually compared with piscivores like Northern Pike, Walleye, and Burbot. This more gradual increase in mercury muscle tissue concentrations in the benthic species is attributed to their diet with greater proportions of plant material and invertebrates with lower mercury concentrations compared with diets composed of other fishes containing higher mercury concentrations. As can be seen in **Figure 5**, with for Lake Whitefish from Dog Lake exhibits a linear relationship between mercury muscle tissue concentration and total length ($Y = 0.004x - 0.04$, $R^2 = 0.515$).

3.4 Comparison of findings for recent fish monitoring with current Ontario online consumption guidelines

Inspection of fish length:mercury relationships in MFN territory reported herein led to the observation that the Ontario online consumption guide generally reflects

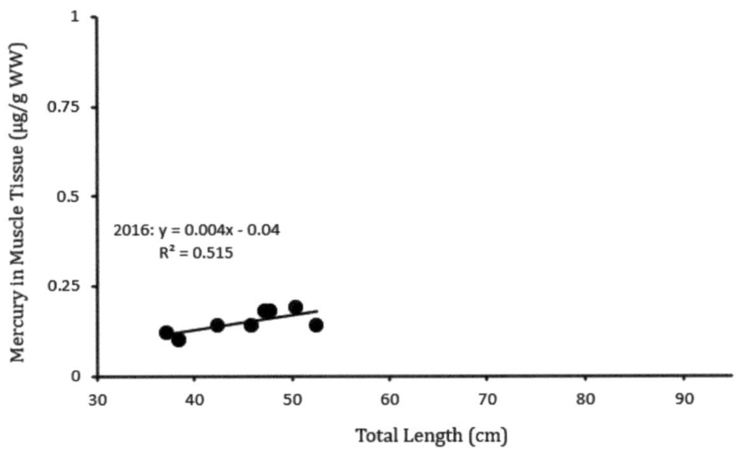

Figure 5.
Relationship between total length and mercury in muscle tissue for Lake Whitefish sampled within Dog Lake during 2016 (●). For these Lake Whitefish in Dog Lake, the methylmercury fish tissue concentration in muscle equals ~0.06 µg/g WW (Ontario recommendation for sensitive population wishing to consume 32–16 meals per month) is near 35 cm and 0.5 µg/g (Ontario "do not eat" recommendation for sensitive populations) occurs with fish lengths beyond 90 cm (i.e., beyond length distribution in sample).

fish length when ~0.06 µg/g WW is first evident in the tissue monitoring data. For this analysis, the length (+5 cm) at ~0.06 µg/g WW was estimated from the Model-1 length:mercury regression relationships for fish from the habitats under study. Also, the length for initial consumption restriction from the Ontario online guide for each species was identified and both sets of information were compared. This comparison revealed that, for most species, when mercury tissue residue observations are available, the online consumption guidelines frequently started near or at the length where mercury tissue monitoring data concentrations were ~0.06 µg/g WW.

4. Discussion

Available historical public and private studies of fish in MFN territory were assessed. For each study, the data was verified as validated before this assessment. This verification and validation process resulted in the identification of data sets often not readily available in the public domain. As a complement to this survey of past studies, fish sample collections were also completed by MFN in 2021, and resulted in the independent estimate of current concentrations of COCs including mercury in Northern Pike from Dog Lake and other habitats.

The 2021 data collected by MFN also confirmed that only mercury consistently exceeded human health consumption guides while other chemicals such as metals and metalloids were usually present at concentrations in tissue considered to be of low risk to fish consumers; these findings were consistent with information reported in 2020 by MECP. Before these studies, the information available to MFN residents via public sources was considered out-of-date with the community's perception that these consumption guidelines were not useful.

The analyses herein also identified a spatial pattern of increased mercury concentrations in all habitats across watersheds with larger and older fishes across seven species, and this suggests a regional process is shaping increases in mercury concentrations. These results collectively indicate the older (and larger) the fish, the higher mercury tissue concentration.

The identification of higher concentrations over time can be partially interpreted for this study to also represent the presence of smaller and younger fish in the more recent samples for some habitats compared with larger and older specimens assessed in the earlier samples. This time series pattern of smaller fish captured during the most recent sample collections may reflect the presence of over-fishing in some habitats as well as the confounding effect of excessive sample collections in monitoring programs.

Where monitoring data is available, larger and older fish in a habitat consistently have higher mercury concentrations compared with smaller specimens. This pattern was consistent for fish species that primarily eat other fish, such as Northern Pike, Walleye, Burbot, and Lake Trout. Species with mixed diets like Smallmouth Bass also demonstrated elevated mercury tissue content, likely reflecting the pattern of larger bass showing a diet preference for other fish while small bass often consume predominantly invertebrates; an interesting facet for Smallmouth Bass is they do not appear to have achieved large population sizes since introductions in the early 1900s in these northern habitats [29]. In contrast, species that consume mud, plants, and invertebrates such as Lake Whitefish, Common White Sucker, and Longnose sucker demonstrated lower mercury tissue concentrations.

This assessment, for the last 15 or so years, identified that mercury concentrations were reported as higher during past years compared with more recent years. Similar results were reported previously for species such as Northern Pike, Walleye, and Lake Trout mercury concentrations with increased mercury concentrations of >20% from 1985 to 2005 [4, 5].

With these general findings, no matter the cause of the smaller fish lengths during recent years, a regional pattern is evident represented by the presence of a mercury: body length relationship where mercury consistently increases over time in lakes and rivers. This study generated evidence of this pattern from seven species where monitoring information was available. Hence, mercury accumulation is largely a species-independent pattern across MFN territory with higher concentrations in predatory fishes such as Northern Pike, Walleye, Burbot, and Lake Trout with lower concentrations in benthic species such as Lake Whitefish, White Sucker, and Longnose Sucker. The observations also revealed intermediate mercury levels for smaller Smallmouth Bass with elevated concentrations in larger Smallmouth Bass that likely consume other fish. Thus, local sources of disturbances in combination with regional processes are likely acting together, leading to the observed increase in mercury in all fish, with implications for human and wildlife consumption of fish.

Water mercury concentrations are not a key metric to track changes in mercury content in freshwater lakes and rivers, as the key driver of the rate of methylation of mercury is increased plant nutrient loading [3, 18–20]. In aquatic ecosystems, modest increases in annual plant nutrient loadings, particularly TP, are sufficient to drive increases with *in situ* methylation of existing low inorganic mercury concentrations in water or sediment to produce the necessary amounts of very bioaccumulative methylmercury. A key diagnostic parameter for eutrophication stress in such situations is elevated mercury levels in fish.

Hence, an important first step for ecosystem assessment is to assess fish tissue concentrations involving species representing different feeding guilds for chemicals such as mercury, to quantify local patterns of risk to consumers of fish within sport or subsistence fisheries. The next important step is to minimize local anthropogenic sources of increased nutrient loading into sensitive watersheds. This approach is justified, as subsistence fisheries include individuals who often rely on other country foods, and can have high proportions of food types with elevated mercury residues compared with consumers who do not regularly consume country foods.

4.1 Management of nutrients that cause eutrophication in Ontario

In Ontario, the PWQO for TP addresses the aesthetic, algal bloom issue with the 10, 20, and 30 μg/l objectives for lakes and rivers. Although these concentrations for TP are referred to as water quality objectives, they represent interim guidelines developed in 1979 as a strategy to control eutrophication. Referring to these objectives, Ontario ([24], pp. 13–14) states: "*Current scientific evidence is insufficient to develop a firm Objective at this time. Accordingly, the following phosphorus concentrations should be considered as general guidelines which should be supplemented by site-specific studies ...* ".

Thus, the current site-specific Ontario PWQO for TP is provided in the Lakeshore Capacity Assessment Handbook with both the rationale and application methodology. Specifically, Ontario ([25], pp. 13–14) states: "*The revised PWQO for lakes on the Precambrian Shield allows a 50 percent increase in phosphorus concentration from a modeled baseline of water quality in the absence of human influence.*" It is clear that the PWQO for TP in water bodies on the Precambrian Shield according to current Ontario regulations, is not the aesthetic-based objectives of 10, 20, or 30 μg/l. Rather, site-specific objectives for lakes and rivers must be developed using the methods presented in Ontario [25].

An additional key policy objective in the PWQO process is included in Ontario ([24], p. 6): Policy 2 states, "*Water quality which presently does not meet the Provincial Water Quality Objectives shall not be degraded further and all practical measures shall be taken to upgrade the water quality to the Objectives.*"

The fish tissue methylmercury data presented herein confirms that various water bodies in MFN's traditional territories are currently subject to significant eutrophication stress. The confirmation of eutrophication extends from elevated methylmercury in fish tissue to observations reported by MFN residents of periodic algal blooms and more extensive submerged aquatic vegetation during recent years compared to the past for these habitats and is consistent with other technical studies that have also reported the presence of eutrophication in these different areas [1, 14, 15, 18, 19].

Thus, site-specific TP PWQOs are required to be developed to limit TP loadings to local watersheds due to the regional pattern of evidence of eutrophication [17]. As eutrophication is also caused/enhanced by nitrogenous substances, [ammonia, nitrate, nitrate, Total Kjeldahl nitrogen (TKN)], which are also plant nutrients, loadings of these substances should also be limited. As well, total organic carbon (TOC) is a parameter reflecting general aquatic nutrient status. This data suggests a safe rule of thumb is most surface waters are suffering from excessive nutrient loads, consistent with observations for the Lake Superior watershed [30, 31], and from other parts of Ontario [21, 32–34].

Studies since 2018 led by MFN for proposed and current large-scale activities (e.g., mines) within MFN traditional territory identified evidence that demonstrates the release of nutrients such as phosphorus and nitrogen above government guidelines. Evidence of eutrophication has also been documented or reported by others for habitats across Ontario that receive this effluent with elevated phosphorus and nitrogen [3, 4, 16–20]. These examples where large-scale activity is associated with increased nutrient loads followed by eutrophication are also now associated with elevated concentrations of mercury in fish tissue [2, 3, 6, 11, 20]. Such observations of linkages involving nutrient management, eutrophication, and mercury tissue concentrations confirm the importance and need to carefully manage nutrients as well as fish consumption in these habitats.

Considerations such as land use, nutrient management, vegetation replanting, and gravel road construction are topics of concern to MFN and are identified within the current land use planning document under preparation. In the future, MFN plans to manage environmental settings with a science-based approach intended to retain and protect existing natural heritage and cultural values. With this long-term view, it is hoped that enhanced environmental management will be achieved, and result in lower nutrient loadings to surface waters, including wetlands, creeks, rivers, and lakes. The completion of land-use planning within MFN's traditional territory is expected to result in the achievement of more balanced environmental management and benefit MFN residents as well as terrestrial and aquatic habitats.

4.2 Comparison of findings for recent fish monitoring with current Ontario online consumption guidelines

Available information from verified historical studies indicates that mercury fish tissue concentrations for fish species in lakes and rivers within MFN territory were largely below ~0.06 μg/g WW prior to 1982. Studies completed after the mid-1980s indicate increased mercury concentration in fishes and the need for fish consumption guidelines. Such increases in methylmercury over time identifies the need for comparison with Ontario's fish consumption guidelines within this extended study period as a strategy to place the current guidelines in context. This approach is justified, given observed increases in methylmercury over the study period, across lakes and rivers in MFN territory.

Observed fish length:mercury relationships in MFN territory reported herein identified the Ontario online consumption guide generally follows fish length when ~0.06 μg/g WW is first evident in the tissue monitoring data. However, there was some variability in the comparisons among fish species. Specifically, Lake Whitefish, White Sucker, and Longnose Suckere that consume detritus, plants, mud, and invertebrates demonstrated very good concordance between the initial length in Ontario's consumption guide for fish length with mercury at ~0.06 μg/g WW. For species that are predators on other fishes like Northern Pike, the initial length for consumption restriction was larger, suggesting the Ontario online consumption guide was less restrictive than suggested from available monitoring data. For other predatory fishes, such as Walleye, Lake Trout, Burbot, and Smallmouth Bass, initial lenght for consumption restriction at ~0.06 μg/g WW was smaller, suggesting the Ontario consumption guide was more conservative.

These modest differences for initial length of consumption restriction for a species are likely related to differing mercury accumulation in the various habitats, small or size-limited fish sample sizes, and/or skewed population distributions related to local factors such as high fish harvest rates in fisheries. Such uncertainty and variability represent some of the challenges in developing of fish consumption guidance for highly mobile fishes. Nevertheless, the general pattern is consistent with the understanding increasingly elevated mercury concentrations within older/longer fish compared with smaller/younger fish for a species and mercury tissue concentration of ~0.06 μg/g WW is the threshold that identifies when consumption should be reduced, to reduce risk for consumers.

As previously noted, MFN residents are considered sensitive consumers of fish resources, due to a relatively high frequency for consumption of wild fish. Hence, when any interested person consults this Ontario online resource, it is readily feasible to ascertain what the risk of consumption of a particular fish species exists through simply knowing the length for the fish species where the consumption restrictions start. In addition, this approach also does not require the sex of the fish species to be known, as the general accumulation pattern of mercury in fishes is often comparable between male and female fishes, as reported previously [6, 12], and a pattern also observed across species in this study. Therefore, understanding the basis of the consumption guidelines for fish provides a strategy to inform all individuals that harvest fish concerning the appropriate quantities across fish species and fish lengths.

4.3 Recommendations for fish consumption guidance for study lakes

With the available monitoring information from public and private sources, all habitats assessed within this study included fish specimens that exceeded the generic MOECC and Health Canada threshold for consumption of 0.06 μg/g and 0.5 μg/g mercury, respectively. Those fish specimens with the highest mean concentrations of tissue mercury were predators that consumed other fishes such as Northern Pike, Walleye, Burbot, and Lake Trout.

In contrast, omnivore species that consume mud, detritus, plants, and invertebrates such as Lake Whitefish, White Sucker, and Longnose Sucker had consistently lower mean concentrations of tissue mercury. Also, species such as Smallmouth Bass that consume invertebrates at small lengths and fish at larger lengths demonstrated tissue residues that were often intermediate of the predators and omnivores.

To review, the mean tissue residues ranked from highest to lowest were as follows: Northern Pike, Walleye, Lake Trout, Burbot, Smallmouth Bass, Lake Whitefish, White Sucker, and Longnose Sucker. From this group, Northern Pike, Walleye, Lake Trout, Burbot, and Smallmouth Bass all had specimens where tissue concentrations ≥0.5 μg/g and also frequently with specimens with concentrations ≥1.0 μg/g.

Given that MFN residents are classified as a sensitive population, due to frequent consumption of fish, the consumption guideline to lower risk is set at 0.2 μg/g mercury within tissue. Considering MFN residents as a sensitive population and using a threshold consumption guide of 0.2 μg/g, the recommended consumption rates of Northern Pike are:

- For fish with a length of 15–35 cm: child: one meal per week; all adults: two meals per week;
- For fish 35–50 cm: child: zero meals per week; all adults: one meal per week;
- For fish with length >50 cm: child: 0 meals per week; most adults: zero meals per week; adults with weight equal to or greater than 250 lbs. or 113.6 kg: one meal per week.

For the general population, the recommended consumption of Northern Pike are:

- For fish with a length of 15–35 cm: child: two meals per week; all adults: three meals per week;

- For fish with length of 35–50 cm: children are allowed one meal per week; for most adults: two to three meals per week, and for adults with weight equal to or greater than 250 lbs. or 113.6 kg: four meals per week;

- For fish with length >50 cm: children allowed zero meals per week; most adults: one to two meals per week; adults with weight equal to or greater than 250 lbs or 113.6 kg: two meals per week.

These recommendations for consumption guidelines, based on available fish monitoring information, indicate all members of the sensitive MFN population need to be cautious with consumption of Northern Pike, Walleye, Lake Trout, Burbot, and Smallmouth Bass across habitats. The consumption recommendations for Lake Whitefish, Common White Sucker, and Longnose Sucker are less restrictive due to the consistently lower mercury tissue concentrations in these latter species due to diets of detritus, plants, and invertebrates.

This recommended consumption approach for wild fish in subsistence fisheries uses the precautionary principle. Specifically, the study recommends consumption rates that are expected to avoid health consequences, if they are followed. It is not feasible to forecast possible health outcomes for individuals who do not follow the recommendations provided from the guide and/or the online recommendations from Ontario. It is well-established science that excessive consumption of wild fish can lead to undesirable human health outcomes and this predicate justifies the need for a guide involving the consumption of wild fishes.

5. Conclusions

This study has allowed for an update on the COCs in fishes harvested within MFN territory. The study approach also included the allowance for the analysis of past public and private studies to represent a wider set of observations across lake and river habitats. Time trends in mercury in fish tissue suggest that eutrophication is evident across the four lakes and rivers and these locations extend west to east across MFN territory. These forested watersheds appear to have been experiencing the effects of eutrophication since at least 2009, from an unknown combination of diffuse and defined nutrient sources.

An initial finding is that mercury is the primary contaminant of concern evident in fish tissue. The analyses included herein also identified that sport fish consumption guidelines in Ontario are first initiated for a fish species in a habitat when mercury concentrations in muscle tissue at a given length exceed about 0.06 μg/g. Then for this species, consumption recommendations are severely limited at 0.5 μg/g in tissue at a given length, with no consumption recommended for fish specimens with lengths associated with $\geq$1.0 μg/g.

Analysis of available historical and recent information indicates the larger specimens in many fish populations show the highest concentrations of mercury and should be generally avoided for consumption. Comparisons across habitats suggest consistent concentrations of methylmercury above 0.06 μg/g since at least 2009 with the likely cause of widespread eutrophication.

This study also confirmed that consumption guidance is required for nearly all habitats across MFN's traditional territory confirming this is a regional pattern due to pervasive contamination by mercury. A topic of concern that requires further study that was not addressed concerns the possible consequences on wildlife (e.g., birds, small mammals) that also consume fish with elevated mercury. Similarly, other wildlife (e.g., Moose *Alces alces*) may be at risk from exposure through the consumption of sediment and aquatic plants with elevated mercury. Exploration of consequences on birds, small mammals, and large mammals is a warranted study.

Acknowledgements

This study was supported through all phases by Chief Patricia Tangie of MFN. Chief Tangie embraced the need for careful environmental management and a greater understanding of the lakes, rivers, and traditional foods used by MFN residents across their territory. During the completion of the study, Aaron Bumstead and Stephen Murphy coordinated information collection during the challenges of the Covid-19 Pandemic. The fish sample team was led by Paul Jaques with assistance from Luc Jacques, Kim Rodrigue, Bob Elliot, Danny Couture, Bernard Debassige, Noodin Stone, Cassius Stone, Matthew Stone, and Ryan Lesage. In addition, Jessica Labranche, Christopher Buckle, and John Kim Bell also assisted with other aspects of this study.

Author details

Dean G. Fitzgerald[1]* and Lynn S. McCarty[2]

1 Integrated Ecosystem Solutions, Cambridge, Ontario, Canada

2 L.S. McCarty Scientific Research and Consulting, Ontario, Canada

*Address all correspondence to: deanfishy@hotmail.com

References

[1] Schindler DW. Carbon, nitrogen, and phosphorus and the eutrophication of freshwater lakes. Journal of Phycology. 1971;7(4):321-329. DOI: 10.1111/j.1529-8817.1971.tb01527.x

[2] Rudd JW. Sources of methyl mercury to freshwater ecosystems: A review. Water, Air, & Soil Pollution. 1995;**80**:697-713. DOI: 10.1007/BF01189722

[3] Lavoie RA, Jardine TD, Chumchal MM, Kidd KA, Campbell LM. Biomagnification of mercury in aquatic food webs: A worldwide meta-analysis. Environmental Science & Technology. 2013;**47**(23):13385-13394. DOI: 10.1021/es403103t

[4] Gandhi N, Tang RW, Bhavsar SP, Arhonditsis GB. Fish mercury levels appear to be increasing lately: A report from 40 years of monitoring in the province of Ontario, Canada. Environmental Science & Technology. 2014;**48**:5404-5414. DOI: 10.1021/es403651x

[5] Gandhi N, Bhavsar SP, Tang RWK, Arhonditsis GB. Projecting fish mercury levels in the province of Ontario, Canada and the implications for fish and human health. Environmental Science & Technology. 2015;**40**:14494-14502. DOI: 10.1021/acs.est.5b03943

[6] Environment and Climate Change Canada (ECCC). Canadian Mercury Science Assessment Report. EN84-130/3-2016F-PDF. Gatineau PQ. Environment and Climate Change Canada; 2016. 793 p

[7] Gentès S, Löhrer B, Legeay A, Mazel AF, Anschutz P, Charbonnier C, et al. Drivers of variability in mercury and methylmercury bioaccumulation and biomagnification in temperate freshwater lakes. Chemosphere. 2020;**267**:128890. DOI: 10.1016/j.chemosphere.2020.128890

[8] Blanchfield PJ, Rudd JWM, Hrenchuk LE. Experimental evidence for recovery of mercury-contaminated fish populations. Nature. 2022;**601**:74-78. DOI: 10.1038/s41586-021-04222-7

[9] Health Canada (HC). Human health risk assessment of mercury in fish and health benefits of fish consumption. Ottawa Ontario Health Canada; 2007. 37 p. plus appendices

[10] Charette T, Kaminski G, Rosabal M, Amyot M. Effects of speciation, cooking and changes in bioaccessibility on methylmercury exposure assessment for contrasting diets of fish and marine mammals. International Journal of Environmental Research and Public Health. 2021;**18**:2565. DOI: 10.3390/ijerph18052565

[11] Kidd KA, Muir DC, Evans MS, Wang X, Whittle M, Swanson HK, et al. Biomagnification of mercury through lake trout (*Salvelinus namaycush*) food webs of lakes with different physical, chemical and biological characteristics. Science of The Total Environment. 2012;**438**:35-143. DOI: 10.1016/j.scitotenv.2012.08.057

[12] Ministry of Environment (MOE). Guide to Eating Ontario Sport Fish - 1985. Toronto, Ontario; Ontario Ministry of Environment, 1985. 254 p

[13] Ontario Ministry of Environment, Climate Change, and Parks (Ontario). Guide to Eating Ontario Sport Fish. Toronto, Ontario. Ontario Ministry of Environment, Climate Change, and Parks. 2022

[14] Winter JG, Dillon PJ, Futter MN, Nicholls KH, Scheider WA, Scott LD. Total phosphorus budgets and nitrogen loads: Lake Simcoe, Ontario (1990 to 1998). Journal of Great Lakes Research. 2002;**28**:301-314. DOI: 10.1016/S0380-1330(02)70586-8

[15] Meyers PA. An overview of sediment organic matter records of human eutrophication in the Laurentian Great Lakes region. Water, Air, & Soil Pollution: Focus. 2006;**6**:453-463. DOI: 10.1007/s11267-006-9059-9

[16] Winter JG, Desellas AM, Fletcher R, Heintsch MA, Nakamoto L, Utsumi K. 2011. Algal blooms in Ontario, Canada: Increases in reports since 1994. Lake and Reservoir Management. 2011;**27**(2):107-114. DOI: 10.1080/07438141.2011.557765

[17] Paterson AM, Dillon PJ, Hutchinson NJ, Futter MN, Clark BJ, Mills RB, et al. A review of the components, coefficients and technical assumptions of Ontario's Lakeshore Capacity Model. Lake and Reservoir Management. 2006;**22**:7-18. DOI: 10.1080/07438140609353880

[18] Schindler DW. Lakes as sentinels and integrators for the effects of climate change on watersheds, airsheds, and landscapes. Limnology and Oceanography. 2009;**54**(6part2):2349-2358. DOI: 10.4319/lo.2009.54.6_part_2.2349

[19] Smith VH, Schindler DW. Eutrophication science: Where do we go from here? Trends in Ecology & Evolution. 2009;**24**(4):201-207. DOI: 10.1016/j.tree.2008.11.009

[20] Clayden MG, Kidd KA, Wyn B, Kirk JL, Muir DC, O'Driscoll NJ. Mercury biomagnification through food webs is affected by physical and chemical characteristics of lakes. Environmental Science & Technology. 2013;**47**(21):12047-12053. DOI: 10.1021/es4022975

[21] Schindler DW. The cumulative effects of climate warming and other human stresses on Canadian freshwaters in the new millennium. Canadian Journal of Fisheries and Aquatic Sciences. 2001;**58**:18-29. DOI: 10.1139/cjfas-58-1-18

[22] Rand GM, Wells PG, McCarty LS. Chapter 1: Introduction to aquatic toxicology. In: Rand GM, editor. Fundamentals of Aquatic Toxicology II. Bristol PA: Taylor and Francis; 1995. pp. 3-67

[23] McCarty LS, Borgert CJ, Posthuma L. The regulatory challenge of chemicals in the environment: Toxicity testing, risk assessment, and decision-making models. Regulatory Toxicology and Pharmacology. 2018;**99**:289-295. DOI: 10.1016/j.yrtph.2018.10.001

[24] Ministry of Energy and Environment (Ontario). Water Management Policies. Guidelines: Provincial Water Quality Objectives. PIBS 3303e. Toronto, Ontario: Ontario Ministry of Environment and Energy. 1994. 67 p

[25] Ministry of Environment (Ontario). 2010. Lakeshore Capacity Assessment Handbook. Protecting Water Quality in Inland Lakes on Ontario's Precambrian Shield. PIBS 7642e. Toronto, Ontario: Ministry of the Environment, Ministry of Natural Resources, and Ministry of Municipal Affairs and Housing. 2010. 47 p plus appendices

[26] Fitzgerald DG, and McCarty LS. Fish Biomonitoring Study by Michipicoten First Nation. Ottawa, ON: Submitted to Indigenous Services Canada. December 2022. 81 p

[27] Peterman RM, Bradford MJ. Statistical power of trends in fish abundance. Canadian Journal of Fisheries and Aquatic Sciences. 1987; **44**(11):879-1889. DOI: 10.1139/f87-232

[28] Sokal RR, Rohlf FJ. Biometry. 3rd ed. New York: W.H. Freeman and Co.; 1995. p. 885

[29] Scott WJ, Crossman EB. Freshwater Fishes of Canada. Bulletin 184. Ottawa: Fisheries Research Board of Canada. 1973. 996 p

[30] Dolan DM, Chapra SC. Great Lakes total phosphorus revisited: 1. Loading analysis and update (1994–2008). Journal of Great Lakes Research. 2012; **38**(4):730-740. DOI: 10.1016/J.JGLR.2012.10.001

[31] Chapra SC, Dolan DM. Great Lakes total phosphorus revisited: 2. Mass balance modeling. Journal of Great Lakes Research. 2012;**38**:741-754. DOI: 10.1016/J.JGLR.2012.10.002

[32] Winter JG, Eimers LMC, Dillon PJ, Scott LD, Scheider WA, and Willox CC. Phosphorus inputs to Lake Simcoe from 1990 to 2003: Declines in tributary loads and observations on lake water quality. Journal of Great Lakes Research 2007;33: 381-396. DOI: 10.3394/0380-1330(2007)33[381,PITLSF]2.0.CO;2

[33] Makarewicz JC, Booty WG, Bowen GS. Tributary phosphorus loading to Lake Ontario. Journal of Great Lakes Research. 2012;**38**:14-20. DOI: 10.1016/j.jglr.2012.08.001

[34] Mills EL, Casselman JM, Dermott R, Fitzsimons JD, Gal G, Holeck KT, et al. Lake Ontario: Food web dynamics in a changing ecosystem (1970–2000). Canadian Journal of Fisheries and Aquatic Sciences. 2003;**60**:471-490. DOI: 10.1139/f03-033

Chapter 10

Fish Fauna and Fishery in Ethiopia, Africa

Abenezer Wendimu, Wondimagegnehu Tekalign and Abnet Woldesenbet

Abstract

This chapter highlights the diversity of Ethiopian fish species across 12 drainage basins. These include the Wabishebele-Genale, Abay, Omo-Gibe, Awash, Rift Lakes, and Baro Akobo basins. Ethiopia's lakes, rivers, and reservoirs harbor over 200 fish species, categorized into East African, Nilo Sudanic, and endemic forms. The Nilo-Sudanic species are most diverse, particularly in the Abay, Baro-Akobo, Omo-Gibe, and Tekeze basins. Highland lakes and northern Rift Valley lakes host East African forms. Each of the seven drainage basins holds varying numbers of fish species, with high levels of endemism in the Abay, Rift Valley, and Awash basins. Endemic species counts are as follows: Abay (23), Rift Valley (7), Awash (6), Omo (2), and Baro (1). Rapid population growth, unemployment, and ineffective fisheries management threaten fish diversity. To prevent further degradation, urgent watershed management actions such as forestation, soil conservation, controlled grazing, and banning hillside crop farming are imperative in Ethiopia.

Keywords: drainage basins, fish fauna, ichthyofaunal diversity, Ethiopia, fish species

1. Introduction

Because freshwater is a necessity for all life, it plays a crucial role in socioeconomic development. However, diverse fresh water ecosystems are frequently threatened by activities both on land and in water bodies, putting them at risk. Determining the number of different fish species in fresh water ecosystems is important there for conservation and management. Losses in biodiversity occur at very high rates worldwide, largely because of human activity. Conservationists are concerned about this loss, and they are working hard to protect as much of the remaining diversity as possible.

Knowledge of an area's biodiversity is essential for creating effective conservation strategies [1]. Norris [2] asserted that an ecosystem's species diversity can serve as a barometer of its health. However, a major issue in the modern era is the rapid decline in biological diversity caused by habitat loss and environmental degradation [3]. The inland bodies of water in Ethiopia (Lotic and Lentic) are rich in edible fish resources, which ichthyologists find valuable. This is because Ethiopia's fish fauna is made up of a variety of East African, Endemic, Ethiopian Highlands, and Nilo-Sudanic species [4, 5]. A total of 175 valid fish species and subspecies from 12 orders and 25 families were identified [6, 7]. According to Golubtsov and Mina [8], 168–183 valid fish species have been found in Ethiopia's freshwater bodies. Between 37 and 57 endemic fish species are thought to exist nationwide [8].

2. Aquaculture production and consumption

2.1 Global trend

Aquaculture is a sector of the food industry that expands and produces foods that are high in protein. The total quantity of aquaculture-produced food fish that is consumed by humans, including crustaceans, fish, mollusks, and other aquatic animals, was 52.5 million metric tons in 2008 [9]. The quantity intended for human consumption in 2019 (excluding algae) reached a peak weight of 20.5 kg, and in 2020, it slightly decreased to 20.2 kg per person, which is still more than twice as much per person as the 1960s' average weight of 9.9 kg. Approximately 600 million people's livelihoods are considered dependent on fisheries and aquaculture, at least in part, including workers in the secondary and tertiary sectors as well as their dependents. The number of people working in the primary sector was approximately 58.5 million people. Due primarily to the COVID-19 outbreak in 2020, only around USD 151 billion was produced by fisheries and aquaculture products, less than the US dollar record high of 165 billion in 2018 [10]. A record 180 million metric tons of aquaculture has been produced globally (**Figure 1**), 127 million metric tons more than in 2008, owing to expansion in China, Chile, and Norway [10].

Consumption of aquatic foods is anticipated to increase by 15% to 21.4 kg/person on average in 2030, owing to rising income, urbanization, better post-harvest practices, and dietary trends [10]. Worldwide, people consume more than 100 million tons of fish each year, providing at least 20% of the average amount of animal protein consumed by 2.5 billion people and up to 50% or more in developing nations. Although these levels are still low, fish protein is essential as a significant portion of the animal protein consumed in some regions where food insecurity and malnutrition are most prevalent, such as Asia and Africa. Global fish consumption statistics obscure the significant regional variations between and within nations. As a result of

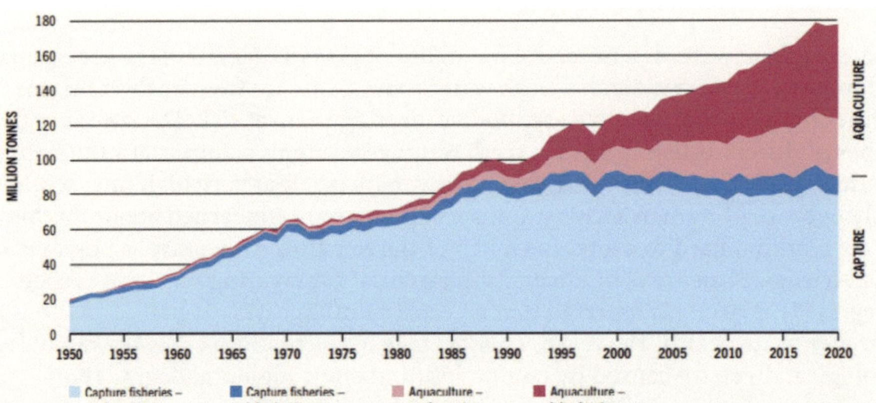

Figure 1.
Capture and aquaculture production on a global scale (excluding aquatic mammals, crocodiles, alligators, caimans, and algae) [10]. Live weight equivalent is the unit of measurement for data.

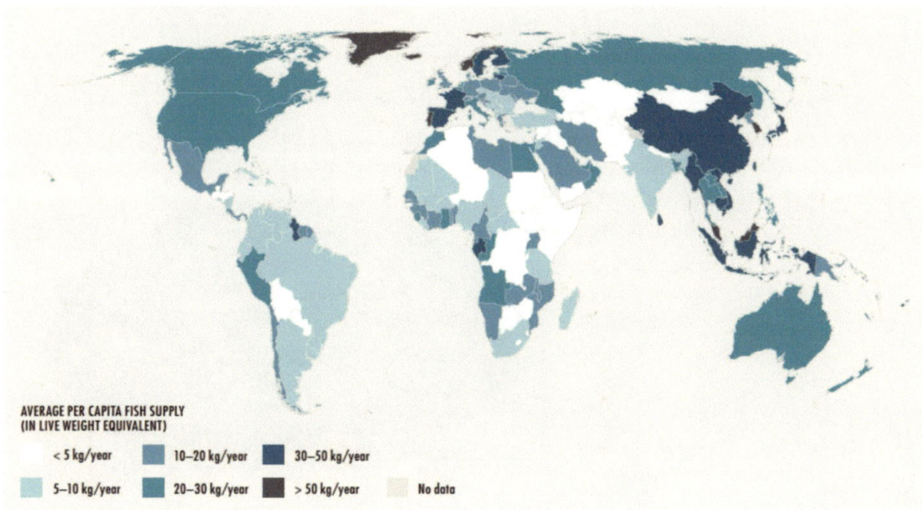

Figure 2.
Estimated amount of fish consumed per person, average 2015–2017 [10].

the impact of geographical, cultural, and economic factors, along with accessibility to and proximity to fish landings and aquaculture facilities, fish consumption varies significantly by individual, from less than 1 kg to 100 kg annually. Despite the fact that there are still observable variations in the quantities of fish eaten between different countries and regions of the world (**Figure 2**), some distinct trends can still be observed.

2.2 Status of aquaculture in Africa

Between 2003 and 2007, aquaculture production in Africa increased by 56% and over 100%, respectively, in both volume and value. This growth was fueled in part by rising aquatic product prices, the emergence and growth of small and medium-sized aquaculture businesses, significant investments in cage culture, and the expansion of larger commercial endeavors, some of which produced high-value goods for markets abroad [9]. Several phylogenetic taxa that are old and isolated (e.g., Lungfish, Polypteridae, Protopteridae, and Bichirs) are common in this region [11]. Although the vast river systems of the continent are also fishery-rich, accounting for up to 50% of the total inland catch, lakes contribute significantly to inland fisheries [12]. The diverse species flocks that the African ichthyofaunal includes as a result of adaptive radiation give it an additional distinction [11]. For instance, flocks of various cichlid species can be found in Malawi, Tanganyika, and the Great Victoria Lakes of East Africa, and *Labeo barbus,* which can be found in Lake Tana (Ethiopia).

According to the FAO [13], Africa contributes very little to the production of fisheries worldwide, with only 7% of the total production coming from regional capture fisheries, becoming the second largest region after Asia in terms of this population. Despite the fact that Africa makes a minimal overall contribution to each of these sub-regions, 3 million tons are produced by inland waters, which account for 25% of all global inland fishery production (**Figure 3**). Because of this comparison, artisanal fishing in Africa is valued, particularly, the significance of continental lakes, such as Lake Victoria. This context is crucial in places with high level of poverty.

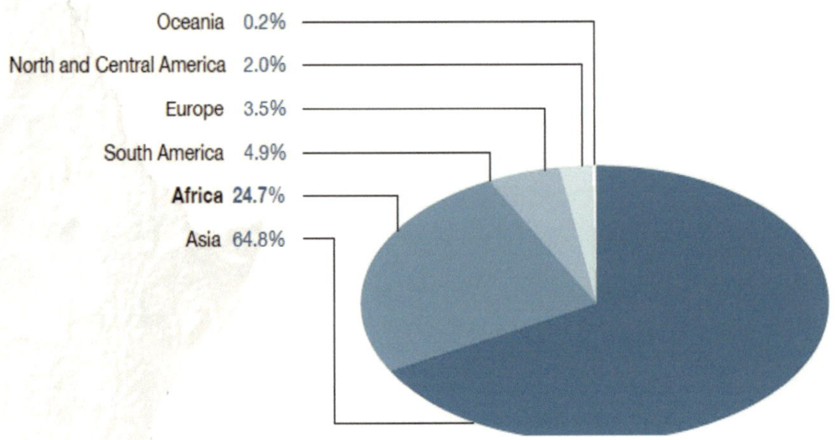

Figure 3.
Inland capture fishing by continent [13].

Africa appears to lag behind global growth in aquaculture. Aquaculture in Africa is relatively rare. There are several reasons for this, but the primary reason is that the industry is not viewed as a commercial enterprise. Even in areas where it is not commercialized, aquaculture in Africa can benefit from sound fisheries management techniques that will help protect these significant food production sectors. Recent data show that the region's aquaculture industry is growing faster than the average for the world [13], and presently makes up 15–19% of all fish produced.

Figures 4 and **5** depict the variations in wild capture, fisheries and aquaculture practices among the five pioneer fish-producing nations in Africa.

Fish consumption in Africa was the lowest at the regional and continental levels, reaching a peak of 10.5 kg in 2014 and starting to fall to 9.9 kg in 2017 (**Table 1**). Nevertheless, demand in Africa varied, with West Africa having a per capita intake of 12 kg and East Africa having a per capita intake of 5 kg. Between 1961 and 2017, North Africa experienced significant growth (per capita from 2.9 kg to 14.7 kg), but sub-Saharan Africa experienced stagnant or declining per capita fish consumption. Several interrelated factors contribute to sub-Saharan Africa's low fish consumption,

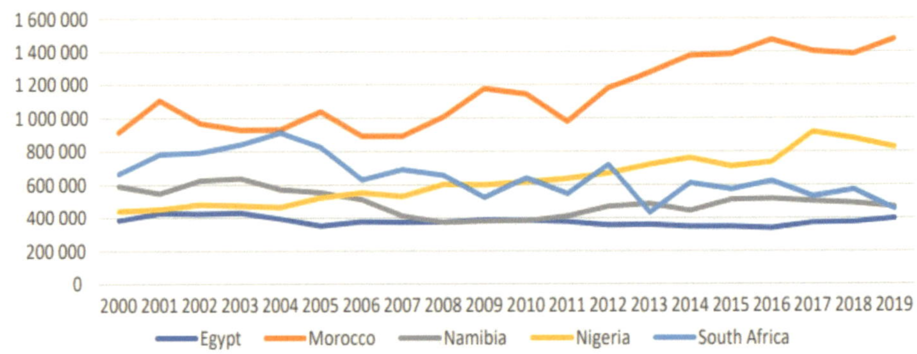

Figure 4.
Tons of wild fish produced overall between 2000 and 2019 in the five selected African nations [13].

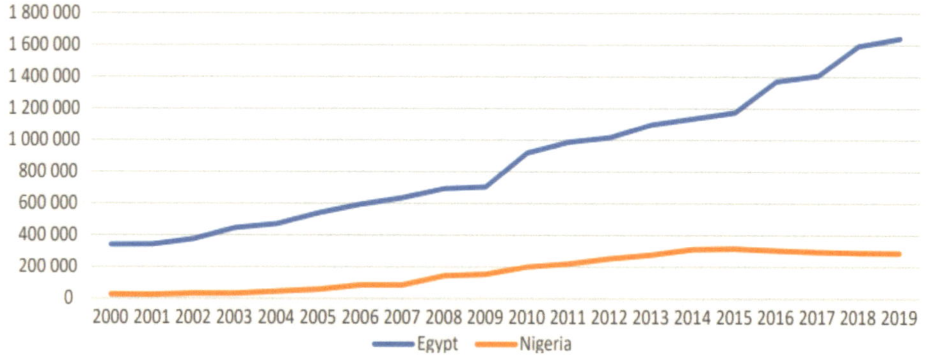

Figure 5.
Total tons produced by aquaculture in Egypt and Nigeria between 2000 and 2019 [13].

Region/economic grouping	Total food fish consumption *(million tonnes live weight equivalent)*	Per capita food fish consumption *(kg/year)*
World	152.9	20.3
World (excluding China)	97.7	16.0
Africa	12.4	9.9
North America	8.1	22.4
Latin America and the Caribbean	6.7	10.5
Asia	108.7	24.1
Europe	16.1	21.6
Oceania	1.0	24.2
Developed countries	31.0	24.4
Least developed countries	12.4	12.6
Other developing countries	109.5	20.7
Low-income food-deficit countries	23.6	9.3

Table 1.
Consumption of fish overall and per person by region and economic grouping, 2017 [13].

including population growth, outpacing the availability of fish for food; stagnation in fish production due to stress on resources from capture fisheries; and an underdeveloped aquaculture industry. Low-income levels, poor infrastructure for landing, storing, and processing fish, a lack of fish product sales channels for marketing and distribution, and all these factors contribute to the low consumption of fish. It should be noted, however, that actual values in Africa are likely greater than official statistics indicate because the contribution of a few artisanal fisheries, subsistence fishing, and unrecorded cross-country trade is underreported.

2.2.1 South Africa

With a 0.5 percent Gross Domestic Product (GDP) contribution and 4000 direct jobs, the fishery industry in South Africa is comparatively small in terms of employment, according to Augustyn *et al.* [14]. Relatively 140,000 jobs depend on these industries. The most recent estimate of the value of all fishery landings is 700 million USD. The biggest and most lucrative commercial sectors are deep-sea trawls, purse seines (small pelagic purse seines), squid jigs, offshore west coast rocks, lobster traps, and line and net fisheries [15]. The government has recognized marine aquaculture as a developing industry that deserves support despite the fact that it remains in its toddlerhood in terms of aquaculture, with a projected worth of USD 34 million [14].

2.2.2 Namibia

After Mauritania, South Africa, and Morocco, Namibia is the continent's fourth-largest nation in terms of catch fisheries. Hake and horse mackerel are the two primary commercial species caught in Namibia. The large pelagic tuna, deep-water crab, monkfish, rock lobster, and orange roughly are different species with more recent significance. The fishing industry exported approximately USD 680 million in 2016, ranking second after mining in terms of foreign exchange earnings. The fishing industry employs 16,800 people directly, with hake fishing and processing accounting for 70% of all jobs. Along with mariculture processing, horse mackerel is a growing industry that creates new jobs [16, 17]. In terms of aquaculture, mariculture is a growing sector primarily based on Lüderitz Bay and Walvis Bay, according to ATFALC [18]. Mariculture production in Walvis Bay and Lüderitz Bays, which totals less than 2000 tons annually, includes onshore abalone production as well as farms for oysters, mussels, and other shellfish.

2.2.3 Morocco

Approximately 3% of Morocco's GDP is derived from fishing. According to Cervantes *et al.* [19], in 2017, 1.4 million tons were thought to have been caught in 2017. Most harvests were made on the Atlantic side. Small pelagic species, including anchovies and sardines, were the primary species caught (approximately 70 percent). The sector of high-sea tuna species is the main target of fishing, which is caught by alien ships, primarily from the fleets of Asia, Russia, and the European Union. When considering the direct and indirect jobs generated by small-scale fisheries, the number of individuals directly employed in Moroccan fishing in 2017 reached 400,000. However, the scarcity of processing, harbors, and infrastructure facilities means that small-scale sectors are still largely underdeveloped. Production from aquaculture is rather meager, with an estimated 1200 tons produced in 2017 and 250 employed people.

2.2.4 Nigeria

The fishing sector is rounded to the nearest half of Nigeria's GDP. It was reported that 734, 731 tons were produced in total in 2016. The fishing business is varied, with 36% coming from marine and wild catch, 33% from inland water systems, and 31% from aquaculture. More than 80% of Nigeria's domestic catch fisheries are produced by artisanal small-scale fishers who live in the coastal, inshore creeks, lagoons, interior rivers, and lakes of the Niger Delta. Nigeria is the top producer of aquaculture in sub-Saharan Africa, specializing in fresh and brackish water aquaculture. Approximately

USD 840 million is the off-farm value of 291,000 tons of this product was produced in 2018 [13]. Production of freshwater is mostly to blame for this. Catfish, which typically grow in ponds and tanks and account for more than 50 percent of all aquaculture production volume, is the most widely farmed species in Nigeria.

2.2.5 Egypt

Over the last ten years, the yield of fish caught has remained largely stable, at 350–400000 tons. In the past 20 years, aquaculture production has increased significantly and as of 2003, it has surpassed the wild catch. Fish production accounted for 75% of the country's total production, or 1.56 million tons and was said to have come from aquaculture in 2018. Tilapia is the most prevalent species in freshwater and inland reservoirs, followed by the flathead gray mullet. The nation's aquaculture ponds cover approximately 115,000 hectares. Sea bass and sea bream are the two most important species in marine culture, although prawns and other species, such as meager (drum) are also produced in modest amounts.

2.3 Overview of fisheries development in Ethiopia

The lentic and lotic inland water bodies in Ethiopia are rich in resources for edible fish, which ichthyologists are particularly interested in. The country has been recognized for having a significant amount of physically accessible freshwater resources, including 122 billion metric cubic (BMC) of mean annual flow distributed among 12 river basins, 4.5 BMC underground, 9 saltwater lakes; 11 freshwater lakes; 12 major marshes; and craters, the majority of which are situated within the Rift Valley Basin [20]. Tesfaye and Wolff [21] stated that most lakes are endorheic because they have no surface water outlets, except for Ziway, Tana, Langano, Abaya, and Chamo. Ethiopia has less potential for ground water than surface water resources. According to research, the nation also has numerous lakes and reservoirs, numerous small water bodies, and expansive floodplain areas that are dispersed throughout the entire nation and cover an estimated 13,637 km^2 of land (**Table 2**), or 1.2% of Ethiopia [23, 24]. There are numerous significant lakes, rivers, and small bodies of water, such as swamps, floodplains, ponds, and canals for irrigation with a smaller than 10 km^2 area [25]. Sadly, many of these water bodies in Ethiopia have not yet been studied, and many of them, because of their wetlands and unlisted locations, are still a mystery to scientists and academics.

The nation's inland waters are home to species that are endemic to the Ethiopian highlands, Nilo-Sudanic, and East African Ichthyofauna [4, 5, 26]. Additionally,

Water resources	Total area (km^2)	Total length (km)	Annual produce estimations (tons/year)	
			Mean	Per unit length
Large lakes	7740	—	39,262	5.8 ± 0.6
Large Reservoirs	1447	—	7879	6 ± 0.6
Small bodies of water	4450	—	25,996	4.1 ± 0.4
Rivers	—	8065	21,405	2.4 ± 0.9
Total	13,637	8065	94,541	

Table 2.
Summary of the various water bodies' production estimates in Ethiopia [22].

approximately ten exotic species of fish have been introduced into the freshwater ecosystems of Ethiopia from abroad for many reasons, including the prevention of malaria and weeds (**Table 3**) [28].

Commercially exploited lakes are about 6500Km^{2,} and yields per annum do not exceed 23,000 (**Table 4**).

Too many fish species can be found in the country's rivers and streams than in lakes and reservoirs. The most significant fish for commerce were *Oreochromis niloticus, Clarias gariepinus, Barbus* species, *Cyprinus carpio,* and *Carassius carassius*. The Main River and miscellaneous small river systems of the country account for approximately 1,135,494Km^2 and more than 20,000 potential yield estimates (**Table 3**).

The majority of fish eaten in Ethiopia are caught from the wild using artisanal (small-scale) techniques. Of the nation's estimated annual fish production, which is approximately 51,481 tons for major water bodies (**Table 4**), only about 38,400 tons have recently been consumed (**Figure 6**) [29].

2.4 Ethiopia's drainage basins and fish diversity

Based on faunal similarities (particularly fish fauna) and the African freshwater ecoregion model, Ethiopia's freshwater systems were conveniently classified into five freshwater ecosystem types in 12 drainage basins (**Figure 7**) [30].

It consists of rivers, streams, and lakes in the highlands of the country, including Lake Tana, lakes in the rift valleys such as Lakes Abaya and Chamo, and Lake Turkana (including the Omo River and its tributaries and Lakes Abaya and Chamo).

Rivers	Catchment area(Km2)	Length within Ethiopia (km)	Potential yield estimates		
			1° Model	2° Model	Weight Per Unit length (ton/km)
Baro	38,400	285	839	232	2.3
Akobo	21,890	203	487	119	2.1
Gillo	13,050	252	295	182	1.7
Alwero	8098	321	185	294	1.6
Pibor	4300	96	100	27	1.5
Blue Nile	176,000	800	3675	1792	3.2
Awash	112,696	1200	2385	3999	3.1
Wabi-Shebele	202,697	1000	4215	2787	3.4
Genale	171,042	480	3575	652	3.5
Omo	79,000	760	1690	1619	2.3
Tekeze	82,350	608	1759	1041	2.2
Mereb	5900	440	136	549	1
Angerib	23,812	220	528	139	1.3
Miscellaneous small rivers	196,259	1400	4085	5426	3.8
Total	1,135,494	8065	23,954	18,855	2.4 ± 0.9

Table 3.
Estimates of the potential fish production in Ethiopian Rivers [21, 27].

Water bodies	Main landing site	Area (Km2)	Fish Yield Possible (ton/ year)
Abaya	Arba Minch	1070	600
Awassa	Awassa	91	611
Chamo	Arba Minch	350	4500
Koka reservoir	Koka	255	700
Langano	Oittu	230	240
Lugo	Lugo	25	400
Tana	BahrDar	3500	10,000
Turkana	Ethiopian (1.3% total area)	94	750
Ziway	Ziway	434	2941
Total		6477	23,342

Table 4.
Commercially exploited lakes, landing areas, and potential fish yield [21, 27].

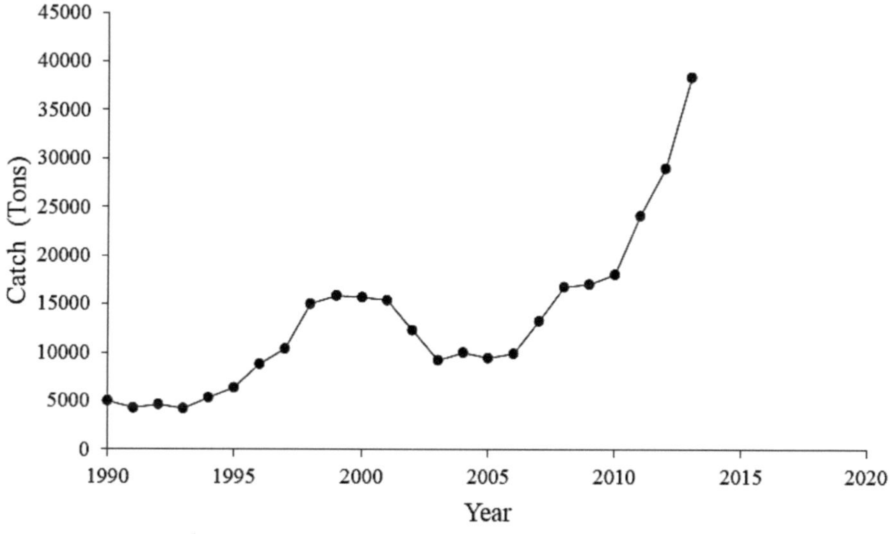

Figure 6.
Trends in the nation's got fish catch from 1990 to 2010 (according to the Federal Ministry of Agriculture and the Central Statistics Office website, both of which were cited by Brook Lemma in 2012 and from 2011 to 2013) [29].

The Wabi-Shebele, Genale, Dawa, and Fafan river basins, as well as the coastal basins of the Red Sea, lie between Shebele and Juba (including the Awash system and the salt lakes of northern Ethiopia, including Lake Abbe, Afambo, Afdera, and Asale).

Golubtsov and Darkov [31] argued that there are six main watersheds that can be used to further subdivide freshwater ecoregions (**Table 5**). These include Rift Valley, Omo-Gibe-Turkana, Shebele-Juba, Blue Nile, White Nile (Baro-Akobo) and Tekeze-Atbara. The country's drainage system is the result of an uplift that occurred in Ethiopia during the tertiary period, forming the Rift Valley and two highlands [33, 34].

The predominant fish species in the Baro, Akobo, Omo, Gibe, Tekeze, and Blue Nile basins (**Table 5**) were Nilo-Sudanic forms. Additionally, some of these fish can be found in the Southern Rift Valley and lakes of the Shebele Genale system

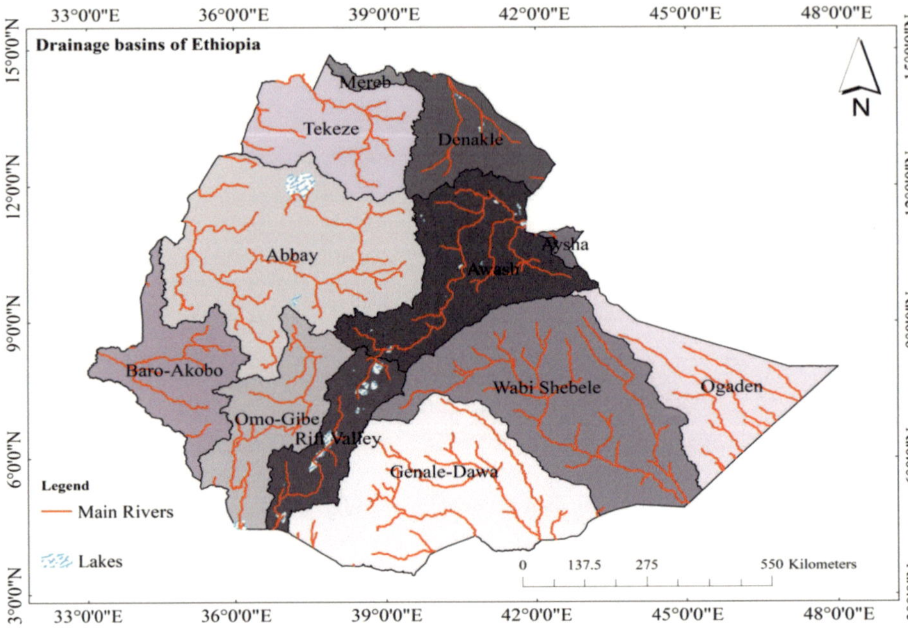

Figure 7.
Drainage basins of Ethiopia (Source: ArcMap by Abenezer Wendimu).

Drainage systems	Number		
	Family	Genera	Species
Wabi-Shebele and Juba	12	21	33
Tekeze-Atbara	10	22	34
Rift Valley	11	18	28–31
Omo-Turkana	20	42	76–79
Baro-Akobo	26	60	113
Abay	16	37	77

Table 5.
Six drainage systems contain a variety of fish [32].

(Abebe, [4]). Due to the historical and continuing connection of the watershed with the river systems of Nile, West, and Central Africa, these systems contain elements of Nilotic fish [26]. The genera represented by these fish were Mormyrus, Malapterurus, Labeo, Hyperopisus, Hydrocynus, Citharinus, Barilius, Bagrus, Alestes, Polypteus, and Protopterus. South of the Rift Valley and in portions of the Shebele–Genale Basin, these forms are present (lakes Abaya and Chamo). These lakes and watersheds are thought to have been connected to the Upper White Nile by Lake Rudolph 7500 years ago [35]. These forms are believed to be related to West African fish, as the Nile was once connected to the river systems in Central and West Africa.

The highland East African forms are found in the lakes of the northern Rift Valley, including Lakes Hawassa, Zuway, and Langano; highland lakes Tana and Hayq; and associated river systems, as well as the Awash Drainage Basin. Among them are the genera

Clarias, Garra, Oreochromis, Barbus, Labeo, Barbus, and Varicorhinus. They are related to fish from Eastern, Northern, and Southern Africa. Some elements are present in the waters of western Africa. For instance, the widely dispersed cyprinid species *G. dembeensis* is found in six distinct nations (Tanzania, Kenya, Cameroun, Ethiopia, Egypt, and Nigeria). The Awash and Northern Rift Valley lakes are almost devoid of Nilotic river fish [30].

Considering the size of the nation, the ichthyofauna of Ethiopia does not appear to be very numerous. Lowland waters have no known endemic species, but the waters of the Ethiopian plateau have a high endemism rate [30]. The fish ichthyofauna of these streams is dominated by two fish genera (Barbus and Garra). *Afronemacheilus abyssinicus* is an odd endemic fish of the Ethiopian plateau [30]. In terms of number, richness, and diversity, the endemic forms have an uneven distribution of fish species among the drainage basins. Since the lakes and rivers are relatively accessible, there is a high chance of discovery, which may contribute to the uneven distribution of species within the major drainage basins.

The Omo-Gibe, Wabi-Shebele, Tekaze, and Rift Valley basins around Lake Abay are next in order of species diversity, then the Baro Basin [31]. The rich and diverse habitats that makeup part of this high diversity are also likely justification for the relatively high level of exploration and collection that has occurred in these easily accessible bodies of water. However, the highest endemism appears to be in the Abay and Awash basins. This is due to the 'species flock' of Lake Tana and endemic fish species that have evolved into special habitats in the creeks of the highlands of northern and central Ethiopia. Lake Tana is home to 28 species and 1 subspecies, 20 of which are endemic to Ethiopia. According to Getahun *et al.* [30], Lake Tana is home to 18 endemic species.

2.4.1 Baro-Akobo Basin

The southwest highlands are relatively small, rounded mountain remnants that are divided by established river valleys and are located south of the Abay [36]. These mountains and hills give rise to a number of the Baro-Akobo basin's tributaries. Major river systems in the basin include the BaroKela, Gilo, Akobo, Baro, Sore, Geba, Birbir, Alwero, Bonga, and Jejebe Rivers. The southern Ethiopian plateau provides the majority of the water for the Sobat, as the so-called Baro-Akobo is known outside of Ethiopia.

The White Nile system in Ethiopia is home to 107 fish species that are divided into 54 genera and 23 families. Only one of the 87 fish species, *Afronemacheilus abyssinicus*, according to Getahun [35], is unique to this basin [8]. More recently, 113 fish species from the same basin were found in 60 genera and 26 families, according to Golubtsov and Darkov [31]. The most varied fish fauna is found in the White Nile system, which is located in Ethiopia. Other drainage systems do not contain the following six families: Protopteridae, Notopteridae, Nothobranchiidae, Cromeriidae, Channidae, and Anabantidae [31]. *Mormirids* sp., *Malapterurus* sp., *Lates niloticus, Labeo horei*, *Hydrocynus* sp., *Heterotis niloticus*, *Gymanrchus niloticus, Bagrus, Barbus* sp., and *Alestes* sp. are the fish species that are most important for commerce. There are about six endemic species in this drainage basin, and there is no information on exotic/alien species of fish. For Ethiopia's development of fish culture, the watershed's diverse fish fauna in low-lying areas is a valuable resource [31].

2.4.2 Blue Nile basin

Lake Tana is the source of the Blue Nile, which flows into central and northwestern Ethiopia. It is Ethiopia's principal river, flowing at about 50 billion cubic meters annually between Lake Tana and the Sudanese border over a distance of 1000 kilometers [36]. Its water system includes the Dinder River, which empties into the Blue Nile far below the Sudan reservoir, and Lake Tana, Ethiopia's largest lake, and its tributaries. It also has several basins, including the Didessa, Beles, Jemma, and Dabus Rivers, as well as the Koga and Fincha basins [37]. Despite having a drainage area of only 324,000 km^2, the entire Nile system receives 58% of its water from this area and the vast bulk of soil particles/sediment from Egypt's deltas and alluvial river valleys [38]. The lower basin, specifically the Jamma, Guder, Didessa, and Dabus Rivers, is where the majority of the Blue Nile flood's water comes from. In the southwest region of the country, where rainfall is heavy, Didessa and Dabus rise on the left bank. Another group of important tributaries include the Beshilo, Dabena, Anger, Mugger, Belessa, and Wonchit. Within Ethiopian borders, the Blue Nile drainage was home to 30 different fish species [39]; JERBE, on the other hand, listed 77 species of fish belonging to 16 families and 37 genera. Cyprinidae is the most varied family of fish. The Blue Nile drainage system has a large number of endemic species (a total of at least 24 endemic species). The cyprinids that are unique to the Lake Tana sub-basin made up one-fourth (19 species) of the total species count.

Three fish species were introduced into Ethiopia and were shown by Golubstov and Mina [40] to be a component of the drainage network for the Blue Nile. One species from each of the three families serves as representation. The *Afronemacheilus abyssinicus, Clarias gariepinus*, and *Oreochromis niloticus* species are the representatives of the Cichlidae, Claridae, and Balitordae, respectively. Cyprinidae, the family of fish with the most members in the lake, is portrayed by the following genera such as Varicorhinus, Labeo, barbus, Barbus, and Garra [30]. Barbus is the genus that houses the *Labeo barbus* species found in Lake Tana. However, the name Labeobarbus is now used for the large hexaploid African barbus [41]. The brand-new genus name more accurately conveys their phylogenetic separation from other Barbus species, which were previously grouped together. Different Labeobarbus species have different reproductive methods in addition to resource allocation (feeding) [42]. The "small" barbs are categorized under the genus Barbus, which has three species: *B. pleurograma*, *B. tanapelagius,* and *B. humilis* [43]. There is only one species of Varicorhinus, *V. beso*. According to Stiassny and Getahun [44], four species make up the genus Garra: *G. tana*, *G. dembecha, G. dembeensisis,* and *G. regressus*. After the cyprinid species in Lake Lanao disappeared due to overexploitation, 15 different species of Labeobarbus now form the world's sole cyprinid species congregating in Lake Tana.

2.4.3 Tekeze-Atbara Basin

This basin includes the rivers that drain the northeast of the country. The Tekeze River, which merges with Sudan's White Nile and Blue Nile rivers to form the Nile, as well as the Guang River, also known as the Atbara River, are included in the drainage system [37]. The sources of its tributaries are on the Ethiopian High Plateau, which is close to the Blue Nile and lies to the east and west of Lake Tana [36].

The fish fauna of the Tekeze-Atbara drainage system was unknown prior to the JERBE survey of this area, according to Tedla [39]. Three endemic species, 2 introduced (exotic) species, 10 families, 22 genera, and 34 fish species were recorded by JERBE from the Tekeze-Atbara drainage within the limits of Ethiopia.

2.4.4 Omo-Turkana Basin

Kafa is the source of the Omo River, which flows south into Lake Turkana [45]. The Omo River basin stretches from the west highlands of southern Kenya to the semi-arid Omo lowlands. Ninety percent of the lake's yearly inflow goes into the richly diverse Omo delta, and 14% of Ethiopia's yearly runoff comes from there, changing in response to changes in lake level [46]. There is evidence that the Nile and this basin have previously connected more than once when it was wet and pale climatic fluctuations [47]. In Kenya and Ethiopia, there are 130,860 km of Lake Turkana watershed. The lake is the fourth-largest lake in Africa and the largest desert lake in the entire world. Ethiopia's largest river system is the Blue Nile, and the second largest in terms of runoff volume is the Omo basin. As a closed basin, Lake Turkana inflows gradually evaporate, leaving the lake's waters nearly salty, unfit for human consumption, and unusable for agriculture.

The Omo River empties into Lake Rudolf, also known as Lake Turkana, on the Kenyan border. The southwest portion of the country's western highlands is drained by a number of rivers, including the Gibe River in the Omo River watershed [26]. Before the JERBE study, only 13 species had been identified within Ethiopia's borders of the Omo drainage system [39]. The Omo-Turkana Basin is home to between 76 and 79 fish species, which belong to 20 families and 42 genera. The Omo River system contains up to eight endemic fish species, or about a quarter of the fish fauna of the system. As of yet, no introduced species have been found [31].

2.4.5 Shebele-Juba Basin

Originating in the Bale Mountains and Ahmar, respectively, Wabi-Shebele and Dawa Genale Gastro flow southeasterly toward Somalia. Genale-Dawa River is the Ethiopian name for the Juba River, claimed by Basnyat and Gadain [48]. In its upper catchment, the Juba River's principal tributaries are the southeast-flowing Wabi Dawa, Genale, and Wabi Gastro. Fourteen fish species were listed as coming from this drainage system before JERBE, according to Tedla [39]. The Juba and Wabi-Shebele Drainage Basins are among the basins in the nation with the largest catchment areas and the least amount of research done on their fish populations. Thirty-three species of fish from 21 genera and 12 families have been described in the works of the JERBE group [31]. The most distinctive ichthyofaunal species of the Nilotic and East African fish taxa are found here, including the cichlid Alestesaffinis and the characin Alestesaffinis (*Oreochromis spilurus*). According to Golubstov and Mina [40], the eel Anguilla sp., a diadromous fish, is only found there in Ethiopia.

2.4.6 Rift Valley Basin

The main Ethiopian rift in the country's center is the Afar Rift System, and the largely rifted zones of southwest Ethiopia are the three main geological zones that make up the Ethiopian Rift Valley, which is the northernmost segment of the East African Rift system [49, 50]. The lakes in the southern part of Ethiopia are Lake Abaya and Lake Chamo. These lakes are located in the north: Zuway, Abijata, Shala, Langano, and Awassa. The salty northern lakes include Afdera, Gamari, Asale, Afambo, and some of Abbe. The crater lakes (Lake Arenguade, Bishoftu, and Hora) and the Bishoftu group and Chitu are all in the Rift Valley Basin of the country [4].

The most diverse fish fauna is found in the southern Ethiopian Rift Valley (Lake Abaya and Chamo), where there are 20 different species of fish, and the river basin in the Awash watershed has 11 different species of fish, which is about 37% of the overall fish weight in great in the Rift Valley of Ethiopia [8]. Taxonomically, the valley holds more than 30 diver species of fish in 18 genera and 11 families, with four introduced and five endemics [31]. Ethiopia's larger cities and administrations in small and large towns provided edible fish species largely from Tana Lake (more than 50%) in the northern part of the country and other Rift valley lakes (40%), according to Tesfaye Wudneh [51].

2.5 Fishery activities

Fishing is a dynamic and expanding industry, according to FAO (2004), because it employs labor-intensive harvesting, processing, and distribution methods to take advantage of marine and freshwater fish resources. This subactivation sector's, whether done part-time, full-time, or only seasonally, regularly aims to provide fish and fishery produces to home consumption as well as regional markets. However, over the last decade or two, market consolidation and increased globalization have increased export-oriented production in many small-scale fisheries.

Various levels of organization operate on a small scale, including sole proprietorships, formal economic enterprises, and informal microenterprises engaged in fish production. The lack of uniformity within and across nations and regions must be taken into account when developing strategies and policies to increase this subsector's contribution to food security and poverty reduction. The fisheries that still operate today typically involve fishing households (as opposed to commercial enterprises), have little money and effort, few (if any) small fishing vessels, and are primarily driven by local consumption. It involves a short trip near the intended coast. The most frequently used terms, either singly or in combination, when referring to small-scale fishing are "subsistence," "traditional," "peasant," "artisanal," and "inshore." According to Kurien [52], small-scale fishing differs from medium- and large-scale fishing in the following ways (**Table 6**).

Characteristics	Small-scale	Medium-scale	Large-scale
Range of value of output per crew per annum (USD)	200 to 1500	Over 8000	Over 15,000
Range of fuel consumption per unit per annum (tones)	1–60	400–450	1600–1800
Range of fish harvest per unit per annum (tones)	2–100	200–1200	5000–8000
Range of fish harvest per ton of fuel (tones)	2–3	2–3	3–4
Investment range per unit	1–80	300–4000	10,000–40,000
Estimated number of units	3,200,000–3500,000	30,000–32,000	5000–5500
Crew range per unit	1–5	25–30	40–60

Table 6.
Rough estimates of the characteristics of different scales of operation in fishery sources: Kurien [52].

2.6 Fishing gear technology

Long lines, hooks, gillnets, beach seines, and cast nets are among the equipment frequently used in Ethiopian fisheries [21]. In particular, the rivers of Ethiopia use a variety of traps, baskets made of plant materials, nets, wires, and scoops [53]. Nearly all of Ethiopia's lakes are fished with gillnets, which account for most of the country's commercial production (**Table 4**). Commercial catches are made with beach seines in the northern Rift Lakes, like Koka Reservoir, Lake Langano, and Lakes Ziway and Langano. Bagrus and Nile perch are caught in Abaya Lake using both surface and bottom longlines, respectively [54].

Additionally, it is employed in Lakes/Chamo to catch Nile perch [21]. Hook and line usage is frequently prohibited when fishing for subsistence [55]. On various lakes and rivers in the nation's drainage basins, many other conventional tools are also used. For instance, in the Gumara River of Lake/Tana, scoop nets and fences are utilized [56]. The majority of the fisheries in the Baro-akobo Basin of the Gambella region use traditional gear [53]. Additionally, there are applications for poisons that are extracted from a variety of plants, such as Millettia ferruginea [53, 57].

2.7 Challenges of small-scale fishery development and prospects

2.7.1 Post-harvest losses

High temperatures cause fish to spoil quickly, which causes post-harvest loss. Getu *et al.* [58] estimated that 10–12 million tons of fish worldwide per annum, or 10 percent of total catch, are lost due to spoilage. According to a study report published by Ayalew *et al.* [59], post-harvest fish losses in Lake Hayq and Lake Tekez, Northern Ethiopia, were estimated to have cost 10,934,000 ETB in losses over a 6-year period. Post-harvest fish losses are attributed to a variety of factors, including limited access to markets, size, and species preferences, inadequate infrastructure for handling, processing, storage, and transportation of fish, and distance to major markets.

2.7.2 Limitations on marketing, inadequate infrastructure, and access to fishing gear

The insufficient processing and marketing of small-scale fisheries is caused by a number of things, including inadequate infrastructure, a lack of standard processing methods, and a lack of financial support [60]. Small-scale fishermen use their production inputs like feed and fingerlings and hire labor inefficiently from an economic standpoint [61]. Poor transportation and preservation facilities have an impact on Ethiopia's fish marketing as well. According to a study by Sairam [62], the main marketing issue in the Lake Hawassa area is the absence of sufficient transportation, fish markets that are permanent processing and storage facilities, and customer perception.

According to Alemu *et al.* [63], there are significant issues with the necessary transportation and infrastructure in the Gidabo River and Lake Abaya fishery production systems. Fishermen are compelled to use hefty motorcycles and donkey backs to transport their goods in order to supply the market. Due to a lack of infrastructure, including electricity, fish handling, storage, and preservation techniques are not used. The fishermen offer discounted dried fish products to consumers, hotel owners, and fish traders in Gololcha or Dilla. Due to a lack of contemporary fishing gear and difficult access to markets, fishermen in all landing areas face unique challenges [62]. Additionally hard to come by in Ethiopia are the lead rope and floats needed for nets.

2.7.3 Overfishing

Due to fishermen's lack of knowledge regarding the timing of first sexual maturity, overfishing may become a problem. The majority of fishermen (50.6%), according to Muluye *et al.* [64], were unable to determine yet if the fish they captured were mature or immature. *Labeo barbus* intermediuss caught in the Koka reservoir, according to Tesfahun [65], were undersized when they reached first maturity. In Lake Hawassa, juvenile fishing of a similar nature was seen (77.6% for *Clarias gariepinus* and 23.0% for *Oreochromis niloticus*) and *Labeo barbus* species (15%) for Lake Tana [66]. Overfishing and resource exploitation are also caused by the type and mesh sizes of fishing gear. The research conducted in Lake Ziway showed that the use of small mesh sizes was the main problem, which resulted in 43.33% overexploitation of the lake's fish stock [64]. Poor legal and policy frameworks, as well as the improper application of current fishery laws and regulations, cause poor exploitation of fishery resources in all areas.

2.7.4 Urbanization, agricultural expansion, and wetland degradation

Small-scale fisheries' ability to produce sustainably is being threatened globally by the destruction of aquatic ecosystems brought on by unsustainable fishing methods [67]. In Ethiopia, wetlands accounted for about (22,600 km^2) of the country's total land area. These wetlands areas have helped to protect a variety of pollutants, including sediments, chemicals, fertilizers, sewage from people and animals, animal waste, pesticides, and heavy metals [68]. Wetland areas are used by fish species like *Labeo barbusnedgia, Garadembecha, Clarias gariepinus,* and *Labeobarbus intermedius*, and for their breeding processes; however, in various Ethiopian lakes, the water shade was deteriorating more quickly [69].

Wondie [70] asserted that shoreline wetland stability was significantly threatened by industrial pollution, agricultural expansion, different activities around drainage systems, and removing wetlands' trees for one's own benefit and financial gain. Because fish breeding grounds were destroyed, Lake Ziway's actual production fell from 2300 tons annually in 2003 to 1127 tons annually in 2011 [71]. FAO [29] stated that the main pollutants harming Ethiopian fisheries and water bodies come from agriculture and industrial sewage. Similar to how the effluents from the textile industries in Awassa and Arba Minch and the tannery at Koka Reservoir can affect the status of the fisheries, fish stocks might suffer as a result of Lake Abijata's mineral extraction. It uses the lake water for its own purposes and discharges various pollutants and nutrients into the catchment, which has an impact on various Lake Biodiversity.

2.7.5 Climate change

Due to insufficient rains and droughts, which were made worse by El Nio in 2015, Ethiopia is dealing with a crisis of food insecurity and severe drought [53]. Certain country's fisheries are declining due to climate change [72]. Higher inland water temperatures decrease available fish stocks by altering the trophic status and water quality of a particular aquatic ecosystem. The effects of climate change on agricultural crops can make fishing households more vulnerable; therefore, fishermen around Lake Langeno were forced to catch any fish, regardless of size [73]. The highest runoff occasionally occurred in various areas due to rainfall vibration, adding to the sediment load in the water bodies. Current issues in Lake Tana include sediment load and siltation [74]. Similar to this, changes in fish species diversity, size, and composition have been observed in Lake Ziway as a result of climate variability and change, as have changes in species distribution [75], potential species extinction [76], and decreased productivity [77].

2.7.6 Fish diseases

Diseases have an impact on fish production as well. According to Meko *et al.* [78], one issue facing the nation's fishery industry is fish diseases. Fish disease and parasite conditions reduce the potential for fish production. It should be noted that poor water quality is the primary cause of the majority of parasitic diseases. The majority of parasitic organisms are opportunistic, potentially persistent in the tank or in small numbers on the fish, and only make fish sick when they are stressed [79]. Fish health is influenced by a variety of factors, including the type of filtration system, water temperature, the number of fish in a tank, lighting, pH, and chemistry of the water. It is the primary issue affecting both aquaculture and catch-based fishing around the globe. It might result in high mortality in a body of water or a fishing location. Aside from post-harvest production loss, ailments have been shown to be the cause of death in aquaculture and capture fisheries, and some are also the root cause of zoonotic human diseases in many parts of the world [24]. Overfishing and parasitic infections, for example, are causing a decline in *Labeobarbus intermedius*, according to Mengesha [80] and Dadebo *et al.* [79] reports; as a result, fish are becoming less available to local fish markets. According to the study assessment findings published by Bekele and Hussien [81], The parasite that caused fish the most trouble in Lake Ziway was Contracaecum. In the fish's digestive tract, parasites like nematodes also made a contribution of 19.02% for *Clarias gariepinus* and 8.60% for *Oreochromis niloticus* [81].

2.7.7 Water hyacinth

Water hyacinths (*Eichhorni acrassipses*) are considered the worst invasive weeds due to their negative impact on aquatic ecosystems, fisheries, transportation, agriculture, living conditions, and social structures [82]. Fish kills brought on by dissolved oxygen depletion were caused by water hyacinth's heavy use of and reduction of dissolved oxygen [83]. These weeds are now primarily to blame for the lakes' declining fish production levels. Approximately 34,500 ha (or 15% of the northern shore) of Lake Tana have been reported as being infested with water hyacinth, according to Wassie *et al*. [84]. Because the expansion of water hyacinth interferes with their ability to fish, all fishermen changed their landing location. According to the same research report, water hyacinth caused a decline in *Labeo barbus'* catch per unit of effort (CPUE), which fell from 63 kg/trip in 1991–1993 to 6 kg/trip in 2010. In relation to this, high water hyacinth infestation levels have also been noted in a number of other Ethiopian rift valley lakes, particularly Lake Koka, Lake Ellen, Aba-Samuel Dam, and Lake Wonji [85], which had a similar impact on the various sectors.

2.7.8 Prospects

The nation has started a number of initiatives to boost the economy and improve the standard of living for its citizens. Large dams and reservoirs would be built, as specified in the GTP plan, with the primary goal of producing electricity. Others are dammed to hold water for agriculture that depends on irrigation. These bodies of water can be used to integrate cultured fisheries and increase fishing output nationally. Some success stories include fish production in Koka, Fincha, Gilgelgibe I, Melkawakena, and Tendaho reservoirs. However, in addition to recently constructed sugar cane factories and small to medium-sized irrigation facilities, more reservoirs will be used for fish production without affecting the reservoirs' primary function. Along with other hydroelectric dams in the Genale and Omo-Gibe basins, the massive Ethiopian Renaissance Dam on the Abay (Nile) River will soon be constructed in areas of the nation with high fish production. In order to stock such large bodies of water, this technology needs facilities for transportation, fish seed, and feed production. Therefore, the concerned ministries and institutions will need to make preparations in advance for adequately supplying and stocking water bodies with the appropriate fish species [78, 86, 87]. Ethiopia needs a multi-sectoral strategy and well-coordinated collaborative efforts from all stakeholders to change the rate at which aquaculture advances. Therefore, a critical task for the future is to increase the participation of producers and relevant public authorities in the distribution and management of aquatic resources and land use. For the aquaculture sector to successfully develop in Ethiopia, it was essential to continue building capacity through universities to graduate skilled labor as well as to train, extend, and educate fish producers at a higher level through appropriate research centers [59, 86].

2.8 Fisheries as a means of ensuring food security

Fish is frequently an undervalued but crucial component in securing access to food and nutrition for all [88]. Fish is a particularly good source of omega-3 fatty acids, eicosapentaenoic acid (EPA) and docosahexaenoic acid (DHA), which is high in nutrients and provides high-quality, low-saturated-fat protein [89]. At both the national and international levels, the importance of fish protein in general and small-scale fisheries in particular to food security and nutrition has been acknowledged and advanced [90]. The Food and Agriculture Organization (FAO) claims that all fishing grounds in West Africa are overfished or exploited [91]. Since the 1950s, a large number of development initiatives aimed at increasing fishing activity in West Africa have frequently failed [92]. However, it is impossible to significantly increase either the domestic fish stocks or the sources of imported fish to meet at least the average global consumption of 20 kg per capita in a short amount of time.

Over the past 30 years, aquaculture has grown steadily and quickly, and more than 40% of all fish consumed today comes from aquaculture. Even though aquaculture harvest is increasingly becoming a part of many Asians' diets, it is much less prevalent among people in sub-Saharan Africa. Fisheries and aquaculture can make contributions through both direct and indirect channels. The contribution of the direct consumption to the production is the direct mechanism. For instance, poor people in developing nations typically rely on diets high in carbohydrates to meet their nutritional needs. However, they are lacking in micronutrients and proteins. Fish can aid in preventing micronutrient deficiencies in this situation because it contains high-quality protein, essential fatty acids, and significant micronutrients like iron, zinc, vitamin A, calcium, and iodine [90].

However, households can use aquaculture to improve their nutritional status directly by eating fish from their own ponds [90]. In India, Kumar and Dey [93] found that houses with ponds for fish farming consume 10.9% more energy than households without ponds but with wage earners and that the proportion of undernourished people in households with fish ponds is 10% lower than in the control population. It has been demonstrated that small fish species are crucial sources of protein for fish ponds owned by low-income families in the Dinajpur District of Bangladesh, particularly during the months when vegetables are scarce or prohibitively expensive. Dey and his coworkers in Malawi compared the consumption of fish between homes with and without fish ponds. They noticed that households with fish ponds consume fresh fish and dried fish more frequently [94].

On the other hand, in addition to its direct impact on dietary intake, fish sales also indirectly improve household food security by raising household income. This can be used to buy additional food supplies, such as less expensive staple foods, and is one of the livelihood strategies that has greatly benefited people in developing nations [95]. It is an essential tactic for helping the poor obtain food, income, and other social benefits [96]. Studies have found that fishing has a positive effect on household income.

According to a 2013 study by Gebremedhin and his colleagues in Lake Tana, fishing generates a sizable portion of the country's income. It contributed 48% of the fishermen's total yearly income. In Malawi, households with fish ponds had incomes that were greater by 1.5 times than those of households without [94]. Rural poor households that engage in aquaculture or capture fishing frequently use the extra money they make from selling fish to buy food. Recent studies have shown that households' increased consumption of staple foods is positively impacted by household income from aquaculture [97].

2.9 Potential of fish feed

Fish feeds, which make up at least 40–60% of the cost of production, affect the fish's ability to survive and be profitable [98]. Even though improvements to the continent's overall production system have the potential to boost aquaculture production, the desired expansion of aquaculture, genetics, and principles of general farm management, which is necessary to satisfy the growing demand for fish, is only possible with reasonably priced, high-quality fish food [99]. The cost of producing animals in developing nations is decreased using conventional feed resources, which provide fish with the protein they need at a lower cost. The most important thing to emphasize is the need to find better and more affordable sources of protein that may not be suitable for human consumption [53]. In Ethiopia, which is in the spotlight for this, there are several agricultural processing byproducts that are not used for human consumption but have great potential as feeds for aquaculture on a small scale. It is acceptable that the production of high-quality fish protein from locally accessible low-protein byproducts can significantly increase the local human population's access to protein [32, 53]. The production potential of a number of Ethiopian agricultural and agroindustrial byproducts for use in poultry and livestock feed has been assessed [100]. However, there is not much information available regarding this resource's suitability as fish feed [101]. As a result, the presence of agriculture is a key determinant of the country's potential for aquaculture because it provides byproducts for fish feed and fertilizer. Small-scale fish farming can use agricultural byproducts to increase yields beyond what the pond's natural production would allow.

2.10 Fish consumption trends

It has been found that fish contains nutrients with a high biological value for human health. Despite the country's strong livestock breeding and meat consumption, trends in fish consumption are modest. In comparison to the rest of Africa, Ethiopia consumes very little fish overall (0.216 kg per year) [53]. However, Wednesdays and Fridays receive a disproportionate amount of consumption, as do those days when people are fasting (15 days in August, 55 days in March/April, and other possibly less widely observed times) [25]. Fish is becoming increasingly more of a luxury good consumed by higher-income groups as a result of rising real prices due to increasing scarcity (apparently reflecting both rising demand and supply constraints) [53, 102].

2.11 Marketing system of fish in Ethiopia

Despite being more expensive and therefore harder to come by, the majority of Ethiopians prefer beef to fish [103]. There are numerous factors that affect fish marketing in Ethiopia as well, the majority of which are common to many developing nations [53, 99].

Ethiopia has a sizable market potential for fish, and there is a demand for fish. However, Ethiopia's fish marketing system is hampered by a lack of a fish trading tradition, an ineffective marketing network, and subpar transportation and preservation facilities. In addition to the aforementioned difficulties, it has been discovered that the marketing situation for fish is harmed by people's selectivity toward certain species. The relationship between fish supply and demand in the nation, which has fluctuated, determines the price of fish. The same research showed that, for instance, in Lake Tana, the cost of whole fish per kilogram nearly doubled in just 5 years. Additionally, the selling price of fish fillets has almost tripled. However, a kilogram of whole or filleted fish now costs between 15 and 20 Birr and 65 to 85 Birr, respectively, in RVA and even more in Addis Abeba [57, 100].

According to Shimada [54], the price of producing one kilogram of fish varies according to the time of year when fish are available near the fishing grounds. Production levels are above average during months like July to September and February to May. Between July and September, primarily in August, production peaks, and this is also the time when production costs are lowest. Consequently, both individual fishermen and fishing cooperatives profit greatly during this season [104]. The amount of produce fell below the mean in the other seasons, although to the point where some months of the year go by without any fish being caught. Commercial fishermen lost money and stopped fishing during this time.

3. Materials and methods

This review used a variety of literature sources, journals, books, book chapters, workshop materials, FAO reports, bulletins, legal documentation, and documents from the Internet to examine Ethiopia's fishery resources. Information on catch composition, seasonal patterns, regional variations in harvest rates, projections of potential output, processing methods, marketing, socioeconomic factors, management, and legal and regulatory frameworks were all reviewed. Using the literature sources, information on the quantity of water resources was also updated. Fisheries data from 42 years (from 1973 to 2023) were analyzed to assess Ethiopia's state of fishery production and aquaculture development.

4. Conclusion

Ethiopia is home to a large variety of ichthyofauna in its lakes, rivers, and reservoirs. Several endemic, Highland East African and Nilo-Sudanic species make up Ethiopia's fish fauna. Ethiopia's inland waterways are home to 168–183 valid fish species, including 37–57 endemics in the nation. Baro, Blue Nile, Wabi-Shebele, and

Omo-Gibe basins are in the order of highest fish species diversity in Ethiopia. But, it appears that the Blue Nile and Awash basins have the highest levels of endemicity. First of all, this is attributed to the flock of a critically endangered species called Labeobarbus that resides in Lake Tana. The Tekeze Basin in Ethiopia's borders has the least variety of fish species when compared to other parts of the Nile basin. Many Nilo Sudanese fishes similar to those found in the southern Rift Valley lakes are thought to inhabit the Wabi-Shebele ecoregion (Lakes Chamo and Abaya). The area with the greatest diversity of introduced fish species is the Rift Valley Basin. Site-specific management is crucial in fishery biology and fish communities because fish communities vary depending on the type of water body.

Authors' contributions

Abenezer Wendimu collected data, organized the data on the computer, identified relevant papers, and wrote the draft manuscript. Dr. Wondimagegnehu Tekalign (Ph.D., Assoc. Prof.) and Dr. Abnet Woldesenebet (Ph.D., Ass. Prof.) advised and reviewed the manuscript and assisted in the language edition. All authors read and approved the final version of the manuscript.

Funding

The study received no funding from government, commercial, or non-profit financing organizations.

Competing interests

There is no conflict of interest between the authors regarding this paper.

Consent for publication

This manuscript does not contain any person's data, and further consent for publication is not required.

Availability of data and materials

The datasets generated and analyzed during the current study are included in the body of this paper.

Author details

Abenezer Wendimu[1,2]*, Wondimagegnehu Tekalign[2] and Abnet Woldesenbet[2]

1 Wolaita Zone Education Bureau, Bilaten Tena Secondary School, Biology Department, Dimtu, Wolaita, Ethiopia

2 Natural and Computational Sciences College, Biology Department, Wolaita Sodo University, Wolaita Sodo, Ethiopia

*Address all correspondence to: wabeniezer@gmail.com

References

[1] Pino-del-Carpio A, Ariño AH, Miranda R. Data exchange gaps in knowledge of biodiversity: Implications for the management and conservation of biosphere reserves. Biodiversity and Conservation. 2014;**23**:2239-2258. DOI: 10.1007/s10531-014-0718-2

[2] Norris RH, Hawkins CP. Monitoring river health. Hydrobiologia. 2000;**435**: 5-17. DOI: 10.1023/A:1004176507184

[3] Vyas D, Krishnayya NSR, Manjunath KR, Ray SS, Panigrahy S. Evaluation of classifiers for processing Hyperion (EO-1) data of tropical vegetation. International Journal of Applied Earth Observation and Geoinformation. 2011;**13**(2):228-235. DOI: 10.1016/j.jag.2010.11.007

[4] Getahun A, Stiassny MLJ. The freshwater biodiversity crisis: the case of the Ethiopian fish fauna. SINET: Ethiopian Journal of Science. 1998;**21**:207-230

[5] Joint Ethio-Russian Biological Expedition (JERBE). An artificial key for fish species of the Gambella Region (The White Nile in the limits of Ethiopia). Addis Ababa, Ethiopia; 1995. p. 45

[6] FishBase. All fishes reported from Ethiopia (landlocked). Paris, France: FishBase; 2023. Available from: https://fishbase.mnhn.fr/country/CountryChecklist.php?what=list&trpp=50&c_code=230&csub_code=&cpresence=present&sortby=alpha&ext_pic=on&vhabitat=all2

[7] Tropical Freshwater Fish. List of Freshwater Fishes for Ethiopia. 2021. Available from: https://tropicalfreshwaterfish.com/data/Ethiopia.htm

[8] Golubtsov AS, Mina MV. Fish species diversity in the main drainage systems of Ethiopia: current state of knowledge and research perspectives. Ethiopian Journal of Natural Resources. 2003;**5**(2):281-318

[9] FAO. The State of World Fisheries and Aquaculture 2010. 2010. Available from: https://www.fao.org/3/i1820e/i1820e.pdf

[10] FAO (Food and Agriculture Organization of the United Nations). *In brief to the state of world fisheries and aquaculture 2022. Towards blue transformation*. Rome, Italy: FAO; 2022

[11] Lundberg JG, Kottelat M, Smith GR, Stiassny MLJ, Gill AC. So many fishes, solitle time: an over view of recent ichthyological discoveries in freshwaters. Annals of the Missouri Botanical Gardens. 2000;**87**(1):26-62

[12] Welcomme RL. The inland fisheries of Africa. Food and Agriculture Organization of the United Nations. Rome: FAO; 1997. committee for inland fisheries of Africa occurrence paper 7, 69

[13] FAO (Food and Agriculture Organization of the United Nations). The state of world fisheries and aquaculture 2020. Sustainability in action. Rome, Italy: FAO; 2020

[14] Augustyn J, Cockcroft A, Kerwath S, Lamberth S, Githaiga-Mwicigi J, Pitcher G, et al. Climate change impacts on fisheries and aquaculture: a Global analysis. In: Phillips BF, Pérez-Ramírez M, editors. Chapter 15: South Africa. Vol. II. Berlin, Germany: Springer 2019. pp. 479-522

[15] DAFF. *Status of the South African marine fishery resources 2016*. Forestry and Fisheries, Cape Town: Department of Agriculture; 2016

[16] MFMR. Fisheries policy 2015 ensuring fisheries and aquatic resources sustainability. Maximising economic and social benefits to all Namibians, Namibia. Windhoek: Ministry of Fisheries and Marine Resources; 2015

[17] MFMR. Strategic plan 2017/18-2021/22. Namibia, Windhoek: Ministry of Fisheries and Marine Resources; 2017

[18] ATFALC. Fisheries and aquaculture industry in Namibia. In: The ministerial conference on fisheries cooperation among African States Bordering the Atlantic Ocean (ATFALCO): Series report no. 2 on the fisheries and aquaculture review in the 22 ATLAFCO member countries. Rabat, Morocco: ATLAFCO -The Ministerial Conference on fisheries cooperation among African States bordering the Atlantic Ocean. 2012

[19] Cervantes A, Faraj A, Enthoven L. Rapport de la Réunion Scientifique Conjointe annuelle relative à l'Accord de pêche signé entre le Royaume du Maroc et l'Union européenne. Madrid, Espagne, 24-26 septembre 2018. Bruxelles: Rapports des Comités Scientifiques Conjoints; 2018

[20] Gordon A, Sewmehon D, Melaku T. Marketing systems for fish from Lake/ Tana, Ethiopia: Opportunities for improved marketing &livelihoods. In: Improving Productivity and Market Success of Ethiopian Farmers. Project Working Paper 2. Nairobi, Kenya: ILRI; 2007

[21] Tesfaye G, Wolff M. The state of inland fisheries in Ethiopia: A synopsis with updated estimates of potential yield. Ecohydrology & Hydrobiology. 2014;**14**(3):200-219. DOI: 10.1016/j.ecohyd.2014.05.001

[22] Kundu R, Aura CM, Muchiri M, Njiru JM, Ojuok JE. Difficulties of fishing at Lake Naivasha, Kenya: Is community participation in management the solution? Lakes & Reservoirs: Research & Management. 2010;**15**(1):15-23. 10.1111/j.1440-1770.2010.00419.x

[23] Tigabu Y, Abera D, Ahmed L. Kinds and levels of post harvest losses of fish and possible ways to reduce loss in Lake Ziway. Proceeding of Ethiopia Society of Animal Production. 2007;**15**:143-150

[24] Wang H, Amesa S, Endebu M, Tirfessa G, Zhong-Yi Z, Zhi-Gang W. Preliminary study on construction of three-in-one fishpond and its effect on aquaculture in Ethiopia. Fisheries and Aquaculture Journal. 2018;**9**(2):2-6

[25] FAO. Information on Fisheries Management in the Federal Democratic Republic of Ethiopia. 2003. Available from: http://www.fao.org/fi/fcp/en/ETH/body.htm

[26] Roberts TR. Geographical distribution of African freshwater fishes. Zoological Journal of the Linnean Society. 1975;**57**:249-319

[27] Alemayehu L. The economic implications of comprehensive approach to learning on fisheries and aquaculture industrial development (policy and managerial capability learning): A case of Ethiopia Research Institute. Japan: Japan International Cooperation Agency (JICA); 2017

[28] Tedla S, Hyelemeske F. Introduction and transplantation of freshwater fish species in Ethiopia. SINET: Ethiopian Journal of Science. 1981;**4**:69-72

[29] FAO (Food and Agriculture Organization). Fishery and aquaculture country profles: the federal democratic republic of Ethiopia. Rome, Italy: FAO Fisheries and Aquaculture Department; 2014

[30] Getahun A, Dejen E, Antehen W. Fishery studies of Ribb River, Lake Tana Basin, Ethiopia. A report submitted to World Bank. 2008;**2**:1573

[31] Golubstov A, Darkov A. A review of fish diversity in the main drainage systems of Ethiopia. In: Dmitry S, Pavlov DY, Andrey A, Mina M, editors. Ecological and Faunistic studies in Ethiopia. Addis Ababa, Ethiopia: JERBE 20 Years Scientific Cooperation, Artistic inc; 2008. p. 173

[32] Vadacchino L, De Young C, Brown D. The fisheries and aquaculture sector in national adaptation programmes of action: importance, vulnerabilities and priorities. In: FAO Fisheries and Aquac No.1064. Rome. Italy: FAO; 2011

[33] Tewabe D. Spatial and temporal distributions and some biological aspects of commercially important fish species of Lake Tana, Ethiopia. The fourth annual conference of the Ethiopian fisheries and aquatic sciences association. Ethiopia: (EFASA); 2012

[34] Westphal E. Agriculture systems in Ethiopia. Joint publication of the college of Agriculture, Haile Selassie I University, Ethiopia and the Agricultural University. Wageningen, the Netherlands: Center for Agricultural Publishing and Documentation; 1975

[35] Getahun A. An overview of the diversity and conservation status of the Ethiopian freshwater fish fauna. Journal of Afrotropical Zoology. 2007:87-96

[36] Getahun A. The Nile in the Ethiopian territory: riverine fish and fisheries. In: A paper presented at the Food and Water challenge international workshop that took place in Addis Ababa, Ethiopia. Addis Ababa, Ethiopia: Addis Ababa University; 2003

[37] Habteselassie R. Fishes of Ethiopia. Annotated checklist with pictorial identification guide. Addis Ababa: Ethiopian fisheries and aquatic science association; 2012

[38] Rzoska J. The Blue Nile system. In: Rzoska J, editor. The Nile, Biology of An Ancient River. The Hague: DR. W. Junk B.V. Publishers; 1976. pp. 219-232. Available from: https://www.rcin.org.pl/dlibra/publication/109546/edition/155646/content?ref=struct

[39] Tedla S. Freshwater fishes of Ethiopia. Addis Ababa, Ethiopia: Department of Biology, Addis Ababa University; 1973. p. 101

[40] Golubstov, A. and, Mina, M. 2003. Fish species Diversity in the main drainage system of Ethiopia: current state of knowledge and research perspectives Ethiopia. Journal of Natural Resources 5:281-318.

[41] Skelton PH. A complete guide to the freshwater fishes of southern Africa. 2nd ed. South Africa: Southern Book Publishers; 2001. p. 388

[42] De Graaf M, Nentwich E, Osse J, Sibbing F. Lacustrine spawning, a new reproductive strategy among 'large'African cyprinid fishes,Journal of Fish Biology. 2005;**66**:1214-1236. DOI: 10.1111/j.0022-1112.2005.00671.x

[43] De Graaf M, Dejen E, Sibbing F, Osse, J. Barbus tanapelagius, a new species from Lake Tana (Ethiopia): Its morphology and ecology. Environmental Biology of Fishes. 2000;**59**:1-9. DOI: 10.1023/A:1007608208630

[44] Stiassny M, Getahun A. An overview of Labeo in relationship and the phylogenetic placement of the Afro-Asian genus *Garra* Hamilton,

1922 (Teleostei: Cyprinidae), with the description of five new species of Garra from Ethiopia, and a key to all African species. Zoological Journal of the Linnean Society. 2007;**150**:41-83

[45] De Graaf M. Lake Tana's piscivorous barbus (Cyprinidae, Ethiopia). Ecology, evolution exploitation [PhD thesis]. The Netherlands: Wageningen Agricultural University; 2003. Available from: https://edepot.wur.nl/121409

[46] Avery ST. Lake Turkana and the lower Omo: The impact of hydrological impacts of major Dam and irrigation Development on the world's largest desert lake. African studies center school of Interdisciplinary Area; 2013

[47] Beadle LC. The Inland Waters of Tropical Africa: An Introduction to Tropical Limnology. 2nd ed. Longman, New York. 1981. 475 p. DOI: 10.1002/iroh.19830680114

[48] Basnyat D, Gadain H. Hydraulic behavior of the Juba and Shebelle Rivers: basic analysis for irrigation and flood management purposes. Technical report No W-13. Nairobi, Kenya: FAO-SWALIM; 2009

[49] Bonini M, Corti G, Innocenti F, Manetti P, Mazzarini F, Abebe T, et al. Evolution of the Main Ethiopian Rift in the frame of Afar and Kenya rifts propagation. Tectonics. 2005;**24**:TC1007

[50] Gabriel G. The main Ethiopian Rift System: an overview on and sedimentation processes. In: Tudorancea C, Taylor WD, editors. *Ethiopian Rift Valley Lakes*. Leiden, Netherlands: Backhuys Publishers; 2002. pp. 13-43

[51] Wudneh T. *Biology and management of fish stocks in Bahir Dar Gulf, Lake Tana, Ethiopia*. Ph.D. dissertation,. Wageningen: Wageningen Agricultural University; 1998. 142 pp

[52] Kurien J. Small-scale fisheries in the context of globalization. In: Tromso. Trivendrum, India: Centre for Development Studies; 1998. pp. 1-47

[53] FAO (Food and Agriculture Organization). The State of Food Security and Nutrition in the World. Rome, Italy: Building Climate Resilience for Food Security and Nutrition; 2016

[54] Shimada G. The Economic Implications of Comprehensive Approach to Learning on Industrial Development (policy and Managerial capability learning) (No. 1001). A Case of Ethiopia Research Institute. JICA (Japan International Cooperation Agency); 2013. Available from: https://ideas.repec.org/p/jic/wpaper/1001.html

[55] Assefa Mitike Janko AM. Fish production, consumption and management in Ethiopia. International Journal of Economics & Management Sciences. 2014;**3**(3):25-39. DOI: 10.4172/2162-6359.1000183

[56] Kefale M, Chinnan KPM. Employment growth and challenges in small and micro enterprises Woldiya, north east Amhara Region, Ethiopia. Educational Research and Essays. 2012;**1**(2):21-22. Available from: https://api.semanticscholar.org/CorpusID:20194198

[57] Daw K, Brown C, Badjeck M. Climate change and capture fisheries. Norwich, World Fish Center, Penang: University of East Anglia; 2012

[58] Getu A, Misganaw K, Bazezew M. Post-harvesting and major related problems of fish production. Fisheries and Aquaculture Journal. 2015;**6**:154

[59] Ayalew A, Fufa A, Wubet B, Samson L. Assessment of post-harvest fish losses in two selected lakes of Amhara Region, Northern Ethiopia.

Heliyon. 2018;**4**(11):e00949. DOI: 10.1016/j.heliyon.2018.e00949

[60] Agboola W. Improving Fish farming Productivity towards Achieving Food Security in Osun State, Nigeria: A Socioeconomic Analysis. Annals of Biological Research. 2011;**2**:62-74

[61] Njagi KA, Njati IC, Smaio GH. Factors affecting profitability of fish farming under economic stimulus programme in Tigania East District, Meru County, Kenya. IOSR Journal of Business and Management. 2013;**15**(3):25-36

[62] Sairam BP. Assessment of fish production and marketing system in Lake Hawassa at Amora Gedel, Ethiopia. Ethiopia: Behara Sairam Patnaik; 2016. https://www.researchgate.net/publication/301888630 [Accessed: December 2022]

[63] Alemu LA, Assefa MJ, Tilahun GA. Fishery production system assessment in different water bodies of Guji and Borana zones of Oromia, Ethiopia. International Journal of Fisheries and Aquatic Studies. 2014;**2**(2):238-242

[64] Muluye T, Tekle-Giorgis Y, Tilahun G. The extent of immature fish harvesting by the commercial fishery in Lake Hawassa, Ethiopia. Momona Ethiopian Journal of Science. 2016;**8**(1):37-49

[65] Tesfahun A. Some biological aspects and immature fishing of the African Big Barb *Labeobarbus intermedius* (R.) in Lake Koka, Ethiopia. A part of MSc thesis submitted to Hawassa University. Hawassa, Ethiopia; 2011

[66] Heide F. Feasibility Study for a Lake Tana Biosphere. United Nations Educational, Scientific and Cultural Organization (UNESCO). Reserve, Ethiopia; 2012

[67] Eggert H, Greaker M. Effects of global fisheries on developing countries possibilities for income and threat of depletion. Environment for Development Initiative. 2009;**393**:27

[68] Gebretsadik T, Mereke K. Threats and opportunities to major Rift Valley Lakes wetlands of Ethiopia. Journal of Agriculture Research and Technology. 2017;**9**(1):0010012

[69] Tessema A, Mengist A, Mejen EA. Survey on fisheries in Chefa wetland and around Kemissie, Oromia Zone Ethiopia. Direct Research Journal of Agriculture and Food Science. 2014;**2**(3):28-32

[70] Wondie A. Improving management of shoreline and riparian wetland ecosystems: the case of Lake Tana catchment. Bahir Dar University, Bahir Dar, Ethiopia. 2010;**10**(2-4):123-132

[71] Hailu M. Ecosystem structure, trophic link and functioning of a shallow rift valley lake: the case of Lake Ziway (Ethiopia). Ethiopia: Addis Ababa University; 2011

[72] Temesgen M, Getahun A. Fishery management problems in Ethiopia: natural and human induced impacts and the conservation challenges. Reviews in Fisheries Science and Aquaculture. 2016;**24**(4):305-313

[73] Temesgen M. Length-weight relationship and condition factor of fishes in Lake Langena, Ethiopia, PhD Dissertation. Ethiopia: Addis Ababa University; 2017

[74] Awoke T, Melaku M. Challenges and possible mitigation of Ethiopia fishery: A review. International Journal of Fisheries and Aquatic Studies. 2017;**5**(1):241-246

[75] Perry RI, Ommer RE, Allison E, Badjeck MC, Barange M,

Hamilton L. The human dimensions of marine ecosystem change: interactions between changes in marine ecosystems and human communities. England: Oxford University; 2009

[76] Cheung WWL, Lam VWY, Sarmiento JL, Kearney K, Watso R, Pauly D. Projecting global marine biodiversity impacts under climate change scenario. Fish and Isheries. 2009;**10**(3):235-251

[77] Abera L, Getahun A, Lemma B. Composition of commercially important fish species and some perspectives into the biology of the African Catfish *Clarias gariepinus* (Burchell), Lake Ziway, Ethiopia. International Journal of Advanced Research. 2014;**2**(1):864-871

[78] Meko T, Kebede A, Hussein A, Tamiru Y. Review on Opportunities and Constraints of Fishery in Ethiopia. International Journal Poultry Fish Science. 2017;**1**(1):18

[79] Dadebo E, Tesfahun A, Teklegiorgis Y. Food and feeding habits of African big barb *L. intermedius* (Rüppell, 1836) (Pisces: Cyprinidae) in Lake Koka, Ethiopia. Journal of Agricultural Research and Development. 2013;**3**:49-58

[80] Mengesha M. Heavy metal pollution in the rift valley Lakes of Awassa and Koka, M.Sc Thesis. Germany: University of Bremen; 2009

[81] Bekele J, Hussien D. Prevalence of internal parasites of *Oreochromis niloticus* and *Clarias gariepinus* fish species in Lake Ziway. Ethiopian Journal of Aquaculture Research Development. 2015;**6**:308

[82] Bhattacharya A, Haldar S, Chatterjee P. Geographical distribution and physiology of water hyacinth (*Eichhornia crassipses*)—The invasive hydrophyte and a biomass for producing xylitol. International Journal of ChemTech Research. 2015;**7**(4):1849-1861

[83] Waithaka E. Impacts of water hyacinth (*Eichhornia crassipes*) on the fishing communities of Lake Naivasha, Kenya. Journal of Biodiversity & Endangered Species. 2013;**1**:2

[84] Wassie A, Minwuyelet M, Ayalew W, Dereje T, Woldegebrael W. Water hyacinth coverage survey report on Lake Tana. Technical Report Series. 2014

[85] Firehun Y, Struik PC, Lantinga EA, Taye T. Water hyacinth in the Rift Valley water bodies of Ethiopia: its distribution, socioeconomic management. International Importance Journal Agricultural Research. 2014;**3**(5):067-075

[86] Abera L. Fisheries production system scenario in Ethiopia. International Journal of Fisheries and Aquatic Studies. 2017;**5**(1):79-84

[87] Lakew A, Tadesse Z, Dagne A. Technologies suitable for aquaculture development in Ethiopia. Ethiopia: Ethiopia Fish and Fisheries Association; 2016

[88] HLPE. Sustainable fisheries and aquaculture for food security and nutrition. A report by the high level panel of experts on food security and nutrition of the committee on world food security, P118. 2014 www.fao.org/3/a-i3844e.pdf [Accessed 23 December 2022]

[89] Hu FB, Bronner L, Willett WC, Stampfer MJ, Rexrode KM, Albert CM. Fish and omega-3 fatty acid intake and risk of coronary heart disease in women. Journal of the American Medical Association. 2002;**287**:1815-1821

[90] Kawarazuka N. The contribution of fish intake, aquaculture, and small-scale

fisheries to improving nutrition: a literature review. In: The world fish center working paper No. 2106. Penang, Malaysia: The world fish center; 2010

[91] FAO (Food and Agriculture Organization). Contribution of fisheries to national economies in West and Central Africa—policies to increase the wealth generated by small-scale fisheries, new directions in fisheries e a series of policy briefs on development issues. Rome, Italy: Food and Agriculture Organization of the United Nations; 2006a

[92] Belhabib D, Sumaila UR, Pauly D. Feeding the poor: contribution of West African fisheries to employment and food security. Ocean and Coastal Management. 2014;**111**:72-81

[93] Kumar P, Dey MM. Nutritional intake and dynamics of undernourishment of farm households in rural India. Indian Development Review. 2006;**4**(2):269-284

[94] Dey MM, Kambewa M, Prein D, Jamu FJ, Paraguas E, Pemsl E, et al. Impact of development and dissemination of integrated aquaculture—agriculture (IAA) technologies in Malawi. NAGA World Fish Centre Quarterly. 2006;**29**(1 & 2):28-35

[95] Aiga H, Matsuoka S, Kuroiwa C, Yamamoto S. Malnutrition among children in rural Malawian fishfarming households. Transactions of the Royal Society of Tropical Medicine and Hygiene. 2009;**103**:827-833

[96] Manasi S, Latha N, Raju KV. Fisheries and livelihoods in Tungabhadra Basin, India: current status and future possibilities. Working Paper 217. Bangalore: The Institute for Social and Economic Change; 2009, 24 pp

[97] Jahan KM, Ahmed M, Belton B. The impacts of aquaculture development on food security: lessons from Bangladesh. Aquaculture Research. 2009;**41**(4):1-15

[98] Gabriel UU, Akinrotimi OA, Bekibele DO, Onunkwo DN, Anyanwu PE. Locally produced fish feed: potentials for aquaculture development in subsaharan Africa. African Journal of Agricultural Research. 2007;**2**(7):287-295

[99] Hirpo LA. Fisheries production system scenario in Ethiopia. International Journal of Fisheries and Aquatic Studies. 2017;**5**:79-84

[100] Asfaw Z. The future of wild food plants in southern Ethiopia: ecosystem conservation coupled with enhancement of the roles of key social groups. Acta Horticulturae. 2008;**806**:701-708

[101] Craig S, Helfrich LA. Understanding fish nutrition, feeds and feeding. Cooperative Extension Service. 2012;**13**(1):420-256

[102] FAO (Food and Agriculture Organization). Fishery (http://www.fao.org/faoterm/en/?defaultCollId=21) (Entry: 98327). Rome, Italy: Food and Agriculture Organization of the United Nations; 2009a. p. 24. Accessed: 21 December 2022

[103] Adugna T. Feed resource for producing export quality meat and livestock in Ethiopia from selected woredas in Oromia and SNNP regional states, Ethiopia. 2007. Available from: http://hdl.handle.net/123456789/3038

[104] EIDP. Evolution and current performance. In: Presentation on "Learning to Compete: Accelerating Industrial Development in Africa" Conference Organized by UNUWIDER; 24/25 June; Helsinki, Finland. 2013. Available from: https://www.wider.unu.edu/expert/finn-tarp?page=30

Chapter 11

Timber and Trout: An Examination of the Logging Legacy and Restoration Efforts in Headwater Streams in New England (USA)

Joshua C. Hoekwater, Amy M. Villamagna, Brigid C. O'Donnell, Tyson R. Morrill, Ben J. Nugent and Jared B. Lamy

Abstract

The forested landscape of New England (USA) was dramatically altered by logging during the nineteenth and early twentieth centuries. Although the northern temperate forests of the region have largely regenerated, the streams and rivers remain impacted. The loss of terrestrial wood, organic material, and nutrient inputs during the forest regeneration period has affected habitat quality and biotic communities, most notably in small headwater streams. The same waterways are further impacted by now undersized stream crossings, mostly culverts associated with old infrastructure that alter hydrology and sediment transport; moreover, these culverts have created barriers to the movement of riverine organisms. We synthesize literature on headwater stream wood additions and culvert removal in North America and discuss observed patterns in organic matter, benthic macroinvertebrates, and Brook Trout (*Salvelinus fontinalis*) from before and after wood additions and stream-crossing enhancements in a previously logged watershed in New England. There were minimal changes to habitat and substrate two years after restoration efforts. However, streams with wood additions retained a higher density of rafted organic matter and had significantly higher benthic macroinvertebrate density. Additionally, two years after restoration, one year-old Brook Trout were significantly longer in restored streams than prior to restoration. Collectively, these results document a relatively rapid increase in organic matter retention, macroinvertebrates, and Brook Trout size, soon after restoration efforts.

Keywords: freshwater streams, Brook Trout, habitat fragmentation, restoration, benthic macroinvertebrates, organic matter

1. Introduction

The forested landscape of the New England region of the United States was dramatically altered in the nineteenth and early twentieth century [1]. Old-growth forests were largely cleared, first for agriculture and then in response to the growing demand for wood products [2]. By 1839, New England and New York accounted for

41% of the nation's timber harvest [3] and by the mid-1880s approximately 80% of New Hampshire forests had been cleared [4]. Since this time of dramatic disturbance, logging within riparian areas has decreased mature forested cover, increased stream temperatures and stream siltation, and reduced wood inputs that provide critical habitat [1, 5–11].

Allochthonous coarse particulate organic matter (CPOM), commonly in the form of autumn-shed leaves, supplies most of the energy to first order streams in eastern US deciduous forests [12, 13]. Retention of this energy in headwaters is often driven by large woody debris from riparian forests that trap CPOM and sediments, creating wood jams that operate as bioenergetic hotspots [8, 11, 14] and influence nutrient cycling, microbial populations, and macroinvertebrate densities locally and downstream [15–20]. Retained CPOM can be transformed into dissolved organic matter (DOM) by leaching [21] or broken down into fine particulate organic matter (FPOM) by mechanical or biological processes. CPOM, DOM, and FPOM become the vectors for increased microbial and invertebrate biomass that influence food webs from the bottom up [22–25]. The lack of such inputs has been shown to dramatically reduce invertebrate productivity [13, 26]. Wood jams also form pools, create protective overhead cover, and diversify habitat [27–29], providing fish in headwater streams with refuge from predation [30] and high flow events [31]. Several studies have suggested that wood additions benefit lotic fish populations [29, 32–34]. However, others suggest the impacts are lessened in headwater streams where boulder substrates and high gradients create habitat complexity [32, 35]. Headwater streams represent 79% of river length in the conterminous United States, and so clarifying the impacts of restoration is a worthwhile goal [36].

Although the northern temperate forests of the region have largely regenerated with the decrease in logging, the suppression of natural fires, and land conservation (e.g., the White Mountains National Forest), the waterways that crisscross the landscape remain impacted. Mature forests contribute less wood, organic material, and nutrients to streams and rivers than their old-growth counterparts [8, 37]. The absence of these terrestrial inputs has a documented effect on habitat quality, species assemblages, and species densities [1, 38, 39]. The same systems remain impacted by dams and stream crossings (mostly culverts) associated with the timber industry. The artificial water conduits of the intensive logging period were constructed for short-term economic gain, not long-term resilience. Although some stream crossings were replaced during the development of the modern automobile corridor, those that remain within the reforested areas are undersized for current hydrologic flows or poorly fit, resulting in altered hydrologic and sediment flows and substantial impediments to the movement of aquatic species [40, 41]. Culverts can alter sediment transport, organic matter retention, and hydrology [41]; some evidence shows culverts may negatively impact benthic macroinvertebrate communities [18, 42–44]. Impassable road-stream crossings also limit the ability of fish to thermoregulate, forage efficiently, and spawn effectively [45–48].

Headwater streams maintain robust populations of Brook Trout (*Salvelinus fontinalis*) throughout their historic range in the eastern USA [49]. However, this species is highly vulnerable to habitat fragmentation and degradation within headwater streams [46, 50]. Brook Trout require high-quality and heterogeneous habitat to thermoregulate, avoid competition, access spawning habitat, and to find refuge during seasonal droughts and extreme flow events [47, 51, 52]. Therefore, habitat fragmentation and degradation attributed to impassable stream crossings and reduced woody inputs pose significant challenges to headwater Brook Trout populations and increase their risk of extirpation [47, 53–56]. These threats are exacerbated by warming waters and the loss of suitable habitat [57–61].

Stream restoration efforts in historically logged watersheds often focus on wood additions [11], but documented responses across trophic levels are limited for headwater systems [32]. To date, few studies have documented a positive fish response to culvert removal or wood additions in headwater streams, and fewer still have tracked multitrophic responses to such restoration efforts [32, 62–64]. Researchers who study stream restoration efforts have emphasized a need for more monitoring before and after intervention, over longer time periods, and with detailed measurements of abiotic and biotic factors [29, 32, 33, 65]. While long and controlled experimental designs (e.g., Before-After-Control-Impact, or BACI) are generally preferred, the conditions for such studies are inherently rare given the limited and opportunistic nature of restoration (i.e., financial investment, time, and access to equipment).

Despite the prevalence of movement barriers and the lack of wood recruitment impacting streams, few studies have been published regarding Brook Trout populations in New Hampshire [49, 66–68]. Herein, we describe an adaptive sampling approach focused on evaluating early fine-scale abiotic and multitrophic responses to culvert removal and upstream wood additions in headwater streams within a watershed in central New Hampshire. We discuss patterns of wood accumulation, organic matter retention, benthic macroinvertebrates, and Brook Trout before and after restoration treatments which suggest rapid local and downstream responses across trophic levels. For a discussion of the impacts of habitat fragmentation and recent reconnection on the population genetic structure of wild Brook Trout in the watershed, we direct the reader to Lamy et al. [69]. In closing, we discuss the value of our work in the context of environmental management and conservation.

2. Site history and description

The Beebe River watershed (USGS Hydrologic Unit Code: 01070010403; area = 20,187 acres/81.8 sq. km) is located in central New Hampshire (USA). This river is a tributary to the Pemigewasset River that flows south through New Hampshire into Massachusetts where it continues to the Atlantic Ocean. Like most forested areas in central and northern New Hampshire, the Beebe River watershed has a long history (>150 years) of logging, including the more selective timber harvests of the current era to accommodate power lines that run parallel to a section of the river's mainstem. Indications of logging within the watershed were first documented on a map produced in 1860 when a completed road paralleled the Beebe River toward Sandwich and a sawmill was constructed on the upper river [70]. The ~42 km Beebe River Railroad was constructed between 1917 and 1921 [71]. The establishment of this access route initiated extensive logging in the region. Toward the end of WWI, virgin Red Spruce (*Picea rubens*) from the Beebe River area contributed one quarter of the timber used in construction of US military aircraft [72]. All the spruce was reportedly removed by 1920, but the railroad remained as one of three active routes through the White Mountains [71]. In 1923, a fire, likely attributed to forest slash ignited railside, burned more than 14 sq. km (3500 acres) of the Beebe headwater forest [72]. Much of the regional softwood had been harvested and other large expanses burned, while patches of hardwood forests remained. Logging continued throughout the region for 40–50 years, depleting the area of usable timber. As a result, the current forest stand age largely ranges between 80 and 100 years old [4].

Between 1977 and the early 2000s, ownership changed several times and the property was broken into smaller parcels [73], including 1960 ha purchased by the US Forest Service to expand the White Mountain National Forest. In 2014, The Conservation Fund (TCF), a US-based land conservation organization, purchased 22 sq. km (5441 acres), containing 27 percent of the Beebe River watershed, and led stream restoration and land protection efforts (in cooperation with Redstart Forestry in 2017, the town of Sandwich, NH in 2017, and the town of Campton, NH in 2018). Such efforts included the removal and replacement by steel and plank bridges of five undersized culverts and bridges that impeded stream crossings. This restoration was funded by the TCF's Working Forest Fund as well as the U.S. Department of Agriculture's Regional Conservation Partnership Program (RCPP). The RCPP was an initiative of the 2014 Farm Bill and was funded from the U.S. Forest Service's Forest Legacy Program through the Land and Water Conservation Fund; this enabled TCF to permanently protect the land with working forest conservation easements that maintain public recreational access in perpetuity. In total, TCF reconnected nearly six miles of wild Brook Trout spawning and rearing habitat before selling the easements in 2023 (The Conservation Fund 2023).

The study described herein arose as an opportunity to document the legacy effects of logging on the watershed and monitor the effects of restoration efforts. Data collection was conducted through a partnership between Plymouth State University, the NH Fish and Game Department, and the Pemigewasset Chapter of Trout Unlimited. Three tributaries to the Beebe River, denoted as GR3, GR4, and ECR1 (**Figure 1**) were the focus of this work. These streams were selected for this study based on their similarity in channel width (2–3 m), gradient (4–10% within each), substrate (boulder dominant; **Table 1**), canopy cover, proximity to one another (**Figure 1**), but also differences relating to the extent of habitat fragmentation, degradation, and restoration efforts. Mean daily maximum water temperatures in July ranged from 18.2°C to

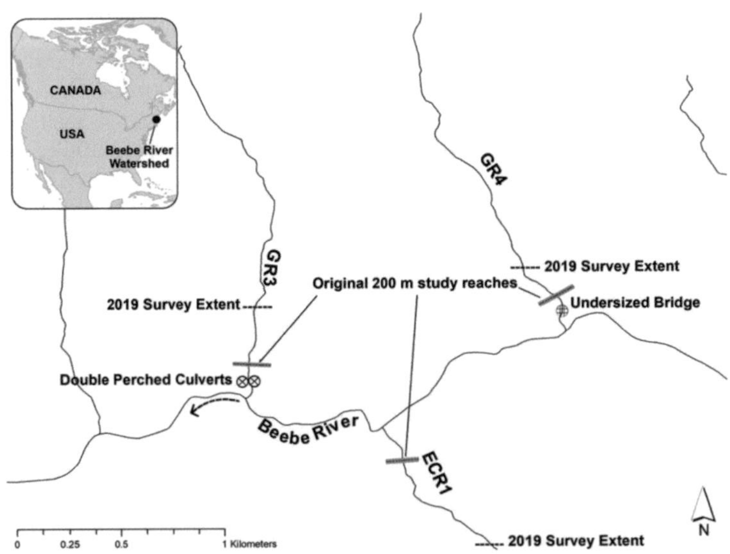

Figure 1.
Map of three study streams within the Beebe River watershed in central New Hampshire (USA). GR3 had two perched culverts replaced by a bridge and wood additions. GR4 had a passable but undersized bridge replaced by new bridge with wood additions. ECR1 was not affected by restoration efforts.

	ECR1			GR3			GR4		
Year	**'17**	**'18**	**'19**	**'17**	**'18**	**'19**	**'17**	**'18**	**'19**
B & C (%)	62	45	65	55	56	70	76	57	88
L & S (%)	36	38	28	44	38	25	24	43	11
F (%)	2	18	6	1	6	6	0	1	2
P (%)	36	33	40	45	42	34	31	28	44
G (%)	10	4	5	21	29	7	22	28	7
R (%)	13	31	14	12	11	30	23	11	26
C (%)	41	32	41	22	18	30	24	33	22
P:R ratio	2.8	1.1	2.9	3.8	3.8	1.1	1.3	2.5	1.7

B & C: Boulder and Cobble; L & S: Large and Small Gravel; F: Fines; P: Pool; R: Riffle; C: Cascade; and P:R: Pool to Riffle.

Table 1.
Substrate and habitat attributes over years (2017–2019) and across the three 200 m study reaches.

20.6°C in GR3 (wood and culvert), 18.0°C to 21.0°C in GR4 (wood), and 17.9°C to 19.5°C in ECR1 (untreated) between 2016 and 2019. Mean daily average water temperatures in July ranged from 15.6°C to 18.0°C in GR3, 16.4°C to 18.2°C in GR4, and 16.3°C to 17.1°C in ECR1 between 2016 and 2019.

A historic logging road parallels the Beebe River and intersects the south-facing study streams (GR3 and GR4) and a similar road intersects the north-facing study stream (ECR1) at a point 500 m upstream from the confluence with the mainstem. Up until 2017, the stream crossing at GR3 contained two perched culverts that were deemed impassable to Brook Trout, while GR4 contained a fully passable bridge that was a pinch point during high stream flow events. In ECR1, a perched culvert remains at the road-stream crossing 300 m above our original 200 m study reaches (**Figure 1**); however, ECR1 is considered the reference or untreated stream because the culvert was not removed and wood was not added over the study period. In August 2017, double perched culverts in GR3 and an undersized bridge over GR4 were replaced with steel stringer bridges that are fully passable by Brook Trout. After the 2017 sampling season, one to two trees per 100 m of stream were added upstream of our original 200 m study reaches in GR3 and GR4. GR3 and GR4 in a perpendicular orientation, with one to two locations having crisscrossed logs and several branches to accumulate leaves and create debris dams, according to federal protocols [74]. In August 2018, four trees were added to GR3 and one tree to GR4 within 100 m of the stream-river confluence. The 2018 wood additions occurred almost entirely downstream of benthic macroinvertebrate sampling. Therefore, we consider the August 2018 wood additions to have minimal influence on our results and attribute the observed responses to upstream wood additions. In summary, GR3 was treated with wood additions and culvert removal, GR4 was treated with only wood additions, and ECR1 was untreated.

3. Study design and data collection

In the summers of 2017–2019, we conducted habitat and benthic macroinvertebrate surveys once a season beginning at the stream-river confluence and continuing upstream to 200 m, including side channels (hereafter referred to as the "original 200 m reach"). We conducted Brook Trout surveys throughout the same reaches in July of 2016–2019. Surface areas within the 200 m reach varied over time with changes in water level and side channels (GR3: 371–587 m^2, GR4: 444–687 m^2, ECR1: 473–532 m^2). All surveys were extended in 2019 to span the full upstream extent of Brook Trout (GR3 culvert and wood = 650 m, GR4 wood = 450 m, and ECR1 untreated = 900 m). Natural vertical barriers (≥1 m hydraulic jump) delimited the upstream extent of Brook Trout in the study streams. This expansion of our survey areas was intended to capture the full scope of available benthic macroinvertebrates and habitat for Brook Trout following changes in upstream habitat.

3.1 Habitat surveys

We surveyed stream habitat following a modified version of NH Fish and Game's *Rapid Habitat Assessment* (New Hampshire Fish and Game, personal communication) and measured wood following *TWF Monitoring Program: Large Woody Debris Survey* [75]. Habitat and wood surveys were conducted annually in late July to mid-August, during base flow conditions (determined by USGS station #01016500). Habitat assessments included characterization of habitat unit type, habitat unit dimensions (length, width, average depth), channel characteristics (wetted and bankfull width), dominant substrate type, stream gradient, hydraulic jump height, stream bank stability, and riparian forest type. Surveys of large wood included characterization of tree species, wood origin, wood decay state, and wood dimensions. Large wood included any piece of wood that was ≥1 m length and ≥10 cm diameter [76] and located at least partially within the bankfull prism, as defined by water levels assumed during bankfull flow [77]. The locations and dimensions of smaller organic material moved by flowing water (i.e., rafted organic matter [ROM]) within the bankfull prism were recorded and associated with measured wood [78, 79]. We summarized relative changes in large wood density, dry and wet ROM, habitat availability, and substrate type in focal streams over time (2017–2019). Water temperature in the streams was continuously monitored every 30 minutes with HOBO sensors (U22-0001; Onset Computer Corp., Bourne Massachusetts) distributed longitudinally to measure thermal variability within and across streams [78, 79].

3.2 Benthic macroinvertebrate surveys

We collected benthic macroinvertebrates in June of 2017, 2018, and 2019 using Surber samplers and D-frame nets (sample area = 0.09 m^2, net mesh size = 500 μm) from randomly selected riffles and pools in proportion to their occurrence (i.e., count) throughout the original 200 m study reaches. Six to seventeen D-frame samples were collected from the reaches in 2017 and 2018. To increase the efficiency of sampling and identification in 2019, we pooled seven samples from each stream in the field (**Table 2**). Macroinvertebrates were stored in 70% ethanol and transported to the lab for family-level identification.

To test for significant changes in total and order-level benthic macroinvertebrate densities within a given stream between 2017 and 2018, we used a Pairwise Wilcoxon test with a Benjamini-Hochberg (BH) p-value adjustment. Considering samples were grouped by sampling zone in the field before laboratory identification in 2019, we conducted a one-sample T-test with a BH p-value adjustment to compare 2019 to 2018 and 2017. We adopted a non-parametric approach because the data did not conform to

	ECR1			GR3			GR4		
Year	**'17**	**'18**	**'19**	**'17**	**'18**	**'19**	**'17**	**'18**	**'19**
N	7	11	1*	17	15	1*	13	6	1*
Density (#/m^2)									
Tot	286	361.1	221.4	48.1	139.2	459.8	185.5	188.4	513.6
D	28.6	68.6	28.8	4.3	16.7	147.1	13	71.6	205.4
E	91.5	155.3	90.8	21.6	64	110.4	61.2	32	30.8
P	120.1	72.2	37.6	9.6	39	119.5	72.3	22.6	148.9
T	34.3	39.7	31	9.6	16.7	69	13	41.4	87.3
Biomass (g/m^2)									
Tot	0.1	0.166	0.319	0.017	0.06	0.141	0.137	0.056	0.264
D	0.009	0.007	0.006	0.002	0.002	0.03	0.001	0.006	0.037
E	0.032	0.091	0.061	0.006	0.033	0.057	0.023	0.014	0.013
P	0.032	0.043	0.217	0.005	0.008	0.021	0.019	0.012	0.129
T	0.026	0.015	0.016	0.004	0.005	0.027	0.005	0.017	0.045

**Seven samples from the original 200 m were pooled in the field in 2019.*
Tot: Total; D: Diptera; E: Ephemeroptera; P: Plecoptera; and T: Trichoptera.

Table 2.
Density and biomass of benthic macroinvertebrates in original 200 m reaches by stream, year, and order: GR3 (wood and culvert), GR4 (wood), and ECR1 (untreated).

parametric assumptions and non-parametric methods are considered suitable, if not better, for environmental data that is not normally distributed [80, 81]. Dry weight biomass was estimated for each year using random sub-samples at the family-level and summarized at the order-level (**Table 2**). Family-level pooled samples were dried in the winter of 2019 in an oven (Precision Econotherm Lab Oven, Thermo Scientific) with forced ventilation for 16 h at 105°C, cooled for 24 h, and weighed with an electronic balance [82].

3.3 Brook Trout surveys

We surveyed fish using two-pass depletion methods [83] with Smith-Root model LR-24 DC (Vancouver, WA, U.S.A.) backpack electrofishing units at 600–800 volts in July of 2016–2019 to ensure age 0 (young of year) Brook Trout had reached catchable size. In addition to Brook Trout, we occasionally captured *Rhinichthys atratulus* (Blacknose Dace) and *Rhinichthys cataractae* (Longnose Dace) near the stream-river confluence. Non-target species were returned to the stream without measurement. Estimated Brook Trout capture probabilities were calculated from multi-pass removal sampling as described in Carle and Strub [84] and averaged between 84 and 100% across sampling events, with an overall average estimated capture probability of 95%. High capture probability justified our use of two-pass depletion methods. Upon capture, the location of each Brook Trout was recorded to ensure fish were returned to the same subsection after data collection. All fish were measured for weight (g) and total length (mm) and each untagged Brook Trout that was ≥ 60 mm total length and ≥ 2g was sampled for scales above the lateral line and posterior to the dorsal fin [85]. Scales were brought back to the lab for age estimation, which enabled us to assess and compare length at specific ages over time. Brook Trout scale samples were dry mounted between two glass microscope slides and viewed under 10X magnification to identify periods of winter growth and summer growth [85]. Age was determined with knowledge of the sampling month, but without knowledge of fish length or location. Thirty percent of age assignments were reviewed by a third party. For the few discrepancies noted between assessors, scales were reassessed. The age structure of Brook Trout captured in July from the original 200 m reach of each tributary was compared over time using a two-sample Kolmogorov-Smirnov test (**Figure 2**).

Biomass and density values were calculated by summing the individual weights and number of captured Brook Trout in each tributary and standardized by area. We used a Pairwise Wilcoxon Rank Sum test with a BH p-value adjustment to compare Fulton's Condition Factor [86], total length of 1 year old (hereafter, age 1) Brook Trout, and density and biomass of Brook Trout within streams over time. We

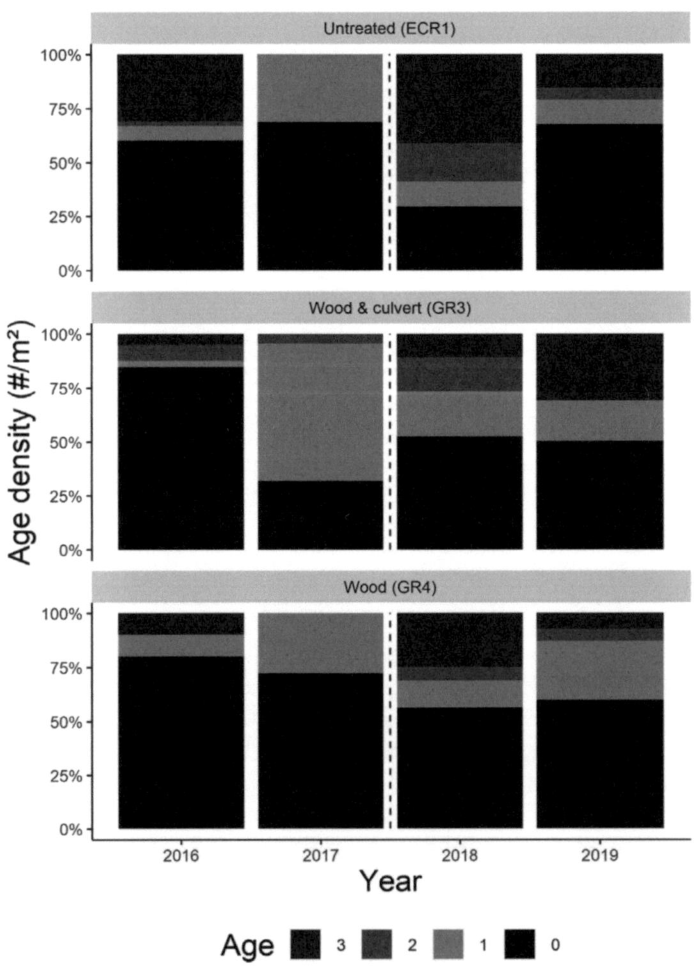

Figure 2.
Proportional abundance of Brook Trout by age, based on July captures in the original 200 m reaches in GR3 (wood and culvert), GR4 (wood), and ECR1 (untreated) from 2016 to 2019. Hashed lines mark when habitat restoration efforts occurred.

constrained our total length analyses to age 1 Brook Trout because this class was the most abundant in our samples (**Figure 2**), and younger individuals are more responsive to prey availability [87].

4. Results and discussion

4.1 Habitat

There were no notable changes in substrate composition or habitat types in the original 200 m reaches after perched culvert removal (GR3) or wood additions (GR3 and GR4), except immediate downstream of the culvert in GR3 where a large scour pool was replaced with a long riffle. Wet wood density (i.e., wood that is fully immersed in the stream) and ROM density in the original 200 m reaches varied over time, but there was no clear relationship with stream modifications in either case (**Figure 3A** and **B**). There were no consistent changes in habitat or substrate types 2 years after wood additions in the original 200 m reaches (**Table 1**), likely because less than 6% of total wood additions occurred there. Roughly half of the 2019 wet wood density across the full survey extent in GR3 and GR4 was associated with wood additions; however, ~94% of this added wood was situated upstream of the original 200 m

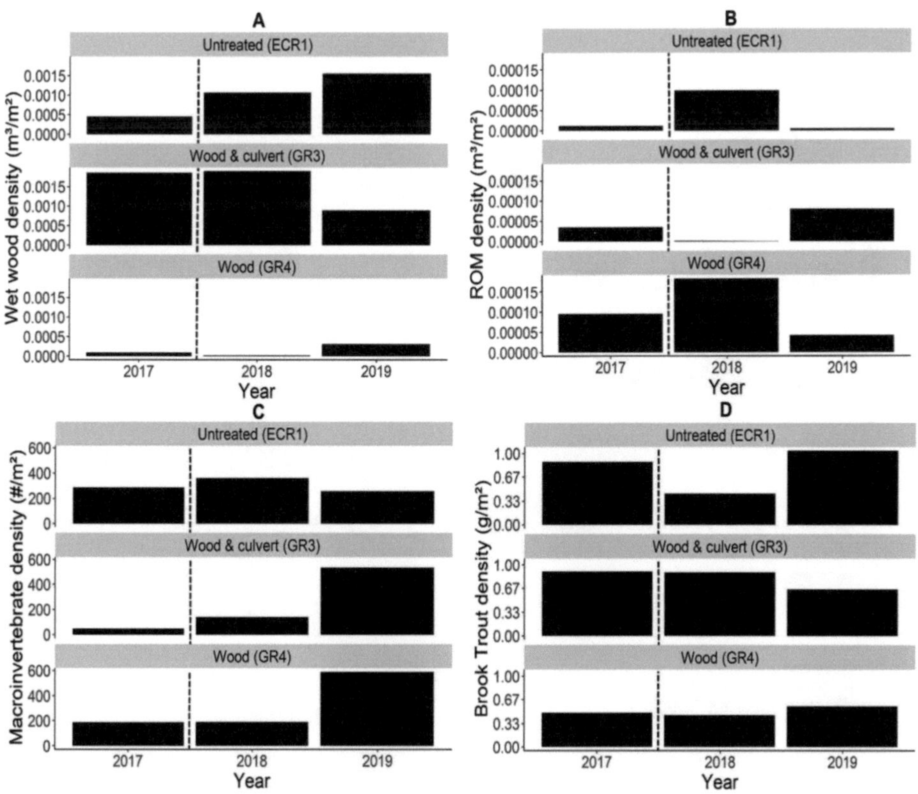

Figure 3.
Density of (A) wet wood, (B) rafted organic matter (ROM), (C) macroinvertebrates, and (D) Brook Trout in GR3 (wood and culvert), GR4 (wood), and ECR1 (untreated) from 2017 to 2019 in the original 200 m reaches. Hashed lines mark when habitat restoration efforts occurred.

reach (**Figure 4A**), and ROM was predominantly retained upstream of the original 200 m reach (**Figure 4B**). Moreover, GR3 held roughly four times and GR4 two and a half times greater ROM density than ECR1 in 2019. Of that ROM, 53% in GR3 and 95% in GR4 was associated with added wood (**Figure 4B**).

4.2 Benthic macroinvertebrates

Macroinvertebrate density roughly tripled 1 year after perched culvert removal and wood additions in GR3 ($P = 0.015$; **Figure 3C**), then tripled again, 2 years after restoration efforts ($P < 0.001$; **Figure 3C**). These increases were not proportional across orders (**Figure 5**) but were significant for all orders after 2 years ($P = 0.001$–0.020; **Table 2**). In GR4, macroinvertebrate density did not significantly differ 1 year after wood additions ($P = 0.903$; **Figure 3C**) but tripled after 2 years ($P < 0.001$; **Figure 3C**), with significant increases observed across all orders ($P = 0.000$–0.039; **Table 2**), except Ephemeroptera ($P = 0.116$; **Table 2**). Meanwhile, in our untreated stream (ECR1), we did not observe a significant change in macroinvertebrate density 1 year after restoration efforts ($P = 0.358$; **Figure 3C**), or 2 years after restoration efforts in any of the orders ($P = 0.173$–1; **Table 2**). We observed an increase in benthic macroinvertebrate biomass between 2017 and 2019 in all streams, but these values were partially influenced by the presence of a few large individuals (Plecoptera and Odonata) that often accounted for half of the biomass in a sample (**Table 2**). After

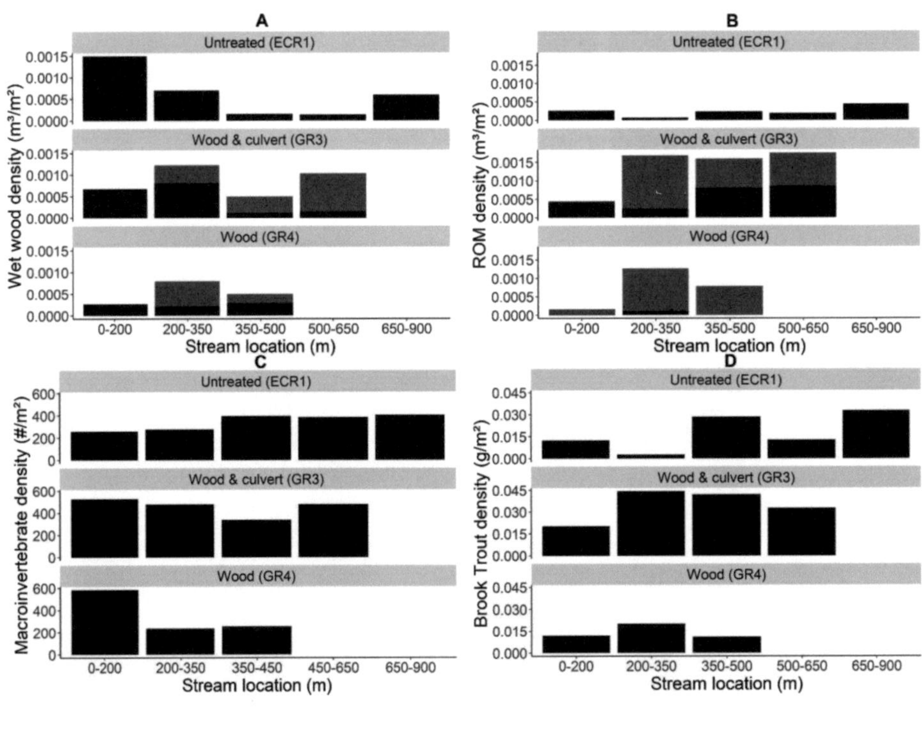

Figure 4.
Density of (A) wet wood, (B) rafted organic matter (ROM), (C) macroinvertebrates, and (D) Brook Trout in 2019, following culvert removal and wood additions, in the extended survey reaches: 650 m in GR3 (wood and culvert), 450 m in GR4 (wood), and 900 m in ECR1 (untreated). In A and B, black indicates natural/existing wood and gray indicates added wood.

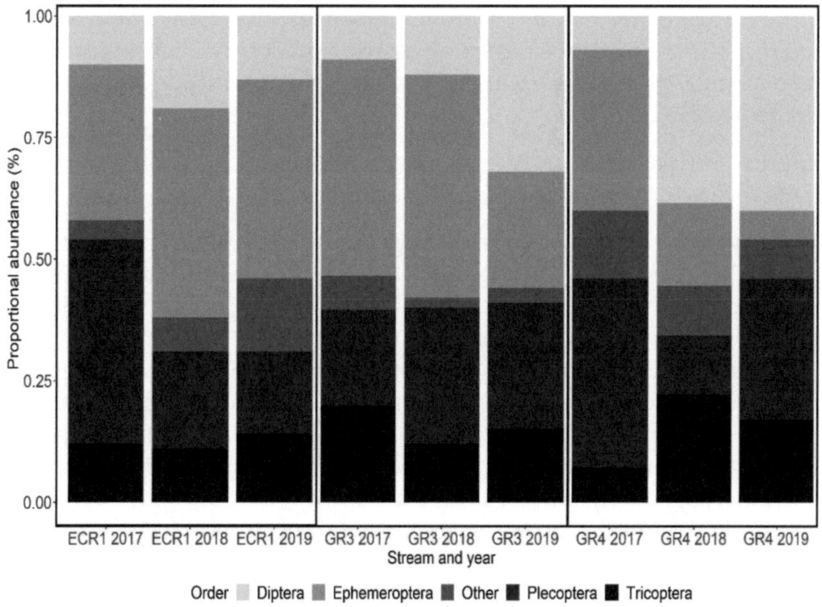

Figure 5.
Proportional abundance of benthic macroinvertebrate orders sampled from 2017 to 2019 in the original 200 m reaches of GR3 (wood and culvert), GR4 (wood), and ECR1 (untreated).

excluding very large individuals, the trends in macroinvertebrate biomass were similar with those for density.

Prior to restoration efforts, benthic macroinvertebrate density in GR3 was one third that of GR4. Given the proximity of these streams, their shared underlying geology, and similar channel morphology and temperature regimes, this large difference in density prior to restoration was unexpected. One year and 2 years after restoration efforts, GR3 and GR4 macroinvertebrate densities were roughly equivalent (**Figure 3C**); therefore, we suspect that the impassable culvert in GR3 may have constrained benthic macroinvertebrate density prior to restoration efforts. This observation is consistent with studies that have found significant differences in abundance and taxonomic richness between benthic macroinvertebrate communities residing in habitats upstream and downstream of culverts [18, 42, 43]. In addition, these findings align with a study that observed significant increases in invertebrate richness in headwater streams following culvert removals [44]. However, to date, no studies have reported significant differences in benthic macroinvertebrate abundances before and after culvert removal [88].

We posit that increased ROM retention from wood additions in the upper reaches contributed to increased benthic macroinvertebrate density and possibly led to a bottom-up response following habitat restoration. Our observations of increased organic matter retention following wood additions in headwater streams corresponding with downstream increases in invertebrate secondary production are consistent with previous studies [25, 89–91].

The dramatic change in habitat and substrate within the area directly impacted by culvert removal and bridge installation may have influenced community composition and density. However, since we did not examine the post-restoration effects of culvert removal independent of wood additions in GR3, we cannot fully discount the role of other factors influencing benthic macroinvertebrates in these study streams. Future studies examining benthic macroinvertebrate functional feeding groups may provide more insight into culvert and wood impacts on stream abiotic processes [90].

4.3 Brook Trout

While we did not detect significant changes in Brook Trout population density ($P = 0.243$–0.438) (**Figure 3D**) or biomass ($P = 0.280$–0.480) in the original 200 m reaches in response to wood additions or culvert removal, previous studies have documented trout population increases one to 2 years after habitat modifications [62, 63, 92]. However, these same studies noted that rapid increases in trout density and biomass may be driven by immigration [62, 63, 92]. For example, Kratzer [63] noted an increase in Brook Trout abundance in treated stream sections 1 year after wood additions. This increase initially coincided with a decrease in biomass in nearby control sections, suggesting immigration from control sections into treated sections. However, within 2–4 years, biomass in control sections rebounded and then exceeded previous levels while treated sections maintained initial biomass increases, suggesting an overall increase in carrying capacity over time. A recent meta-analysis of salmonid responses to stream restoration efforts across 100 studies showed a significant increase in abundance and biomass but noted that monitoring time (<5 years) factors into variation in impacts following restoration [34].

Despite no significant change in density or biomass, we detected a significant increase in age 1 Brook Trout length within the original 200 m reaches of treated streams 2 years after habitat modifications (GR3 culvert and wood: 2016–2019 $P = 0.015$, 2017–2019 $P = 0.031$; GR4 wood only: 2016–2019 $P = 0.011$; **Figure 6**). We found no significant difference in age 1 Brook Trout length in the untreated stream (ECR1 $P = 0.433$). Fulton's condition did not differ within the original 200 m reaches over time ($P = 0.17$–0.99; **Figure 7**), with the exception of a significantly higher Fulton's condition in GR3 in 2018 compared to 2016 ($P = 0.038$).

Within our study, age 1 Brook Trout represented between 5 and 60% of the observed populations in July between 2016 and 2019 (**Figure 2**). We detected a significant length increase in age 1 Brook Trout 2 years after habitat modifications (**Figure 6**), without concomitant significant increases in population density and biomass within the original 200 m reaches (**Figure 2D**). Notably, increases in Brook Trout size have not been detected in previous wood addition studies [62]. We hypothesize that age 1 stream-dwelling Brook Trout were able to utilize increased benthic macroinvertebrate density for growth, which could explain observed increases in length. The hypothesis is supported by Keeley [93] who found that smaller Brook Trout rely more heavily on invertebrate prey than larger trout that become piscivorous. Additionally, previous studies have shown that headwater Brook Trout growth and population density are regulated by density dependence and macroinvertebrate production [94, 95]. Although we cannot factor out the possibility that some larger age 1 Brook Trout moved into GR3 and GR4 from the mainstem, GR4 has had no impediments to fish passage and our PIT tag mark-recapture data supports the contention that age 1 Brook Trout largely remain in their respective headwater streams during the sampling season [78, 79]. Ultimately, we anticipate that increased food availability and Brook Trout size may translate into higher fecundity, population density, and biomass over multiple generations.

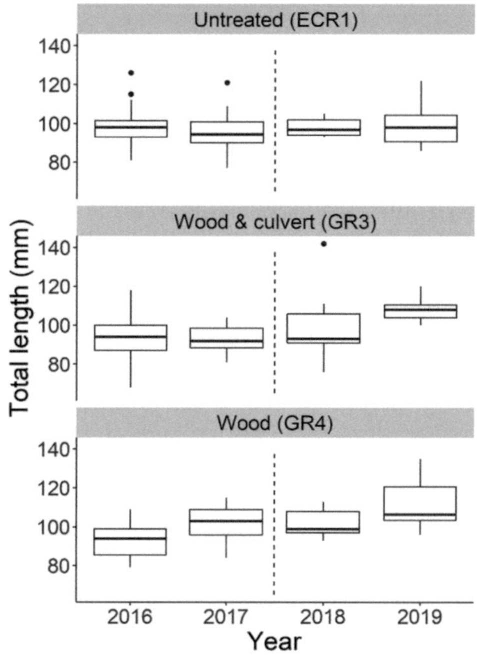

Figure 6.
Total length of age 1 Brook Trout captured in July in the original 200 m reaches in GR3 (wood and culvert), GR4 (wood), and ECR1 (untreated) from 2016 to 2019. The thick line indicates the median, boxes show the interquartile range. Hashed lines mark when habitat treatments occurred.

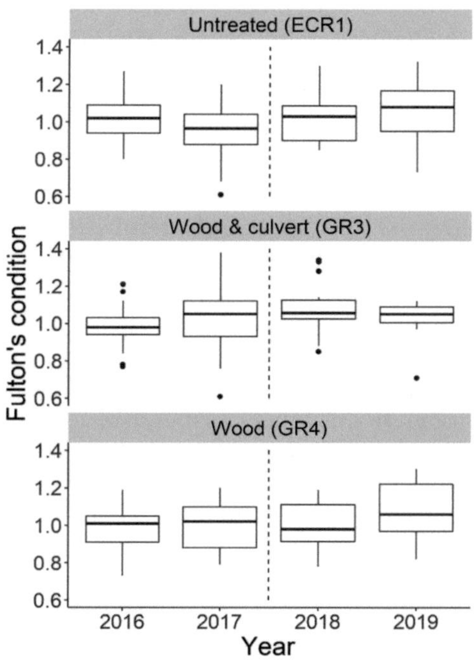

Figure 7.
Fulton's condition for Brook Trout captured in July in the original 200 m reaches in GR3 (wood and culvert), GR4 (wood), and ECR1 (untreated) from 2016 to 2019. The thick line indicates the median, boxes show the interquartile range. Hashed lines mark when habitat treatments occurred.

5. Conclusions and management relevance

Our results suggest that macroinvertebrates experienced a relatively rapid response to wood additions, but potential positive effects from culvert removal still need further exploration. Downstream of wood additions, we did not observe substantial short-term habitat changes or increases in Brook Trout density or biomass. However, we did observe significant increases in benthic macroinvertebrate density and age 1 Brook Trout length downstream of wood additions that retained the majority of ROM. Previous studies have shown that headwater Brook Trout populations are regulated by density dependence and invertebrate production [94, 95]. In light of this, we may see increases in Brook Trout biomass or density downstream of wood additions with continued years of monitoring.

Our results reflect short-term responses to restoration efforts in high gradient headwater streams of historically logged but still young forests where in-stream wood recruitment is limited. Broadly, researchers and managers should consider the ecological benefits of in-stream restoration efforts from a bottom-up perspective of habitat creation and energy capture toward increased secondary production of macroinvertebrates and fish [96]. While a long-term ecological monitoring plan that collects extensive data before and after manipulations across broad upstream and downstream perspectives is ideal [34, 90], it is rare given resource limitations and the opportunistic nature of many restoration endeavors. We were fortunate that our partners valued the data to inform decisions and invested in a pre-restoration assessment and post-restoration monitoring; however, given the resources available at the onset, we were limited in the extent of our monitoring. Early on our partners presented research and restoration plans widely, building community support for the endeavor. With an unanticipated infusion of funding in 2018, we were able to expand sampling efforts to capture the habitat characteristics associated with wood additions and track a suite of variables potentially driving our observed downstream responses. Rather than ignore the upstream extent of the focal streams to preserve the original reach of 200 m, we opted to expand our scope to better inform management and conservation decisions. It was through the consistent communication of our seasonal data collection efforts, observations and data analyses with our partners that prompted them to establish additional environmental protection by writing in a mandatory 15 m "no harvest" riparian buffer into the conservation easements that would conserve the property in perpetuity. To our knowledge, this was the first explicit insertion of such strict stream protection policy into a conservation easement in New Hampshire where the law states that no more than 50% of the basal area of trees can be cut within 15 m of a stream (Basal Area Law, 2019; RSA 227-J:9). This study, examined as a whole, demonstrates the importance of engaging collaborative and diverse partners (management, conservation, and scientific) in collaborative restoration and monitoring decisions and effective communicating scientific observations with non-scientific audiences to encourage data-driven conservation action.

Acknowledgements

This study was a collaborative effort among faculty and students at Plymouth State University, the New Hampshire Fish and Game Department (B. Nugent, D. Timmins, J. Magee), the Pemigewasset chapter of Trout Unlimited (#726), and The Conservation Fund. We thank numerous research assistants for their contribution to field surveys and lab analysis: M. Collins, A. Dunlap, D. Tarr, J. Carboneau, C. Barnes, K. Crowley, S. Bevier, S. Welch, J. Corbet, N. Roberts, S. Bevier, N. Scola, and G. Nyhan. We also thank E. Laflamme for statistical advice and comments on earlier drafts. Sampling was permitted through New Hampshire Fish and Game Department Scientific Licenses F2016-101, F2017-097, F2018-94, and F2019-102 and Plymouth State University IACUC approval. Financial support included funding from the National Fish and Wildlife Foundation "Bring Back the Natives" program, New Hampshire Water Resources Research Center through the US Geological Survey National Institutes for Water Resources Grants Program (Section 104 g) [grant number G16AP00070], Tourism, Environment, and Sustainable Societies Cluster and the Lovett Fund at Plymouth State University, graduate assistantship funding from Plymouth State University, and undergraduate research funding from the Center for the Environment and the Student Research and Creativity Fund at Plymouth State University.

Author details

Joshua C. Hoekwater[1], Amy M. Villamagna[1]*, Brigid C. O'Donnell[2], Tyson R. Morrill[2], Ben J. Nugent[3] and Jared B. Lamy[2]

1 Environmental Science and Policy, Center for the Environment, Plymouth State University, Plymouth, New Hampshire, USA

2 Biological Sciences, Center for the Environment, Plymouth State University, Plymouth, New Hampshire, USA

3 New Hampshire Fish and Game Department, New Hampton, New Hampshire, USA

*Address all correspondence to: amvillamagna@plymouth.edu

References

[1] Nislow KH, Lowe WH. Influences of logging history and riparian forest characteristics on macroinvertebrates and Brook Trout (*Salvelinus fontinalis*) in headwater streams (New Hampshire, U.S.A.). Freshwater Biology. 2006;**51**:388-397

[2] Foster DR. Land-use history (1730-1990) and vegetation dynamics in Central New England, USA. Journal of Ecology. 1992;**80**(4):753-771

[3] Williams M. Clearing the United States forests: Pivotal years 1810-1860. Journal of Historical Geography. 1982;**8**(1):12-28

[4] Langley-Turnbaugh SJ, Keirstead DR. Soil properties and land use history: A case study in New Hampshire. Northeastern Naturalist. 2005;**12**(4):391-402

[5] Hayes DB, Taylor WW, Drake MT, Marod SM, Whelan GE. The value of headwaters to brook trout (*Salvelinus fontinalis*) in the Ford River, Michigan, USA. In: Haigh MJ, Krecke J, Rajwar GS, Kilmartin MP, editors. Headwaters: Water Resources and Soil Conservation. Brookfield, Vermont: Balkeman, A. A; 1998. pp. 175-185

[6] Marschall EA, Crowder LB. Assessing population responses to multiple anthropogenic effects: A case study with brook trout. Ecological Applications. 1996;**1**:152-167

[7] Beschta RL, Bilby ER, Brown WG, Holtby LB, Hofstra DT. Stream temperature and aquatic habitat: Fisheries and forestry interactions. In: Salo EO, Cundy TW, editors. Streamside Management: Forestry and Fishery Interactions. Seattle, Washington USA: University of Washington, Institute of Forest Resources. Contribution 57; 1987. pp. 191-232

[8] Keeton WS, Kraft CE, Warren DR. Mature and old-growth riparian forests: Structure, dynamics, and effects on Adirondack stream habitats. Ecological Applications. 2007;**17**(3):852-868

[9] Macdonald JS, MacIsaac EA, Herunter HE. The effect of variable-retention riparian buffer zones on water temperatures in small headwater streams in sub-boreal forest ecosystems of British Columbia. Canadian Journal of Forest Research. 2003;**33**(8):1371-1382

[10] Wheeler AP, Angermeier PL, Rosenberger AE. Impacts of new highways and subsequent landscape urbanization on stream habitat and biota. Reviews in Fisheries Science. 2005;**13**(3):141-164

[11] Warren RD, Kraft EC, Keeton SW, Nunery SJ, Likens GE. Dynamics of wood recruitment in streams of the northeastern U.S. Forest Ecology and Management. 2009;**258**(5):804-813

[12] Wallace JB, Whiles RM, Eggert S, Cuffney TF, Lugthart GJ, Chung K. Long-term dynamics of coarse particulate organic matter in three Appalachian mountain streams. Journal of the North American Benthological Society. 1995;**14**(2):217-232

[13] Wallace JB, Eggert S, Meyer J, Webster J. Stream invertebrate productivity linked to forest subsidies: 37 stream-years of reference and experimental data. Ecology. 2015;**96**(5):1213-1228

[14] Bilby RE, Likens GE. Importance of organic debris dams in the structure and

function of stream ecosystems. Ecology. 1980;**1980**(61):1107-1113

[15] Battin TJ, Kaplan LA, Findlay S, Hopkinson CS, Martin E, Packman AI, et al. Biophysical controls on organic carbon fluxes in fluvial networks. Nature Geoscience. 2008;**1**:95-100

[16] Beckman ND, Wohl E. Carbon storage in mountainous headwater streams: The role of old-growth forest and logjams. Water Resource Research. 2014;**50**:2376-2393

[17] Elosegi A, Díez J, Pozo J. Contribution of dead wood to the carbon flux in forested streams. Earth Surface Processes and Landforms. 2007;**32**:1219-1228

[18] Khan B, Colbo MH. The impact of physical disturbance on stream communities: Lessons from road culverts. Hydrobiologia. 2008;**600**:229-235

[19] Smock LA, Metzler GM, Gladden JE. Role of debris dams in the structure and functioning of low-gradient headwater streams. Ecology. 1989;**70**:764-775

[20] Valett HM, Crenshaw CL, Wagner PF. Stream nutrient uptake, forest succession, and biogeochemical theory. Ecology. 2002;**83**:2888-2901

[21] Petersen RC, Cummins KW. Leaf processing in a woodland stream. Freshwater Biology. 1974;**4**:343-368

[22] Rau GH, Anderson NH. Use of 13 C/12 C to trace dissolved and particulate organic matter utilization by populations of an aquatic invertebrate. Oecologia. 1981;**48**:19-21

[23] Wallace JB, Webster JR, Cuffney TF. Stream detritus dynamics: Regulation by invertebrate consumers. Oecologia. 1982;**53**:197-200

[24] Peters GT, Benfield EF, Webster JR. Chemical composition and microbial activity of seston in a southern Appalachian headwater stream. Journal of the North American Benthological Society. 1989;**1**:74-84

[25] Entrekin SA, Rosi EJ, Tank JL, Hoellein TJ, Lamberti GA. Quantitative food webs indicate modest increases in the transfer of allochthonous and autochthonous C to macroinvertebrates following a large wood addition to a temperate headwater stream. Frontiers in Ecology and Evolution. 2020;**8**:114

[26] Wallace JB, Eggert SL, Meyer JL, Webster JR. Multiple trophic levels of a forest stream linked to terrestrial litter inputs. Science. 1997;**277**(5322):102-104

[27] Champoux O, Biron MP, Roy GA. The long-term effectiveness of fish habitat restoration practices: Lawrence Creek, Wisconsin. Annals of the Association of American Geographers. 2003;**93**(1):42-54

[28] Opperman JJ, Merenlender AM, Lewis D. Maintaining wood in streams: A vital action for fish conservation. In: Division of Agriculture and Natural Resources. Davis, California, USA: University of California; 2006. DOI: 10.3733/ucanr.8157

[29] Whiteway SL, Biron PM, Zimmermann A, Venter O, Grant JWA. Do in-stream restoration structures enhance salmonid abundance? A meta-analysis. Canadian Journal of Fisheries and Aquatic Sciences. 2010;**67**(5):831-841

[30] Harvey BC, Stewart AJ. Fish size and habitat depth relationships in headwater streams. Oecologia. 1991;**87**:336-342

[31] Lewis SL. Physical factors influencing fish populations in pools of a trout stream. Transactions of

the American Fisheries Society. 1969;**98**:14-19

[32] Nagayama S, Nakamura F. Fish habitat rehabilitation using wood in the world. Landscape and Ecological Engineering. 2010;**6**:289-305

[33] Roni P, Beechie TJ, Pess GR, Hanson KM. Wood placement in river restoration: Fact, fiction, and future direction. Canadian Journal of Fisheries and Aquatic Sciences. 2015;**72**(3):466-478

[34] Foote KJ, Biron PM, Grant JW. Impact of in-stream restoration structures on salmonid abundance and biomass: An updated meta-analysis. Canadian Journal of Fisheries and Aquatic Sciences. 2020;**9**:1574-1591

[35] Warren RD, Kraft EC. Brook trout (*Salvelinus fontinalis*) response to wood removal from high-gradient streams of the Adirondack Mountains (N.Y., U.S.A.). Canadian Journal of Fisheries and Aquatic Sciences. 2003;**60**(4):379-389

[36] Colvin SA, Sullivan SMP, Shirey PD, Colvin RW, Winemiller KO, Hughes RM, et al. Headwater streams and wetlands are critical for sustaining fish, fisheries, and ecosystem services. Fisheries. 2019;**44**(2):73-91

[37] Bilby RE, Ward JW. Characteristics and function of large woody debris in streams draining old-growth, clear-cut, and second-growth forests in southwestern Washington. Canadian Journal of Fisheries and Aquatic Sciences. 1991;**48**:2499-2508

[38] Benke AC, Wallace JB. Influence of wood on invertebrate communities in streams and rivers. In: Gregory SV, Boyer KL, Gurnell AM, editors. The Ecology and Management of Wood in World Rivers. Symposium 37. Bethesda, MD: American Fisheries Society; 2003. pp. 149-177. DOI: 10.47886/9781888569568

[39] Dolloff CA, Warren ML Jr. Fish relationships with large wood in small streams. In: Gregory SV, Boyer KL, Gurnell AM, editors. The Ecology and Management of Wood in World Rivers. Symposium 37. Bethesda, MD: American Fisheries Society; 2003. pp. 179-193. DOI: 10.47886/9781888569568

[40] Poplar-Jeffers IO, Petty JT, Anderson JT, Kite SJ, Strager MP, Fortney RH. Culvert replacement and stream habitat restoration: Implications from brook trout Management in an Appalachian watershed, U.S.a. Restoration Ecology. 2009;**17**(3):404-413

[41] Wellman JC, Daniel LC, Bradford SC. Long-term impacts of bridge and culvert construction or replacement on fish communities and sediment characteristics of streams. Journal of Freshwater Ecology. 2000;**15**(3):317-328

[42] Peterson T. The effect of road culverts on the benthic macroinvertebrate community in wadeable lotic ecosystems. PSU McNair Scholars Online Journal. 2010;**4**:52-74

[43] Gál B, Weiperth A, Farkas J, Schmera D. The effects of road crossings on stream macro-invertebrate diversity. Biodiversity and Conservation. 2020;**29**:729-745

[44] Omoniyi GE, Piscart C, Pellan L, Bergerot B. Responses of macroinvertebrate communities to hydromorphological restoration of headwater streams in Brittany. Water. 2022;**4**:553

[45] Kentaro M, Shoichiro Y. Effects of habitat fragmentation by damming on

the persistence of stream-dwelling charr populations. Conservation Biology. 2002;**16**(5):1318-1323

[46] Letcher BH, Nislow KH, Coombs JA, O'Donnell MJ, Dubreuil TL. Population response to habitat fragmentation in a stream-dwelling brook trout population. PLoS One. 2007;**11**:1-11

[47] Petty JT, Hansbarger JL, Huntsman BM, Mazik PM. Brook trout movement in response to temperature, flow, and thermal refugia within a complex Appalachian riverscape. Transactions of the American Fisheries Society. 2012;**141**(4):1060-1073

[48] Warren ML, Pardew MG. Road crossings as barriers to small-stream fish movement. Transactions of the American Fisheries Society. 1998;**127**:637-644

[49] Hudy M, Thieling TM, Gillespie N, Smith EP. Distribution, status, and land use characteristics of subwatersheds within the native range of brook trout in the eastern United States. North American Journal of Fisheries Management. 2008;**28**(4):1069-1085

[50] Kratzer JF, Warren DR. Factors limiting brook trout biomass in northeastern Vermont streams. North American Journal of Fisheries Management. 2013;**33**(1):130-139

[51] Hoffman R, Dunham J. Fish-Movement Ecology in High-Gradient Headwater Streams: Its Relevance to Fish Passage Restoration through Stream Culvert Barriers. Virginia: US Geological Survey; 2007. OFR 2007-1140

[52] Magoulick DD, Wilzbach MA. Are native brook charr and introduced rainbow trout differentially adapted to upstream and downstream reaches? Ecology of Freshwater Fish. 1998;**7**:167-175

[53] Thompson DP, Rahel JF. Evaluation of artificial barriers in small Rocky Mountain streams for preventing the upstream movement of brook trout. North American Journal of Fisheries Management. 1998;**18**(1):206-210

[54] Nathan LR, Smith AA, Welsh AB, Vokoun JC. Are culvert assessment scores an indicator of brook trout Salvelinus fontinalis population fragmentation? Ecological Indicators. 2018;**84**:208-217

[55] Torterotot JB, Perrier C, Bergeron NE, Bernatchez L. Influence of forest road culverts and waterfalls on the fine-scale distribution of brook trout genetic diversity in a boreal watershed. Transactions of the American Fisheries Society. 2014;**6**:1577-1591

[56] Wood DM, Welsh AB, Todd PJ. Genetic assignment of brook trout reveals rapid success of culvert restoration in headwater streams. North American Journal of Fisheries Management. 2018;**5**:991-1003

[57] Battin J, Wiley MW, Ruckelshaus MH, Palmer RN, Korb E, Bartz KK, et al. Projected impacts of climate change on salmon habitat restoration. Proceedings of the National Academy of Sciences of the United States of America. 2007;**104**:6720-6725

[58] Durance I, Ormerod SJ. Climate change effects on upland stream macroinvertebrates over a 25-year period. Global Change Biology. 2007;**13**:942-957

[59] Flebbe PA, Roghair LD, Bruggink JL. Spatial modeling to project southern Appalachian trout distribution in a warmer climate. Transactions of the American Fisheries Society. 2006;**135**(5):1371-1382

[60] Hakala JP, Hartman KJ. Drought effect on stream morphology and brook

trout (*Salvelinus fontinalis*) populations in forested headwater streams. Hydrobiologia. 2004;**515**:203-213

[61] Ouellet V, Gibson EE, Daniels MD, Watson NA. Riparian and geomorphic controls on thermal habitat dynamics of pools in a temperate headwater stream. Ecohydrology. 2017;**10**(8):e1891

[62] Gowan C, Fausch KD. Long-term demographic responses of trout populations to habitat manipulation in six Colorado streams. Ecological Applications. 1996;**6**:931-946

[63] Kratzer JF. Response of brook trout biomass to strategic wood additions in the east branch Nulhegan River watershed, Vermont. North American Journal of Fisheries Management. 2018;**38**:1415-1422

[64] Sweka JA, Hartman KJ. Effects of large woody debris addition on stream habitat and brook trout populations in Appalachian streams. Hydrobiologia. 2006;**559**:363-378

[65] Rubin Z, Kondolf GM, Rios-Touma B. Evaluating stream restoration projects: What do we learn from monitoring? Water. 2017;**9**(3):174

[66] Kelson SJ, Kapuscinski AR, Timmins D, Ardren WR. Fine-scale genetic structure of brook trout in a dendritic stream network. Conservation Genetics. 2015;**16**(1):31-42

[67] Nislow KH, Lowe WH. Influences of logging history and stream pH on brook trout abundance in first-order streams in New Hampshire. Transactions of the American Fisheries Society. 2003l;**132**(1):166-171

[68] Warren DR, Likens GE, Buso DE, Kraft CE. Status and distribution of fish in an acid-impacted watershed of the northeastern United States (Hubbard brook, NH). Northeastern Naturalist. 2008;**15**(3):375-390

[69] Lamy JB, O'Donnell BC, Villamagna AM, Morrill TR, Nugent BJ, Hoekwater JC. Genetic analysis reveals a complex mosaic of admixture in brook trout in a historically fragmented watershed. North American Journal of Fisheries Management. 2023;**43**(4):947-961

[70] Old Maps of New Hampshire. Grafton County NH Maps [Internet]. 2009. Available from: http://www.old-maps.com/nh/nh_CoGrafton.htm [Accessed: December 10, 2023]

[71] Logging in Lincoln: Beebe River [Internet]. 2021. Available from: https://logginginlincoln.com/Beebe_River.html [Accessed: December 10, 2023]

[72] Gove B. Logging Railroads along the Pemigewasset River. Littleton NH, USA: Bondcliff Books; 2006. 166 p

[73] Hagan JM, Irland LC, Whitman AA. Changing timberland ownership in the Northern Forest and implications for biodiversity. Manomet Center for Conservation Sciences, Report # MCCS-FCP2005-1. 2005. Brunswick, Maine, 25 p. Available from: https://www.iatp.org/sites/default/files/181_2_78270.pdf

[74] USDA. Practical Guide for Strategic Wood Additions (SWA) to Streams in Northern New England (ME, NH and VT). 2018. Natural Resources Conservation Services. Available from: https://efotg.sc.egov.usda.gov/api/CPSFile/22821/395_VT_PS_Stream_Habitat_Improvement_and_Management-SWA_Practical_2020

[75] Schuett-Hames D, Pleus AE, War J, Fox M, Light J. TFW monitoring program method manual for the large

woody debris survey. Washington State Dept. of Natural Resources Timber, Fish and Wildlife Agreement. 1999. TFW-AM9-99-004. DNR #106. Available from: https://www.dnr.wa.gov/publications/fp_tfw_am9_99_004.pdf

[76] Richmond DA, Fausch KD. Characteristics and function of large woody debris in subalpine rocky mountain streams in northern Colorado. Canadian Journal of Fisheries and Aquatic Sciences. 1995;**52**(8):1789-1802

[77] Overton KO, Wollrab SP, Roberts BC, Radko MA. R1/R4 (Northern/Intermountain Regions) fish and fish habitat standard inventory procedures handbook. 1997. Gen. Tech. Rep. INT-GTR-346. Ogden, UT: U.S. Department of Agriculture, Forest Service, Intermountain Research Station. 73 p. DOI: 10.2737/INT-GTR-346

[78] Morrill T. Wild Brook Trout (*Salvelinus fontinalis*) Demographics and Movement in the Presence of Undersized Road Crossings in Headwater Streams in Central New Hampshire [Thesis]. Plymouth: Plymouth State University; 2019

[79] Hoekwater J. Brook Trout (*Salvelinus fontinalis*): Wood Additions and Culvert Removal in Headwater Streams: A Trophic Cascade Perspective [Thesis]. Plymouth: Plymouth State University; 2020

[80] Mumby PJ. Statistical power of non-parametric tests: A quick guide for designing sampling strategies. Marine Pollution Bulletin. 2002;**44**(1):85-87

[81] Riaz M, Mahmood T, Arslan M. Non-parametric versus parametric methods in environmental sciences. Bulletin of Environmental Studies. 2016;**1**:128-130

[82] Leuven RSEW, Brock TCM, van Druten HAM. Effects of preservation on dry and ash-free dry weight biomass of some common aquatic macroinvertebrates. Hydrobiologia. 1985;**127**:151-159

[83] Zippin C. The removal method of population estimation. The Journal of Wildlife Management. 1958;**22**(1):82-90

[84] Carle FL, Strub MR. A new method for estimating population size from removal data. Biometrics. 1978;**34**:621-630

[85] Schneider JC, Laarman PW, Gowing H. Manual of fisheries survey methods II: with periodic updates. Michigan Department of Natural Resources: Fisheries Special Report, 25. 2000. Ann Arbor, MI

[86] Carlander KD. Handbook of Freshwater Fishery Biology. Dubuque, Iowa: Brown; 1950. p. 281

[87] Utz R, Hartman KJ. Density-dependent individual growth and size dynamics of central Appalachian brook trout (*Salvelinus fontinalis*). Canadian Journal of Fisheries and Aquatic Sciences. 2009;**66**:1072-1080

[88] Frankiewicz P, Radecki-Pawlik A, Wałęga A, Łapińska M, Wojtal-Frankiewicz A. Small hydraulic structures, big environmental problems: Is it possible to mitigate the negative impacts of culverts on stream biota? Environmental Reviews. 2021;**29**(4):510-528

[89] Flores L, Giorgi A, González JM, Larranaga A, Díez JR, Elosegi A. Effects of wood addition on stream benthic invertebrates differed among seasons at both habitat and reach scales. Ecological Engineering. 2017;**106**:116-123

[90] Al-Zankana A, Matheson T, Harper D. Adding large woody material into a headwater stream has immediate

benefits for macroinvertebrate community structure and function. Aquatic Conservation: Marine and Freshwater Ecosystems. 2021;**4**:930-947

[91] Deane A, Norrey J, Coulthard E, McKendry DC, Dean AP. Riverine large woody debris introduced for natural flood management leads to rapid improvement in aquatic macroinvertebrate diversity. Ecological Engineering. 2021;**163**:106197

[92] White LS, Charles G, Fausch KD, Josh GH, Saunders WC, Rosenfeld J. Response of trout populations in five Colorado streams two decades after habitat manipulation. Canadian Journal of Fisheries and Aquatic Sciences. 2011;**68**(12):2057-2063

[93] Keeley ER. Demographic responses to food and space competition by juvenile steelhead trout. Ecology. 2001;**82**:1247-1259

[94] Grossman G, Ratajczak R, Wagner C, Petty J. Dynamics and regulation of the southern brook trout (*Salvelinus fontinalis*) population in an Appalachian stream. Freshwater Biology. 2010;**55**(7):1494-1508

[95] Waters FT. Annual production by a stream brook charr population and by its principal invertebrate food. Environmental Biology of Fishes. 1982;**7**(2):165-170

[96] Layman CA, Rypel AL. Secondary production is an underutilized metric to assess restoration initiatives. Food Webs. 2020;**25**:e00174

Chapter 12

Wetlands and the Ecological Services that They Provide on Multiple Spatial Scales, from Landscape Down to Soil

Erwin J.J. Sieben and Donovan C. Kotze

Abstract

In this chapter, we present a hierarchical framework to consider wetlands and their ecosystem services in landscape planning. Wetlands are important in a landscape setting as they are intricately linked to the water cycle, and they provide many ecosystem services. Collectively, wetlands can be regarded as wet ecological infrastructure. Wetlands can be categorized as different hydrogeo-morphic types, which all play a different role in the overall hydrology and lead to different ecosystem services. Ecosystem services can act on various spatial levels, and all of these levels need to be considered when conserving wetlands and securing their ecosystem benefits. The levels that can be recognized for this are the catchment level, the individual wetland (hydrogeomorphic unit) level, the wetland habitat level, and at the smallest scale even the soil level, as some of the most important ecosystem services are related to the biogeochemistry associated with wetland soils.

Keywords: ecological infrastructure, ecological functioning, catchments, biogeochemistry, wetland habitats, wetland soils

1. Introduction

Wetlands are vulnerable ecosystems because they find themselves in the center of a wide range of human economic activities. In most cases, historically, the first settlements of humans have always appeared around places that have easy access to water for irrigation. People want access to the resources that wetlands can provide, but they also drain wetlands to transform the land for different uses. For this reason, people have had an ambiguous relationship with wetlands.

Today, in an age of high population densities, intense industrialization and large-scale changes in land use, many wetlands have been lost, and the remainder have been often degraded, with their original hydrology and water quality compromised. This is a problem in its own right, but this loss of wetlands also plays an important role in the degradation of the landscape as a whole. Wetlands provide many ecosystem services and as such they are often regarded as the most valuable ecosystems [1]. Ecosystem services can be defined as the ways in which the presence of ecosystems affects human well-being, and their existence usually becomes evident only after ecosystems degrade or disappear [2].

Another way of describing the utility of ecosystems for human well-being is by stating that wetlands form an important part of the "ecological infrastructure" [3]. Ecological infrastructure is a term that encompasses all the natural elements of a landscape that are vital to its overall functioning and thereby to the provisioning of ecosystem services. Civil infrastructure can be built to enhance the utility of the landscape for human purposes but generally society benefits from leaving critical ecological infrastructure intact as far as is possible and from mitigating eventual losses. Historical or current losses of ecological functioning are often used as a justification for spending resources on ecological restoration [4, 5]. This is because sustainable livelihoods are dependent on the productive capacity of ecosystems and thereby on their natural functioning. This applies particularly to aquatic ecological infrastructure, as many of the most important ecological processes are tied in with the water cycle. The way in which water moves through a landscape and how water gets distributed throughout it determines the ecological functions of the landscape.

In order to optimally utilize the landscape for human well-being, we need to have an overview of the ecosystem services that all wetland ecosystems within it provide. This depends on the type of wetland as well as on the position that that wetland occupies within the broader water catchment. All wetlands together perform certain tasks that account for outcomes at the level of the catchment, such as flood retention and the maintenance of base flow, and that need to be optimized for the sake of water resource planning. A hierarchical view of ecological infrastructure helps in planning land use and civil infrastructure around the existing ecological infrastructure.

The valuation of wetland ecological infrastructure does not only depend on the wetlands themselves but also on the "demand" for that service, which depends on where people are living within the catchment, where are the needs for certain goods and services coming from the wetland, and on where is the industrial activity and other sources of pollution.

In this chapter, we would like to argue how the presence of wetlands in a landscape is an important aspect of the ecological functioning of that landscape as a whole and we would like to present a hierarchical approach into assessing the health of a landscape in terms of its wetland "ecological infrastructure."

2. A hierarchical view of wet ecological infrastructure

The benefits and ecosystem services that wetlands provide can be felt at different spatial scales. The highest spatial scale is that of a river catchment. People depend on rainfall in order to be able to grow sufficient food, but this rain will partially percolate into the soil or drain into channels that carry away the excess water. Sustainable livelihoods depend on the capacity of the wetland infrastructure to retain water during dry periods and to remove excess water quickly during wet periods, as flooding will damage civil infrastructure. In integrated catchment management, this often means that the wetland infrastructure needs to be supplemented with gray infrastructure outside of the floodplain areas to properly spread out the flood events.

For the sake of environmental impact assessments of any human activity, one core activity often revolves around the identifying and mapping of wetlands, as many countries recognize that wetlands require special attention in conservation.

These mapping activities take place by looking at soil and vegetation characteristics which differentiate the wetland from the surrounding uplands [6]. Of course, the first characteristic to look at is the presence of standing water, but since water tables fluctuate in the course of the seasons, soils and vegetation provide a good proxy for the presence of water that can be seen throughout the year [7], but does require some expert judgment and good local knowledge.

Once wetlands are mapped, we need to consider various spatial scales of wetland configuration. Firstly, we need to consider where wetlands are within the context of an entire catchment. The catchment can be defined at many spatial scales, such as the primary catchment, secondary catchment, tertiary catchment, and quaternary catchment, depending on the size of the main drainage channel and the number of tributaries that are contained within the catchment. Where wetlands are located within the overall configuration of drainage channels has an effect on what role these wetlands may play for the flow regime of the river as a whole, but may also have an impact on water quality as water flowing through a wetland may be filtered and nutrients may be removed.

After wetlands are mapped, they should also be typed in order to understand what is going on in the wetland as not all wetland types have the same functions within the overall hydrology of the catchment [8]. When classifying a wetland as belonging to a certain type, individual wetlands should rather be recognized as hydrogeomorphic (HGM) units [9, 10] as the landform and hydrology of a wetland determine the ecological processes that may take place in the wetland. In defining an HGM unit, a researcher has to address the question why it is that water accumulates in that place and what are the sources and losses of water, and how does it flow through the wetland. Each HGM unit can be regarded as a spatial unit that has a uniform hydrological functioning, but it is also possible that wetlands of different HGM units are organized together into a wetland complex whereby water that is released from one HGM unit can flow becoming the source of water for another HGM unit. Larger wetland areas may often actually be wetland complexes even though many characteristics of the wetland such as water flow, vegetation, and soils are probably distinctive between the units that make up the complex [8]. Many HGM units are typically linked to the drainage network, such as floodplains, channeled and unchanneled valley bottom wetlands, and a large part of the seepages. Others are more typically separate from the drainage network or at least the main drainage channels, although they may be connected to it by means of interflow and aquifers: these are the depressions, wetland flats, and some of the seepages [10].

Even while hydrogeomorphic units determine the hydrology and ecological functioning of wetlands as a whole, many ecological processes are also determined at a finer scale. The hydrology of wetlands creates environmental gradients and along these environmental gradients we find many different habitats that create different conditions for plant growth. Within a wetland, we can therefore recognize different vegetation types or also called different habitat types [8]. These habitat types are mostly determined by hydroperiod, which is the fraction of time that a site is inundated with water [11]. Other environmental conditions, such as pH, nutrient status, salinity, soil texture, and organic matter contents, also determine the habitat type. Because wetlands exhibit such wide environmental conditions on a small scale, they play an important role in evolutionary adaptations and speciation in many groups of plants and animals [12]. The categorization of wetland habitat types is often done on the basis of plant communities, which are the most visible and consistent aspect of the overall biotic community in a wetland.

3. The link between wetland typology and ecosystem services

When talking about wetlands, it is useful to categorize wetlands in order to cover the wide variety in ecosystems, the species that occupy them, and the processes that take place within them. Over time, many different wetland classifications have been developed, but they are only considered useful if they make accurate descriptions of ecosystem processes and ecosystem functions within those wetlands [8]. For this reason, classifications based on hydrogeomorphic types (HGM types) are regarded as the most suitable way of classifying wetlands, and the type to which a wetland is allocated already tells us about which ecosystem services the wetland will provide [9, 10].

Hydrogeomorphic types are determined by looking at the source of water in a wetland as well as the shape of the basin in which this water is retained. A wetland is a place where there is a net surplus of water for at least part of the year, which means that the water entering the wetland exceeds the water exiting the wetland. Water can enter or exit the wetland in three different ways, namely the atmosphere (entry by means of precipitation and exit by means of evaporation), surface water (entry by means of flooding and exit by means of drainage), or subsurface to deep groundwater (entry by means of recharge and exit by means of discharge) [13]. The source of water, how this water is retained within the wetland and the way it is released from the wetland determine the role of this wetland within the overall water cycle and thereby the contribution it makes toward ecosystem services (**Table 1**) [8].

Ecosystem services are those ecological processes that take place that directly benefit human well-being and that tend to be taken for granted. In recent times, efforts have been made to attach values to ecosystem services by means of various valuation techniques, and it is consistently found that wetlands are among the ecosystems with the highest economic value. This is because they are intricately linked to the water cycle and humans depend on the water cycle for their food production and hygiene but can be harmed by excessive flooding and the associated erosion. At the same time, the water cycle is strongly tied to other biogeochemical cycles, and wetlands play an important role in the natural cycling of carbon, nitrogen, phosphorus, sulfur, and other minerals. Overall, ecosystem services can be subdivided into four main categories, namely regulating, supporting, provisioning, and cultural services (**Table 2**) [14].

Hydrogeomorphic type	Main water source	Water losses	Main ecosystem services provided
Floodplain	Surface flow	Runoff	Flood attenuation, sediment deposition, and water quality enhancement
Channeled valley bottom	Surface and subsurface flow	Runoff	Erosion control
Unchanneled valley bottom	Surface and subsurface flow	Runoff	Erosion control and water quality enhancement
Seepage	Subsurface flow	Runoff	Sustenance of base flow
Depression	Precipitation	Evaporation	Provisional services
Wetland flat	Subsurface flow	Evaporation	Provisional services

Table 1.
Hydrogeomorphic types, their inputs and outputs of water, and the main services that they typically provide.

Ecosystem services		Spatial level supply of service	Demand for service	Properties that facilitate this service
Regulating and supporting services	Flood attenuation	HGM unit, position in catchment	When there are downstream properties that need protection	Surface roughness is influenced by vegetation structure, longitudinal slope of the wetland, and storage capacity of depression basins.
	Streamflow regulation	HGM unit and position in catchment	When rainfall is episodic downstream	The properties of the vegetation and its influence on evaporative losses, subsurface flow in the wetland
	Sediment trapping	HGM unit and position in catchment	When there is excessive erosion upstream	The same features that assist in flood attenuation: surface roughness and storage capacity, as slowing down of floodwaters facilitates deposition of sediments
	Phosphate assimilation	HGM unit	When there are sources of phosphate upstream	Most phosphate assimilation takes place where sediments as deposited as phosphates are attached to sediments.
	Nitrate assimilation	HGM unit	When there are sources of nitrogen upstream	The presence of vegetation
	Toxicant assimilation	HGM unit	When toxicants are released upstream	Factors that facilitate sediment deposition. Presence of plants capable of assimilating heavy metals (Typha, Pontederia, etc.).
	Erosion control	HGM unit	When the landscape has steep slopes and high runoff	Vegetation and its stem density of vegetation as it holds soil together as well as canopy cover as it protects soils from the impact of drip.
	Carbon storage	Habitat unit	Cumulative effect of all wetlands	Productivity of vegetation, permanence of inundation as anaerobic conditions slow down organic breakdown
Provisioning services	Grazing resources	Habitat unit	When the surrounding landscape is grazed	High productivity of wetland vegetation, especially when it endures during the dry season
	Provisioning of medicinal plants	Habitat unit	When there are few medical services around	Depending on species present in the wetland
	Harvestable thatch and other materials	Habitat unit	When people rely on traditional materials	Depending on species present in the wetland
	Provision of water	HGM unit and position in catchment	When there are no other sources of water in the area	Depending on easy access, presence of open water
Cultural services	Cultural heritage	HGM unit	When there are communities with cultural practices	Presence of places with special cultural significance within the wetland

Ecosystem services		Spatial level supply of service	Demand for service	Properties that facilitate this service
	Tourism and recreation	HGM unit and habitat unit	When the wetland is on touristic routes	Presence of high conservation value and biodiversity, scenic beauty of the wetland, and the surrounding landscape
	Education and research	HGM unit	When there is good access to the wetland	Presence of existing research infrastructure

Table 2.
Various ecosystem services that wetlands provide, the scales at which they are relevant, and the properties that determine their effectiveness.

Regulating ecosystem services refer to the benefits that are derived from the natural functioning of ecosystems, whereby the ecosystem in question ensures the continuation or regulation of a process. Wetlands may play a role in the ensurance of baseflow in rivers by storing water for extended periods of time although the evaporative losses in wetlands may be quite large as well [15], but also in the retention of floodwaters, which lead to a reduction in the risk of flooding for downstream properties. Other regulating services include the removal of nutrients from water that flows through a wetland and erosion control by slowing down the flow of water. From the perspective of the climate, wetlands can help in climate regulation by means of the storage of carbon.

Supporting ecosystem services are the underlying natural processes that take place that support the regulating ecosystem services, such as photosynthesis at the basis of primary productivity and the carbon and water cycles at the basis of climate control and streamflow regulation. These services are often not specifically considered when monetary valuations are calculated, because they benefit humans mostly indirectly by supporting specific regulatory and provisional services. One specific supporting service that may be considered here as well is the support of species diversity, by providing a habitat where many species can thrive and reproduce.

Provisioning ecosystem services refer to the goods and materials that are provided by ecosystems that benefit people directly, either by the collection of construction materials, subsistence farming, or by commercial operations that benefit a wider community of people. In the case of wetlands, this of course includes the provision of water, either for oneself or for domestic animals that graze around the wetlands. Water can also be used for irrigation of nearby farmlands whereby water is pumped out of the wetland. Other goods that people may harvest from wetlands are fish, waterfowl, edible plants, construction materials, or medicine. The agricultural crops that are planted in wetlands or in areas that are irrigated by water from wetlands can also be regarded as goods that the wetland provides.

Lastly, cultural ecosystem services refer to the ways in which ecosystems have influenced our culture and our use of the landscape, either traditionally or in modern times. Across the world, wetlands have often been at the center of our civilization, as people have always settled around areas where easy access to water was guaranteed and successful harvests could be made to support a growing population. For this reason, many cultural and spiritual practices have emerged that have centered around wetlands, such as traditional cleansing and baptizing ceremonies. Today, wetlands are also central to many recreational practices and many wetlands provide opportunities for tourism. Lastly, wetlands can also be important for research and education, and in some places, research centers and education facilities have been centered around wetlands.

The reality of the ecosystem services that wetlands provide means that people will always have to live with wetlands in their environment. In the following section, we will consider the ecological processes that are part of wetland environments at various spatial scales.

4. Ecosystem services at a catchment scale

Wetlands are scattered throughout the landscape and are connected in various ways to the water cycle. Several of the ecosystem services that wetlands are known to provide are directly linked to this water cycle and help to ensure the accessibility of water to many communities. The position of the wetland within the catchment and in relation to residents living within that catchment has a big influence on ecosystem service supply. For example, the location of wetlands within a catchment has an influence on the collective nitrate removal for the catchment as a whole [16]. Some hydrogeomorphic types of wetlands are also typically found in specific areas within the catchment; for example, seepages are often found close to the upper reaches of the catchment, close to the water divide. In some cases, when such ecosystems contain peat, they may slow down the flow of water into the drainage system as peat bodies hold water that trickles gradually into the streams so that water continues to be released into the fluvial system long after it has precipitated down in the form of rain or snow. Snowfall in the high mountains is another mechanism by which the release of water into rivers is delayed as it will only start percolating into the soil after it starts to melt toward springtime.

Flooding risk is ameliorated by various types of wetlands that are found along the stream channel, especially floodplains, that retain water for extended periods of time, thereby protecting properties downstream. Especially floodplains that contain oxbow lakes and abandoned side channels form basins that can store excess water. One of the main reasons to restore floodplain wetlands and to provide rivers with the space to overtop their banks is to provide storage for floodwaters in order to protect towns and farmlands downstream.

The presence of seepage wetlands, floodplain wetlands, and valley bottom wetlands in a catchment is therefore crucial to ensure the resilience of the livelihoods within that catchment. Many of these wetlands combined can be regarded as a form of "ecological infrastructure" that combined with the river channels that connect them ensure that water remains available throughout the catchment. When wetlands are lost by drainage for agriculture, the need arises to mitigate these losses by creating artificial wetlands within the same catchment.

Another way in which wetlands provide services on a landscape scale or catchment scale is in the maintenance of biodiversity. Wetlands are scattered throughout the landscape, often as small pockets of habitat for aquatic biota. In some cases, wetlands are connected by means of fluvial channels, but they can also be cut off from each other by terrestrial habitat. Species that require wetland habitats have to migrate across the landscape to colonize new wetland habitats, and the metapopulation of a species can best be maintained if wetland habitats are large and wetlands are proximate to each other.

5. Ecosystem services at a wetland scale

At the scale of an individual wetland, wetlands serve as sinks for sediments from the surrounding landscape. This leads to conditions at the level at the individual wetland that may create local ecosystem services. It means that many nutrients, but also pollutants can accumulate in the wetland. Wetlands have the capacity to filter the water that flows through them, due to plant uptake and bacterial conversions. Artificial wetlands called helophyte filters are often constructed to process mine effluents, sewage, and other industrially produced pollution. The filtering capacity of the wetland depends on the size of the wetland, the rate of flow through it, and the type of plants that grow in the wetland. In many large wetlands, because of this filtering effect, the water quality is guaranteed for local subsistence farmers who also get their water needs from the wetland.

Wetlands also form a sink for sediment. This can be either clastic sediments, in areas where there is sufficiently strong flow, but this can also be organic sediments in places that are permanently inundated where the water is stagnant. The storage of a large mass of sediment makes the wetlands soak up water after rains, but it also makes them vulnerable for erosion [15]. Wetlands that develop on steep slopes are unstable as excessive surface runoff may wash away all stored sediment in the wetland.

Wetlands keep soil in place and prevent erosion by spreading water over a larger surface area which reduces the energy of concentrated channeled flow. The trade-off between slope and surface area of wetland follows a function where larger wetlands run a larger risk of eroding and turning into a channel, losing all their sediment.

In the case where wetlands store large amounts of peat, wetlands also perform a role in climate regulation. This is a two-edged sword as it also means that degrading wetlands may emit large volumes of carbon dioxide and methane. In order to emphasize their relatively high global importance as C stores, it is important to note that although wetlands occupy only 4–5% of the land area of the globe, and they hold approximately 30% of the carbon in the terrestrial biosphere [17, 18].

The scenic beauty of a wetland determines its use as a tourist destination or even as a place of worship or spiritual inspiration. Larger waterbodies and waterbodies that attract large amounts of waterfowl can be regarded as attractive tourist destinations, and the economic activities that may arise around such wetlands can be regarded as cultural ecosystem services.

6. Ecosystem services at a habitat scale

Many of the provisioning services that a wetland provides depend on there being suitable habitat for the species that is to be harvested. For a wetland to provide sufficient amounts of fish, there needs to be suitable habitat for fish, which is generally the parts of the wetland that are permanently inundated. Some fish may seek shelter in the inundated plains of a floodplain, especially if there are large predators around. Such shallow water habitats are suitable hunting grounds for fishermen.

Reeds and other tall graminoids are regularly harvested for construction materials or for weaving, arts, and crafts. The diversity of plant morphologies of wetland plants contributes to the diversity of different weave types, e.g., the fine culms of *Juncus kraussii* have particular value for twining and straight sewing, while the robust *Cyperus latifolius* leaves are well suited for plaiting [19]. Sometimes, the plants from wetlands can provide materials to thatch roofs. Wood can be harvested from swamp forests, but generally this can also be harvested from upland areas. Swamp forests may become dominant in floodplain areas in tropical climates and may become dominant along rivers in temperate climates, but are rare in many other parts of the world.

Some wetlands are utilized for small-scale agriculture. The crops that are harvested in this manner can also be regarded as provisioning ecosystem services as it is the wetland soils and soil moisture that provide the fertility to grow crops. There are only few crops that can cope with permanent inundation, but it may be quite feasible for farmers to use the temporary or seasonally wet areas of wetlands, whereby water from the wetland can be used to provide water to the crop. Traditional wetland cultivation systems have characteristically been closely coupled with the natural hydrological regimes of wetlands. This is illustrated by the rice farmers of central Sierra Leone, who through their rich knowledge of the soil moisture requirements of different rice varieties are able to match the different rice varieties with the soil moisture variation along the wetland soil catena [20].

Many wetlands are key grazing resources, particularly those located in arid to semi-arid climate, where they are especially valuable in dry years and at the end of the dry season in most years, owing to their residual moisture. Thus, wetlands are often critical forage sources over the dry season for both livestock and wild grazers [21].

Wetlands are also rich in other biodiversity that plays a role in people's livelihoods, for example, in the form of medicinal plants such as sweet flag (*Acorus calamus*) which is widely used for oral hygiene and the treatment of digestive complaints. The strong environmental gradients create many niches for plant species, and a substantial section of all useful plants can be found in wetland habitats [22].

Different wetland habitats do not only lead to different provisioning ecosystem services, but they can alter the hydrological aspects of the wetland and thereby have an effect on regulating services. For example, if the vegetation in certain sections of the wetland consists of shrubby vegetation, it will lead to more resistance to flow, which will impact ecosystem services that have to deal with erosion control or floodwater retention. Such ecosystem services are mainly impacted by larger-scale processes, but are modified on a local scale by different habitat types [23].

The services most strongly associated with different habitats in wetland potentially complement each other, thereby enhancing the wetland's overall supply for a variety of ecosystem services. For example, a temporarily flooded alluvial fan occupying the inflow of a valley bottom wetland may be particularly important for the trapping and retaining of sediment and the phosphates adsorbed to these sediments, given that phosphates are generally strongly absorbed by sediments, and therefore, phosphate removal tends to be strongly associated with the trapping of new sediment. Further downstream in the same wetland, where permanently flooded habitat and associated low redox conditions and high primary productivity of emergent vegetation and abundant organic sediments predominate, the storage of carbon, and the assimilation of nitrates may be particularly important, given the favorable biogeochemical conditions of this habitat for these particular services.

7. Ecosystem processes at a soil or sediment scale

At the most basic level, most of the processes that are unique to wetlands play out within an inundated soil. The reason for this is that wetlands provide a wide variety of metabolic pathways as there is no oxygen available and bacteria will have to find an alternate electron acceptor for the energy-gaining reactions that drive their life-cycle. This reflects the fact that wetlands today are among the habitats where the original metabolic pathways at the basis of life prevail, from the period before oxygen started to accumulate in the Earth's atmosphere as a waste product of photosynthesis. The primary way in which the soil biochemistry is affected by these inundated conditions is by the inhibition of the breakdown of dead organic matter that accumulates in the soil. This organic matter forms the substrate where many of the redox reactions in anaerobic soils take place.

Within these soils with anaerobic conditions, there is also variation in redox potential as water levels fluctuate and additionally wetland plants also have their influence. The root systems of these plants transport oxygen into the inundated soils as it may leak around the roots, creating narrow aerated zones around the smaller roots. The existence of steep gradients in redox potential are the conditions in which important biogeochemical processes such as denitrification take place [24]. This has an important reaction within the nitrogen cycle, but it is only one example of the wide range of redox reactions that can take place in wetland soils.

The wide variety of metabolic pathways arise from the combination of reactants that can exchange electrons in redox reactions that release free energy to be exploited by bacteria. The type of metabolism that is favored depends on the free energy released by the two half-reactions in the redox exchange, on the amount of organic matter that is available to be broken down and on the concentration of products and reactants that is available in the substrate [25]. This situation creates a multitude of possible reactions, and therefore, there is a wide range of niches at various depths of the substrate available to be exploited by different types of bacteria that each utilize a different energy source.

The elements that may become involved in these redox reactions occur in a specific sequence from the surface to the deeper soil layers or through time [25, 26] (see **Table 3**). Firstly organic matter is degraded in aerobic respiration as oxygen is energetically the most favored electron acceptor (reaction 1). When the oxygen is depleted, nitrate (NO_3^-) is depleted as an electron acceptor in which it is either reduced to ammonium (NH_4^+) or released as nitrogen gas in denitrification (reactions 2 and 3). When the nitrate is gone, the next electron acceptor is manganese (Mn^{4+}) that is reduced to Mn^{2+} (reaction 4), followed by Iron (Fe^{3+}) that is reduced to Fe^{2+}, creating a grayish color in the soil (reaction 5). When none of these ions are available, sulfate (SO_4^{2-}) is reduced to form H_2S gas, which can be recognized by the foul odor of rotten eggs (reaction 6). Lastly, when there is still organic matter available, it can

Reaction	Electron acceptor	Reaction	Redox potential
1	Organic matter	$C_6H_{12}O_6 \rightarrow CO_2 + H_2O$	250 mV
2, 3	Nitrate	$C_6H_{12}O_6 + 4NO_3^- \rightarrow 6CO_2 + 2N_2 + 6H_2O$ $C_6H_{12}O_6 + 4NO_3^- + 4H^+ \rightarrow 6CO_2 + 4NH_4^+ + 6H_2O$	250 mV
4	Manganese	$CH_2O + 2MnO_2 + 4H^+ \rightarrow CO_2 + 2Mn^{2+} + 2H_2O$	225 mV
5	Iron	$CH_2O + 2FeOOH + H^+ \rightarrow CO_2 + 2Fe^{2+} + H_2O$	120 mV
6	Sulfate	$2CH_3CHOHCOO^- + 2SO_4^{2-} + 3H^+ \rightarrow$ $4CO_2 + 4H_2O + 2HS^-$	−75 mV
7	Organic matter	$CH_3COO^- + 4H^+ \rightarrow 2CH_4 + 2H_2O$	−250 mV

Table 3.
Sequence of redox reactions taking place in inundated soils.

be broken down in the methanogenesis reaction (reaction 7), which is a very inefficient reaction and therefore happens at a very slow rate [27].

The complexity of the many reactions taking place in the anoxic soils of wetlands showcases the importance of wetlands in global biogeochemical cycles, especially for carbon, nitrogen, manganese, iron, and sulfur as they are directly involved in the redox reactions. Many other nutrients, such as phosphate, potassium, sodium, and chloride may become concentrated in wetlands in specific circumstances, which may provide either opportunities (phosphate and potassium) or limitations for life forms living in the wetland environment (sodium and chloride). The importance of wetlands for the global cycling of many of the elements that are deeply involved in various life cycles means that these ecosystems provide important supporting ecosystem services that extend beyond the border of the wetland itself.

This means that, even though we recognize these processes as taking place at the smallest scale, their impact is actually on a much larger scale, as it is the cumulative effect of all biogeochemical processes in all wetlands in an area that determine the fluxes of exchange between wetland and atmosphere. For these reasons, people living the direct environment may not be able to "value" the services that the wetland provides, especially when it comes to a service such as "climate regulation." Another problem is that these services may easily be altered when wetlands get degraded where they actually make the problem worse, such as with methane emissions [28].

8. Considering ecosystem services in the prioritization of wetlands for restoration or conservation

Historically, humans have always had an ambiguous relationship with wetlands: they acted as the centers of many civilizations where the first cities emerged on riverbanks and around other places with a steady access to water, but they have also always been regarded as a place where diseases spread and as places that need drastic changes by means of drainage before they can be rendered useful as agricultural land. Some of the world's most productive crops, such as rice and taro, can be grown in wetlands, but for many other crops, we need access to wetland water for irrigation, but the actual wetland itself is unsuitable for cultivation as the soil also requires aeration.

The importance of wetlands in ecosystem services and in the overall biogeochemical pathways across our planet means that as humanity we have to aim to design our landscapes in such a way that it provides space for wetlands and where we even may construct artificial wetlands if natural wetlands have disappeared. For this reason, urban planners and landscape architects should always keep considering the wet ecological infrastructure as all of our lives are strongly intertwined with the flow of water through our landscapes. This implies that the supply of ecosystem services must match the demands that society requires at each of the places where wetlands are designated or restored [29].

9. Conclusion

As wetlands are increasingly under threat worldwide, much has been written about the ecological services that they provide and there is a growing realization that there needs to be a presence of wetlands across any living landscape. The mapping of wetland ecosystem services requires us to understand that the different ecosystem services play out on different spatial scales. Some ecosystem services can take place within a single wetland, whereas for most of the hydrological services, the configuration of wetlands across the entire catchment is crucial. All these aspects of wetland ecology need to be considered in any landscape policy plan or natural resource management plan.

Author details

Erwin J.J. Sieben* and Donovan C. Kotze
School of Agricultural, Earth and Environmental Sciences, University of KwaZulu-Natal, Pietermaritzburg, South Africa

*Address all correspondence to: siebene@ukzn.ac.za

References

[1] Russi D, ten Brink P, Farmer A, Badura T, Coates D, Förster J, Kumar R, Davidson N. The Economics of Ecosystems and Biodiversity for Water and Wetlands. Vol. 78. London, Brussels: IEEP; 2013. p. 118

[2] Millennium Ecosystem Assessment. Ecosystems and Human Well-Being. Vol. 5. Washington, DC: Island Press; 2005

[3] Cumming TL, Shackleton RT, Förster J, Dini J, Khan A, Gumula M, et al. Achieving the national development agenda and the Sustainable Development Goals (Sdgs) through investment in ecological infrastructure: A case study of South Africa. Ecosystem Services. 2017;**27**:253-260

[4] Alexander S, Aronson J, Whaley O, Lamb D. The relationship between ecological restoration and the ecosystem services concept. Ecology and Society. 2016;**21**(1):34-42

[5] Benayas JMR, Newton AC, Diaz A, Bullock JM. Enhancement of biodiversity and ecosystem services by ecological restoration: A meta-analysis. Science. 2009;**325**(5944):1121-1124

[6] Tiner RW. Wetland Indicators: A Guide to Wetland Formation, Identification, Delineation, Classification and Mapping. Boca Raton: CRC Press; 2017

[7] Tiner RW. The primary indicators method—A practical approach to wetland recognition and delineation in the United States. Wetlands. 1993;**13**:50-64

[8] Sieben EJJ, Khubeka SP, Sithole S, Job NM, Kotze DC. The classification of wetlands: Integration of top-down and bottom-up approaches and their significance for ecosystem service determination. Wetlands Ecology and Management. 2018;**26**(3):441-458

[9] Brinson MM. A Hydrogeomorphic Classification for Wetlands. U.S. Army corps of Engineers. Washington, D.C.: Technical Report WRP-DE-4; 1993

[10] Ollis DJ, Snaddon K, Job NM, Mbona N. Classification System for Wetlands and Other Aquatic Ecosystems in South Africa: User Manual: Inland Systems. Pretoria: South African National Biodiversity Institute; 2013

[11] Kotze DC, Hughes JC, Klug JR, Breen CM. Improved criteria for classifying hydric soils in South Africa. South African Journal of Plant and Soil. 1996;**13**:67-73

[12] Moor H, Rydin H, Hylander K, Nilsson MB, Lindborg R, Norberg J. Towards a trait-based ecology of wetland vegetation. Journal of Ecology. 2017;**105**(6):1623-1635. DOI: 10.1111/1365-2745.12734

[13] Mitsch WJ, Gosselink JG. Wetlands. Hoboken, N.J: Wiley; 2007

[14] Malinga R, Gordon LJ, Jewitt G, Lindborg R. Mapping ecosystem services across scales and continents—A review. Ecosystem Services. 2015;**13**:57-63

[15] Bullock A, Acreman M. The role of wetlands in the hydrological cycle. Hydrology and Earth System Sciences Discussions. 2003;**7**(3):358-389

[16] Hansen AT, Dolph CL, Foufoula-Georgiou E, Finlay JC. Contribution of wetlands to nitrate removal at the watershed scale. Nature Geoscience. 2018;**11**(2):127-132

[17] Roulet NT. Peatlands, carbon storage, greenhouse gases, and the Kyoto protocol: Prospects and significance for Canada. Wetlands. 2000;**20**(4):605-615

[18] Joosten H, Clarke D. Wise Use of Mires and Peatlands. Totnes, Devon, UK: International Mire Conservation Group and International Peat Society; 2002. p. 304

[19] Kotze DC, Traynor CH. Wetland plant species used for craft production in Kwazulu–Natal, South Africa: Ethnobotanical knowledge and environmental sustainability. Economic Botany. 2011;**65**(3):271

[20] Roggeri H. Tropical Freshwater Wetlands: A Guide to Current Knowledge and Sustainable Management. Dordrecht: Springer; 2013

[21] Fynn RWS, Murray-Hudson M, Dhliwayo M, Scholte P. African wetlands and their seasonal use by wild and domestic herbivores. Wetlands Ecology and Management. 2015;**23**(4):559-581

[22] Cronk JK, Fennessey MS. Wetland Plants: Biology and Ecology. Boca Raton: Lewis Publishers; 2001

[23] Richardson CJ, Vaithyanathan P. Biogeochemical dynamics II: Cycling and storage of phosphorus in Wetlands. In: Maltby E, Barker T, editors. The Wetlands Handbook. Oxford: Blackwell Publishing; 2009

[24] White JR, Reddy KR. Biogeochemical dynamics I: Nitrogen cycling in Wetlands. In: Maltby E, Barker T, editors. The Wetlands Handbook. Oxford: Blackwell Publishing; 2009

[25] Schlesinger WH, Bernhardt ES. Biogeochemistry: An Analysis of Global Change. Waltham, Mass: Academic Press; 2013

[26] Zhang Z, Furman A. Soil redox dynamics under dynamic hydrologic regimes—A review. Science of the Total Environment. 2021;**763**:143026

[27] Inglett KS, Chanton JP, Inglett PW. Methanogenesis and methane oxidation in wetland soils. In: Methods in Biogeochemistry of Wetlands. Vol. 10. 2013. pp. 407-425

[28] Bridgham SD, Cadillo-Quiroz H, Keller JK, Zhuang Q. Methane emissions from wetlands: Biogeochemical, microbial, and modeling perspectives from local to global scales. Global Change Biology. 2013;**19**(5):1325-1346

[29] Wei H, Fan W, Wang X, Lu N, Dong X, Zhao Y, et al. Integrating supply and social demand in ecosystem services assessment: A review. Ecosystem Services. 2017;**25**:15-27